D1573469

Sensors Applications
Volume 1
Sensors in Manufacturing

Sensors Applications

Upcoming volumes:

- Sensors in Intelligent Buildings
- Sensors in Medicine and Health Care
- Sensors in Automotive Technology
- Sensors in Aerospace Technology
- Sensors in Environmental Technology
- Sensors in Household Appliances

Related Wiley-VCH titles:

W. Göpel, J. Hesse, J. N. Zemel
Sensors Vol. 1–9

ISBN 3-527-26538-4

H. Baltes, W. Göpel, J. Hesse
Sensors Update

ISSN 1432-2404

Sensors Applications
Volume 1

Sensors in Manufacturing

Edited by
H. K. Tönshoff, I. Inasaki

Series Editors:
J. Hesse, J. W. Gardner, W. Göpel

WILEY-VCH

Weinheim – New York – Chichester – Brisbane – Singapore – Toronto

the volume will close with a look ahead to likely developments and applications in the future. Actual sensor functions will only be described where it seems necessary for an understanding of how they relate to the process or system. The basic principles can always be found in the earlier series of *Sensors* and *Sensors Update*.

The series editors would like to express their warm appreciation in the colleagues who have contributed their expertise as volume editors or authors. We are deeply indebted to the publisher and would like to thank in particular Dr. Peter Gregory, Dr. Jörn Ritterbusch and Dr. Claudia Barzen for their constructive assistance both with the editorial detail and the publishing venture in general. We trust that our endeavors will meet with the reader's approval.

Oberkochen and Conventry, November 2000   Joachim Hesse
Julian W. Gardner

Preface to Volume 1 of "Sensors Applications"

Manufacturing technology has undergone significant developments over the last decades aiming at improving precision and productivity. The development of numerical control (NC) technology in 1952 made a significant contribution to meeting these requirements. The practical application of NC machine tools have stimulated technological developments that make the tools more intelligent, and allows the machining process to be carried out with higher reliability. Today, thanks to the significant developments in sensor and computer technologies, it can be said that the necessary tools are available for achieving the adaptive control of manufacturing processes, assisted by monitoring systems, which was a dream in the 1950's.

For the following reasons, monitoring technology with reliable sensors is becoming more and more important in modern manufacturing systems:

- Machine tools operate with speeds that do not allow manual intervention. However, collisions or process failures may cause significant damage.
- Manufacturing systems have become larger in scale, and monitoring of such large-scale systems is already beyond the capability of human beings.
- Increase of labor costs and the shortage of skilled operators calls for operation of the manufacturing system with minimum human intervention; this requires the introduction of advanced monitoring systems.
- Ultra-precision manufacturing can only be achieved with the aid of advanced metrology and process monitoring technology.
- The use of sophisticated machine tools requires the integration of monitoring systems to prevent machine failure.
- Heavy-duty manufacturing processes with higher energy consumption should be conducted with minimum human intervention, from the safety point of view.

In addition,

- Environmental consciousness in the manufacturing of today requires monitoring emissions from the process.

This book deals with monitoring technologies in various manufacturing processes, and aims to provide the latest developments in those fields together with the necessary principles behind these developments. We are convinced that the readers of this book, both in research institutes and in industry, can obtain information necessary for their research and developmental work.

The editors wish to thank the specialists who contributed their expertise and forbearance during the various stages of preparation. In addition to the assistance of the authors, we would like to thank the staff of Wiley-VCH for their support.

Hannover and Yokohama, November 2000   Hans Kurt Tönshoff
Ichiro Inasaki

Contents

List of Contributors *XVII*

1   Fundamentals *1*
1.1   Roles of Sensors in Manufacturing and Application Range *1*
        I. Inasaki, H. K. Tönshoff
1.1.1   Manufacturing *1*
1.1.2   Unit Processes in Manufacturing *2*
1.1.3   Sensors *3*
1.1.4   Needs and Roles of Monitoring Systems *4*
1.1.5   Trends *5*
1.1.6   References *6*
1.2   Principles of Sensors for Manufacturing *6*
        D. Dornfeld
1.2.1   Introduction *6*
1.2.2   Basic Sensor Classification *10*
1.2.3   Basic Sensor Types *13*
1.2.3.1 Mechanical Sensors *13*
1.2.3.2 Thermal Sensors *17*
1.2.3.3 Electrical Sensors *17*
1.2.3.4 Magnetic Sensors *18*
1.2.3.5 Radiant Sensors *18*
1.2.3.6 Chemical Sensors *18*
1.2.4   New Trends – Signal Processing and Decision Making *19*
1.2.4.1 Background *19*
1.2.4.2 Sensor Fusion *21*
1.2.5   Summary *23*
1.2.6   References *23*
1.3   Sensors in Mechanical Manufacturing – Requirements, Demands, Boundary Conditions, Signal Processing, Communication *24*
        T. Moriwaki
1.3.1   Introduction *24*
1.3.2   Role of Sensors and Objectives of Sensing *24*
1.3.3   Requirements for Sensors and Sensing Systems *27*

1.3.4	Boundary Conditions 31
1.3.5	Signal Processing and Conversion 32
1.3.5.1	Analog Signal Processing 32
1.3.5.2	AD Conversion 34
1.3.5.3	Digital Signal Processing 36
1.3.6	Identification and Decision Making 39
1.3.6.1	Strategy of Identification and Decision Making 39
1.3.6.2	Pattern Recognition 40
1.3.6.3	Neural Networks 41
1.3.6.4	Fuzzy Reasoning 42
1.3.7	Communication and Transmission Techniques 43
1.3.8	Human-Machine Interfaces 44
1.3.9	References 45

2 Sensors for Machine Tools and Robots 47
  H. K. Tönshoff
2.1 Position Measurement 47
2.2 Sensors for Orientation 58
2.3 Calibration of Machine Tools and Robots 60
2.4 Collision Detection 62
2.5 Machine Tool Monitoring and Diagnosis 65
2.6 References 70

3 Sensors for Workpieces 71
3.1 Macro-geometric Features 71
  A. Weckenmann
3.1.1 Mechanical Measurement Methods 72
3.1.1.1 Calipers 72
3.1.1.2 Protractors 73
3.1.1.3 Micrometer Gages 73
3.1.1.4 Dial Gages 75
3.1.1.5 Dial Comparators 76
3.1.1.6 Lever-type Test Indicators 76
3.1.2 Electrical Measuring Methods 76
3.1.2.1 Resistive Displacement Sensors 77
3.1.2.2 Capacitive Displacement Sensors 77
3.1.2.3 Inductive Displacement Sensors 78
3.1.2.4 Magnetic Incremental Sensors 81
3.1.2.5 Capacitive Incremental Sensors 81
3.1.2.6 Inductive Incremental Sensors 82
3.1.3 Electromechanical Measuring Methods 83
3.1.3.1 Touch Trigger Probe 84
3.1.3.2 Continuous Measuring Probe System 84
3.1.4 Optoelectronic Measurement Methods 86
3.1.4.1 Incremental Methods 86

3.1.4.2 Absolute Measurement Methods 89
3.1.5   Optical Measuring Methods 90
3.1.5.1 Camera Metrology 90
3.1.5.2 Shadow Casting Methods 91
3.1.5.3 Point Triangulation 91
3.1.5.4 Light-section Method 92
3.1.5.5 Fringe Projection 93
3.1.5.6 Theodolite Measuring Systems 93
3.1.5.7 Photogrammetry 94
3.1.5.8 Interferometric Distance Measurement 94
3.1.5.9 Interferometric Form Testing 95
3.1.5.10 Autofocus Method 96
3.1.6   Pneumatic Measuring Systems 96
3.1.7   Further Reading 98
3.2   Micro-geometric Features 98
        A. Weckenmann
3.2.1   Tactile Measuring Method 99
3.2.1.1 Reference Surface Tactile Probing System 100
3.2.1.2 Skidded System 100
3.2.1.3 Double Skidded System 101
3.2.2   Optical Measuring Methods 101
3.2.2.1 White Light Interferometry 102
3.2.2.2 Scattered Light Method 103
3.2.2.3 Speckle Correlation 104
3.2.2.4 Grazing Incidence X-ray Reflectrometry 105
3.2.3   Probe Measuring Methods 106
3.2.3.1 Scanning Electron Microscopy (SEM) 107
3.2.3.2 Scanning Tunneling Microscopy (STM) 108
3.2.3.3 Scanning Near-field Optical Microscopy (SNOM) 110
3.2.3.4 Scanning Capacitance Microscopy (SCM) 111
3.2.3.5 Scanning Thermal Microscopy (SThM) 111
3.2.3.6 Atomic Force Microscopy (AFM) 113
3.2.3.7 Magnetic Force Microscopy (MFM) 117
3.2.3.8 Lateral Force Microscopy (LFM) 118
3.2.3.9 Phase Detection Microscopy (PDM) 119
3.2.3.10 Force Modulation Microscopy (FMM) 120
3.2.3.11 Electric Force Microscopy (EFM) 121
3.2.3.12 Scanning Near-field Acoustic Microscopy (SNAM) 122
3.2.4   Further Reading 123
3.3   Sensors for Physical Properties 123
        B. Karpuschewski
3.3.1   Introduction 123
3.3.2   Laboratory Reference Techniques 125
3.3.3   Sensors for Process Quantities 125
3.3.3.1 Force Sensors 126

3.3.3.2 Power Sensors 128
3.3.3.3 Temperature Sensors 129
3.3.3.4 Acoustic Emission Sensors 131
3.3.4 Sensors for Tools 134
3.3.5 Sensors for Workpieces 136
3.3.5.1 Eddy-current Sensors 136
3.3.5.2 Micro-magnetic Sensors 137
3.3.6 References 141

4 Sensors for Process Monitoring 143
4.1 Casting and Powder Metallurgy 143
4.1.1 Casting 143
H. D. Haferkamp, M. Niemeyer, J. Weber
4.1.1.1 Introduction 143
4.1.1.2 Sensors with Melt Contact 145
4.1.1.3 Sensors without Melt Contact 149
4.1.1.4 Summary 157
4.1.1.5 References 157
4.1.2 Powder Metallurgy 159
R. Wertheim
4.1.2.1 Introduction 159
4.1.2.2 Mixing and Blending of Metal Powders 159
4.1.2.3 Compacting of Metal Powders 162
4.1.2.4 The Sintering Process 166
4.1.2.5 References 171
4.2 Metal Forming 172
E. Doege, F. Meiners, T. Mende, W. Strache, J. W. Yun
4.2.1 Sensors for the Punching Process 172
4.2.1.1 Sensors and Process Signals 173
4.2.1.2 Sensor Locations 174
4.2.1.3 Sensor Applications 176
4.2.2 Sensors for the Sheet Metal Forming Process 181
4.2.2.1 Deep Drawing Process and Signals 182
4.2.2.2 Material Properties 182
4.2.2.3 Lubrication 184
4.2.2.4 In-process Control for the Deep Drawing Process 186
4.2.3 Sensors for the Forging Process 191
4.2.3.1 Sensors Used in Forging Processes 191
4.2.3.2 Sensor Application and Boundaries 195
4.2.3.3 Typical Signals for Forces and Path 198
4.2.3.4 Process Monitoring 200
4.2.4 References 202
4.3 Cutting Processes 203
I. Inasaki, B. Karpuschewski, H. K. Tönshoff
4.3.1 Introduction 203

4.3.2	Problems in Cutting and Need for Monitoring	203
4.3.3	Sensors for Process Quantities	204
4.3.3.1	Force Sensors	204
4.3.3.2	Torque Sensors	209
4.3.3.3	Power Sensors	211
4.3.3.4	Temperature Sensors	211
4.3.3.5	Vibration Sensors	214
4.3.3.6	Acoustic Emission Sensors	215
4.3.4	Tool Sensors	220
4.3.5	Workpiece Sensors	225
4.3.6	Chip Control Sensors	228
4.3.7	Adaptive Control Systems	231
4.3.8	Intelligent Systems for Cutting Processes	233
4.3.9	References	234
4.4	Abrasive Processes	236
	I. Inasaki, B. Karpuschewski	
4.4.1	Introduction	236
4.4.2	Problems in Abrasive Processes and Needs for Monitoring	236
4.4.3	Sensors for Process Quantities	237
4.4.3.1	Force Sensors	238
4.4.3.2	Power Measurement	239
4.4.3.3	Acceleration Sensors	239
4.4.3.4	Acoustic Emission Systems	239
4.4.3.5	Temperature Sensors	241
4.4.4	Sensors for the Grinding Wheel	244
4.4.4.1	Sensors for Macro-geometric Quantities	246
4.4.4.2	Sensors for Micro-geometric Quantities	247
4.4.5	Workpiece Sensors	249
4.4.5.1	Contact-based Workpiece Sensors for Macro-geometry	249
4.4.5.2	Contact-based Workpiece Sensors for Micro-geometry	251
4.4.5.3	Contact-based Workpiece Sensors for Surface Integrity	252
4.4.5.4	Non-contact-based Workpiece Sensors	252
4.4.6	Sensors for Peripheral Systems	256
4.4.6.1	Sensors for Monitoring of the Conditioning Process	256
4.4.6.2	Sensors for Coolant Supply Monitoring	259
4.4.7	Sensors for Loose Abrasive Processes	262
4.4.7.1	Lapping Processes	262
4.4.7.2	Sensors for Non-conventional Loose Abrasive Processes	264
4.4.8	Adaptive Control Systems	265
4.4.9	Intelligent Systems for Abrasive Processes	268
4.4.10	References	271
4.5	Laser Processing	272
	V. Kral, O. Hillers	
4.5.1	Introduction	272
4.5.2	Parameter Monitoring Sensors	273

4.5.2.1 Sensors for Identifying Workpiece Geometry   273
4.5.2.2 Sensors for Identifying Workpiece Quality   273
4.5.2.3 Sensors for Beam Characterization   274
4.5.2.4 Focal Position and Gas Pressure   274
4.5.3   Quality Monitoring Sensors   275
4.5.3.1 Optical Sensors   275
4.5.3.2 Acoustic Sensors   275
4.5.3.3 Visual-based Sensing   275
4.5.4   Conclusion   276
4.5.5   References   277
4.6   Electrical Discharge Machining   277
         T. Masuzawa
4.6.1   Introduction   277
4.6.2   Principle of EDM   278
4.6.3   Process Control   279
4.6.4   Sensing Technology   279
4.6.4.1 Gap Voltage   280
4.6.4.2 Current Through Gap   281
4.6.4.3 Electromagnetic Radiation   283
4.6.4.4 Acoustic Radiation   283
4.6.5   Evaluation of Machinery Accuracy   283
4.6.5.1 VS Method   284
4.6.5.2 Application of Micro-EDM   285
4.7   Welding   286
         H. D. Haferkamp, F. v. Alvensleben, M. Niemeyer, W. Specker, M. Zelt
4.7.1   Introduction   286
4.7.2   Geometry-oriented Sensors   287
4.7.2.1 Contact Geometry-oriented Sensors   287
4.7.2.2 Non-contact Geometry-oriented Sensors   291
4.7.3   Welding Process-oriented Sensors   295
4.7.3.1 Primary Process Phenomena-oriented Sensors   295
4.7.3.2 Secondary Process Phenomena-oriented Sensors   300
4.7.4   Summary   305
4.7.5   References   305
4.8   Coating Processes   307
         K.-D. Bouzakis, N. Vidakis, G. Erkens
4.8.1   Coating Process Monitoring   307
4.8.1.1 Introduction   307
4.8.1.2 Vacuum Coating Process Classification   308
4.8.1.3 Vacuum Coating Process Parameter Monitoring Requirements   309
4.8.2   Sensors in Vapor Deposition Processes   311
4.8.2.1 Vapor Process Parameter Map   311
4.8.2.2 Vacuum Control   311
4.8.2.3 Temperature Control   318
4.8.2.4 Gas Analyzers for Coating Process Control   321

4.8.2.5	Thin-film Thickness (TFT) Controllers for Deposition Rate Monitoring and Control *322*	
4.8.2.6	Gas Dosing Systems and Valves *324*	
4.8.2.7	Other Parameters Usually Monitored During the PVD Process *325*	
4.8.3	References *325*	
4.9	Heat Treatment *326*	
	P. Mayr, H. Klümper-Westkamp	
4.9.1	Introduction *326*	
4.9.2	Temperature Monitoring *326*	
4.9.3	Control of Atmospheres *329*	
4.9.4	Carburizing *329*	
4.9.5	Nitriding *331*	
4.9.6	Oxidizing *332*	
4.9.7	Control of Structural Changes *334*	
4.9.8	Quenching Monitoring *337*	
4.9.8.1	Fluid Quench Sensor *338*	
4.9.8.2	Hollow Wire Sensor *338*	
4.9.8.3	Flux Sensor *339*	
4.9.9	Control of Induction Heating *339*	
4.9.10	Sensors for Plasma Processes *341*	
4.9.11	Conclusions *341*	
4.9.12	References *341*	

5 Developments in Manufacturing and Their Influence on Sensors *343*

5.1 Ultra-precision Machining: Nanometric Displacement Sensors *343*
E. Brinksmeier
5.1.1 Optical Scales *343*
5.1.2 Laser Interferometers *348*
5.1.3 Photoelectric Transducers *351*
5.1.4 Inductive Sensors *352*
5.1.5 Autocollimators *352*
5.1.6 References *353*
5.2 High-speed Machining *354*
H. K. Tönshoff
5.3 Micro-machining *357*
M. Weck
5.4 Environmental Awareness *363*
F. Klocke
5.4.1 Measurement of Emissions in the Work Environment *364*
5.4.1.1 Requirements Relating to Emission Measuring Techniques in Dry Machining *364*
5.4.1.2 Sensor Principles *364*
5.4.1.3 Description of Selected Measuring Techniques *365*
5.4.1.4 Example of Application *366*
5.4.2 Dry Machining and Minimum Lubrication *367*

5.4.2.1 Measuring Temperatures in Dry Machining Operations 367
5.4.2.2 Measuring Droplets in Minimal Lubrication Mode 368
5.4.3 Turning of Hardened Materials 369
5.4.3.1 Criteria for Process and Part Quality 369
5.4.3.2 Sensing and Monitoring Approaches 371
5.4.4 Using Acoustic Emission to Detect Grinding Burn 372
5.4.4.1 Objective 373
5.4.4.2 Sensor System 374
5.4.4.3 Signal Evaluation 375
5.4.5 References 375

List of Symbols and Abbreviations 377

Index 383

List of Contributors

F. v. Alvensleben
Laser Zentrum Hannover e.V.
Hollerithallee 8
30419 Hannover
Germany

E. Brinksmeier
Fachgebiet Fertigungsverfahren
und Labor für Mikrozerspanung
Universität Bremen
Badgasteiner Str. 1
28359 Bremen
Germany

K.-D. Bouzakis
Laboratory for Machine Tool
and Machine Dynamics
Aristoteles University Thessaloniki
54006 Thessaloniki
Greece

E. Doege
Institut für Umformtechnik
und Umformmaschinen
Universität Hannover
Welfengarten 1A
30167 Hannover
Germany

D. Dornfeld
University of California
Berkeley
CA 94720-5800
USA

G. Erkens
CemeCon GmbH
Adenauerstr. 20B1
52146 Würselen
Germany

H. D. Haferkamp
Institut für Werkstoffkunde
Universität Hannover
Appelstr. 11
30167 Hannover
Germany

O. Hillers
Laser Zentrum Hannover e.V.
Hollerithallee 8
30419 Hannover
Germany

I. Inasaki
Faculty of Sciency & Technology
Keio University
3-14-1 Hiyoshi, Kohoku-ku
Yokohama-shi
Japan

B. Karpuschewski
Faculty of Science & Technology
Keio University
3-14-1 Hiyoshi, Kohoku-ku
Yokohama-shi
Japan

F. KLOCKE
Lehrstuhl für Technologie
der Fertigungsverfahren
RWTH Aachen
Steinbachstr. 53
52056 Aachen
Germany

H. KLÜMPER-WESTKAMP
Stiftung Institut
für Werkstofftechnik IWT
Badgasteiner Str. 3
28359 Bremen
Germany

V. KRAL
Laser Zentrum Hannover e.V.
Hollerithallee 8
30419 Hannover
Germany

T. MASUZAWA
I.I.S., University of Tokyo
Center for International Research
on Micromechatronics (CIRMM)
4-6-1 Komaba, Meguro-ku
Tokyo 153-8505
Japan

P. MAYR
Stiftung Institut
für Werkstofftechnik IWT
Badgasteiner Str. 3
28359 Bremen
Germany

F. MEINERS
Institut für Umformtechnik
und Umformmaschinen
Universität Hannover
Welfengarten 1A
30167 Hannover
Germany

T. MENDE
Institut für Umformtechnik
und Umformmaschinen
Universität Hannover
Welfengarten 1A
30167 Hannover
Germany

T. MORIWAKI
Dept. of Mechanical Engineering
Kobe University
Rokko, Nada
Kobe 657
Japan

M. NIEMEYER
Institut für Werkstoffkunde
Universität Hannover
Appelstr. 11
30167 Hannover
Germany

W. SPECKER
Laser Zentrum Hannover e.V.
Hollerithallee 8
30419 Hannover
Germany

W. STRACHE
Institut für Umformtechnik
und Umformmaschinen
Universität Hannover
Welfengarten 1A
30167 Hannover
Germany

H. K. TÖNSHOFF
Institut für Fertigungstechnik
und Spanende Werkzeugmaschinen
Universität Hannover
Schloßwender Str. 5
30159 Hannover
Germany

N. VIDAKIS
Laboratory for Machine Tool
and Machine Dynamics
Aristoteles University Thessaloniki
54006 Thessaloniki
Greece

J. W. YUN
Institut für Umformtechnik
und Umformmaschinen
Universität Hannover
Welfengarten 1 A
30167 Hannover
Germany

J. WEBER
Institut für Werkstoffkunde
Universität Hannover
Appelstr. 11
30167 Hannover
Germany

M. WECK
Laboratorium für Werkzeugmaschinen
und Betriebslehre
RWTH Aachen
Steinbachstr. 53
52056 Aachen
Germany

A. WECKENMANN
Lehrstuhl für Qualitätsmanagement
und Fertigungsmeßtechnik
Universität Erlangen-Nürnberg
Nägelsbachstr. 25
91052 Erlangen
Germany

R. WERTHEIM
ISCAR LTD.
P.O. Box 11
Tefen 24959
Israel

M. ZELT
Institut für Werkstoffkunde
Universität Hannover
Appelstr. 11
30167 Hannover
Germany

1
Fundamentals

1.1
Roles of Sensors in Manufacturing and Application Ranges
I. INASAKI, *Keio University, Yokohama, Japan*
H. K. TÖNSHOFF, *Universität Hannover, Hannover, Germany*

1.1.1
Manufacturing

Manufacturing can be said in a broad sense to be the process of converting raw materials into usable and saleable end products by various processes, machinery, and operations. The important function of manufacturing is, therefore, to add value to the raw materials. It is the backbone of any industrialized nation. Without manufacturing, few nations could afford the amenities that improve the quality of life. In fact, generally, the higher the level of manufacturing activity in a nation, the higher is the standard of living of its people. Manufacturing should also be competitive, not only locally but also on a global basis because of the shrinking of our world.

The manufacturing process involves a series of complex interactions among materials, machinery, energy, and people. It encompasses the design of products, various processes to change the geometry of bulk material to produce parts, heat treatment, metrology, inspection, assembly, and necessary planning activities. Marketing, logistics, and support services are relating to the manufacturing activity. The major goals of manufacturing technology are to improve productivity, increase product quality and uniformity, minimize cycle time, and reduce labor costs. The use of computers has had a significant impact on manufacturing activities covering a broad range of applications, including design of products, control and optimization of manufacturing processes, material handling, assembly, and inspection of products.

1.1.2
Unit Processes in Manufacturing

The central part of manufacturing activity is the conversion of raw material to component parts followed by the assembly of those parts to give the products. The processes involved in making individual parts using machinery, typically machine tools, are called unit processes. Typical unit processes are casting, sintering, forming, material removing processes, joining, surface treatment, heat treatment, and so on. Figure 1.1-1 shows various steps and unit processes involved in manufacturing which are dealt with in this book. The unit processes can be divided into three categories [1]:

- removing unnecessary material (–);
- moving material from one region to another (0);
- putting material together (+).

For example, cutting and abrasive processes are removal operations (–), forming, casting, and sintering are (0) operations, and joining is a (+) operation.

The goal of any unit process is to achieve high accuracy and productivity. Thanks to the significant developments in machine tools and machining technologies, the accuracy achievable has been increased as shown in Figure 1.1-2 [2]. The increase in productivity in terms of cutting speed is depicted in Figure 1.1-3 [2]. The development of new cutting tool materials has made it possible, together with the improvements in machine tool performance, to reach cutting speeds higher than 1000 m/min.

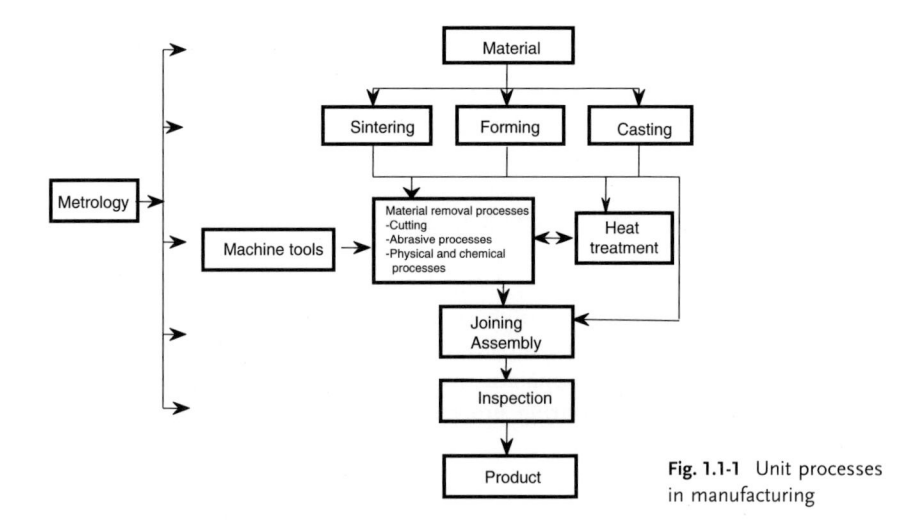

Fig. 1.1-1 Unit processes in manufacturing

Fig. 1.1-2 Achievable machining accuracy [2]

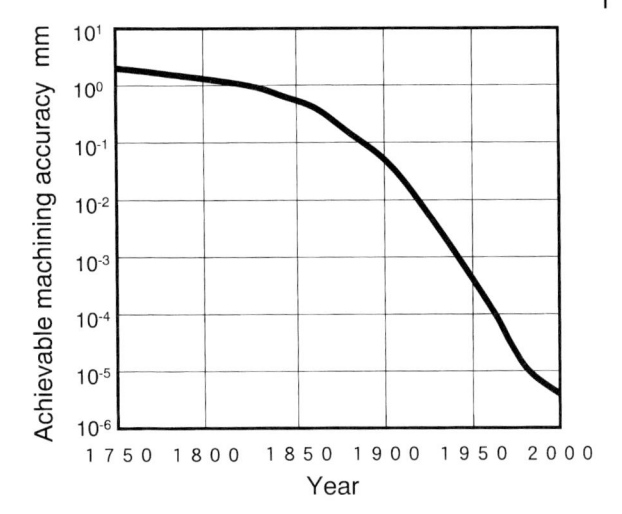

Fig. 1.1-3 Increase of cutting speed in turning [2]

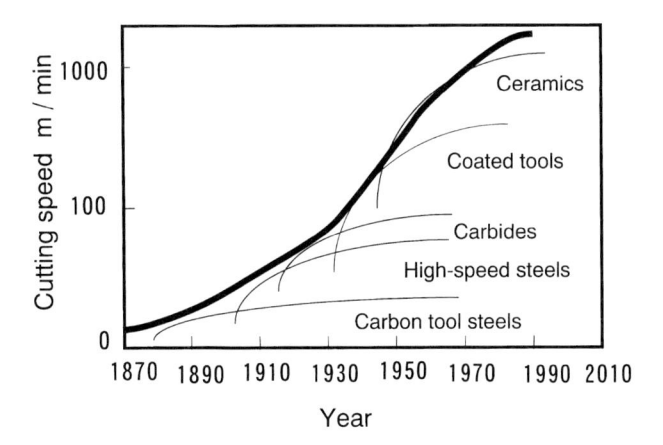

1.1.3
Sensors

Any manufacturing unit process can be regarded as a conversion process of material, energy, and information (Figure 1.1-4). The process should be monitored carefully to produce an output that can meet the requirements. When the process is operated by humans, it is monitored with sense organs such as vision, hearing, smell, touch, and taste. Sometimes, information obtained through multiple sense organs is used to achieve decision making. In addition, the brain as the sensory center plays an important role in processing the information obtained with the sense organs. In order to achieve automatic monitoring, those sense organs must be replaced with sensors. Some sensors can sense signals that cannot be sensed with the human sense organs.

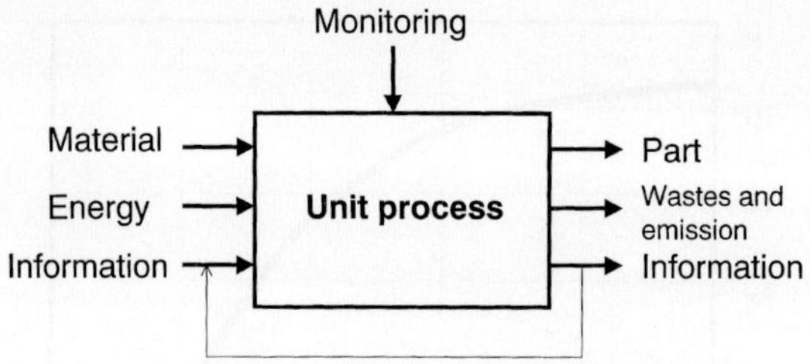

Fig. 1.1-4 Unit process as a conversion process

```
                 Monitoring
                     |
                     v
Material  ----->  ┌─────────────┐  ----->  Part
Energy    ----->  │Unit process │  ----->  Wastes and emission
Information ---->  │             │  ----->  Information
                  └─────────────┘
```

The word sensor came from the Latin *sentire*, meaning 'to perceive', and is defined as 'a device that detects a change in a physical stimulus and turns it into a signal which can be measured or recorded' [3]. In other words, an essential characteristic of the sensing process is the conversion of energy from one form to another. In practice, therefore, most sensors have sensing elements plus associated circuitry. For measurement purposes, the following six types of signal are important: radiant, mechanical, thermal, electrical, magnetic, and chemical [3].

1.1.4
Needs and Roles of Monitoring Systems

Considering the trends of manufacturing developments, the following reasons can be pointed out to explain why monitoring technology is becoming more and more important in modern manufacturing systems:

(1) Large-scale manufacturing systems should be operated with high reliability and availability because the downtime due to system failure has a significant influence on the manufacturing activity. To meet such a demand, individual unit processes should be securely operated with the aid of reliable and robust monitoring systems. Monitoring of large-scale systems is already beyond the capability of humans.
(2) Increasing labor costs and shortage of skilled operators necessitate operation of the manufacturing system with minimum human intervention, which requires the introduction of advanced monitoring systems.
(3) Ultra-precision manufacturing can only be achieved with the aid of advanced metrology and the technology of process monitoring.
(4) Use of sophisticated machine tools requires the integration of monitoring systems to prevent machine failure.
(5) Heavy-duty machining with high cutting and grinding speeds should be conducted with minimum human intervention from the safety point of view.
(6) Environmental awareness in today's manufacturing requires the monitoring of emissions from processes.

Fig. 1.1-5 Roles of monitoring system

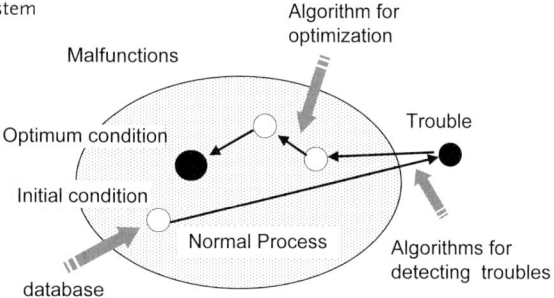

The roles of the monitoring system can be summarized as shown in Figure 1.1-5. First, it should be capable of detecting any unexpected malfunctions which may occur in the unit processes. Second, information regarding the process parameters obtained with the monitoring system can be used for optimizing the process. For example, if the wear rate of the cutting tool can be obtained, it can be used for minimizing the machining cost or time by modifying the cutting speed and the feed rate to achieve adaptive control optimization [4]. Third, the monitoring system will make it possible to obtain the input-output causalities of the process, which is useful for establishing a databank regarding the particular process [5]. The databank is necessary when the initial setup parameters should be determined.

1.1.5
Trends

In addition to increasing needs of the monitoring system, the demand for improving the performance of the monitoring system, particularly its reliability and robustness, is also increasing. No sensing device possesses 100% reliability. A possible way to increase the reliability is to use multiple sensors, making the monitoring system redundant. The fusion of various information is also a very suitable means to obtain a more comprehensive view of the state and performance of the process. In addition, sensor fusion is a powerful tool for making the monitoring system more flexible so that the various types of malfunctions that occur in the process can be detected.

In the context of sensor fusion, there are two different types: the *replicated sensors system* and the *disparate sensors system* [5]. The integration of similar types of sensors, that is, a replicated sensor system, can contribute mainly to improving the reliability and robustness of the monitoring system, whereas the integration of different types of sensors, disparate sensors system, can make the monitoring system more flexible (Figure 1.1-6).

Significant developments in sensor device technology are contributing substantially being supported by fast data processing technology for realizing a monitoring system which can be applied practically in the manufacturing environment.

Fig. 1.1-6 Evolution of monitoring system

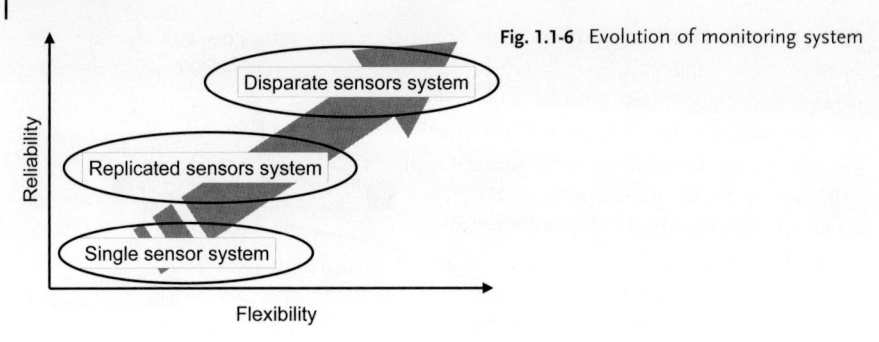

Soft computing techniques, such as fuzzy logic, artificial neural networks and genetic algorithms, which can to some extent imitate the human brain, can possibly contribute to making the monitoring system more intelligent.

1.1.6
References

1 SHAW, M.C., *Metal Cutting Principles*; Oxford: Oxford University Press, 1984.
2 WECK, M., *Werkzeugmaschinen Fertigungssysteme 1, Maschinenarten und Anwendungsbereiche*, 5. Auflage; Berlin: Springer, 1998.
3 USHER, M.J., *Sensors and Transducers*; London, Macmillian, 1985.
4 SUKVITTYAWONG, S., INASAKI, I., *JSME Int.*, Series 3 **34** (4) (1991), 546–552.
5 SAKAKURA, M., INASAKI, I., *Ann. CIRP* **42** (1) (1993), 379–382.

1.2
Principles of Sensors in Manufacturing
D. DORNFELD, *University of California, Berkeley, CA, USA*

1.2.1
Introduction

New demands are being placed on monitoring systems in the manufacturing environment because of recent developments and trends in machining technology and machine tool design (high-speed machining and hard turning, for example). Numerous different sensor types are available for monitoring aspects of the manufacturing and machining environments. The most common sensors in the industrial machining environment are force, power, and acoustic emission (AE) sensors. This section first reviews the classification and description of sensor types and the particular requirements of sensing in manufacturing by way of a background and then the state of sensor technology in general. The section finishes with some insight into the future trends in sensing technology, especially semiconductor-based sensors.

1.2 Principles of Sensors in Manufacturing

In-process sensors constitute a significant technology, helping manufacturers to meet the challenges inherent in manufacturing a new generation of precision components. In-process sensors play different roles in manufacturing processes and can address the tooling, process, workpiece, or machine. First and foremost, they allow manufacturers to improve the control over critical process variables. This can result in the tightening of control limits of a process and as improvements in process productivity, forming the basis of precision machining (Figure 1.2-1). For example, the application of temperature sensors and appropriate control to traditional machine tools has been demonstrated to reduce thermal errors, the largest source of positioning errors in traditional and precision machine tools, and the work space errors they generate. Second, they serve as useful productivity tools in monitoring the process. For example, as already stated, they improve productivity by detecting process failure as is the case with acoustic sensors detecting catastrophic tool failure in cutting processes. They also reduce dead time in the process cycle by detecting the degree of engagement between the tool and the work, allowing for a greater percentage of machining time in each part cycle. As process speeds increase and equipment downtime becomes less tolerable, sensors become critical elements in the manufacturing system to insure high productivity and high-quality production.

With regard to sensor systems for manufacturing process monitoring, a distinction is to be made on the one hand between continuous and intermittent systems and on the other between direct and indirect measuring systems. In the case of continuously measuring sensor systems, the measured variable is available throughout the machining process; intermittently measuring systems record the measured variable only during intervals in the machining process. The distinction is sometimes referred to as pre-, inter-, or post-process measurement for intermit-

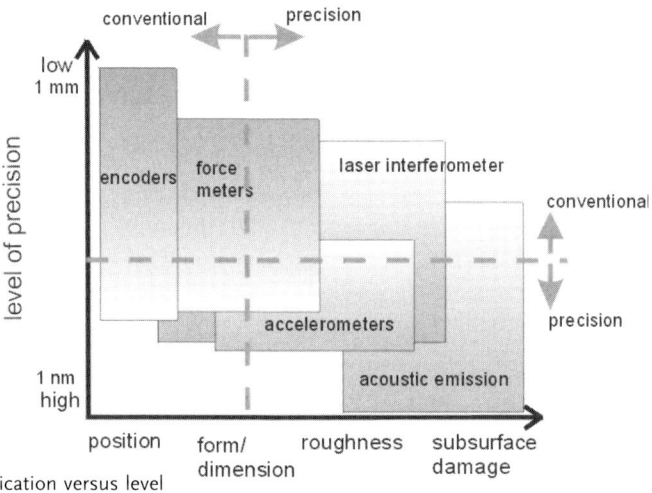

Fig. 1.2-1 Sensor application versus level of precision and error control parameters

tent systems and in-process for continuous systems. Obviously, other distinctions can apply. Direct measuring systems employ the actual quantity of the measured variable, eg, tool wear, whereas indirect measuring systems measure suitable auxiliary quantities, such as the cutting force components, and deduce the actual quantity via empirically determined correlations. Direct measuring processes possess a higher degree of accuracy, whereas indirect methods are less complex and more suitable for practical application. Continuous measurement permits the continuous detection of all changes to the measuring signal and ensures that sudden, unexpected process disturbances, such as tool breakage, are responded to in good time. Intermittent measurement is dependent on interruptions in the machining process or special measuring intervals, which generally entail time losses and, subsequently, high costs. Furthermore, tool breakage cannot be identified until after completion of the machining cycle when using these systems, which means that consequential damage cannot be prevented. Intermittent wear measurement nevertheless has its practical uses, provided that it does not result in additional idle time. It would be conceivable, for example, for measurement to be carried out in the magazine of the machine tool while the machining process is continued with a different tool. Intermittent wear-measuring methods can be implemented with mechanical, inductance-capacitance, hydraulic-pneumatic and opto-electronic probes or sensor systems.

Direct and continuous sensor measuring is the optimal combination with respect to accuracy and response time. For direct measurement of the wear land width, an opto-electronic system has been available, for example, whereby a wedgeshaped light gap below the cutting edge of the tool, which changes proportionally to the wear land width, is evaluated. The wear land width can also be measured directly by means of specially prepared cutting plates, the flanks of which are provided with strip conductors which act as electrical resistors. Another approach uses an image processing system based on a linear camera for on-line determination of the wear on a rotating inserted-tooth face mill. Non-productive time due to measurement is avoided and the system reacts quickly to tool breakage. There are, however, problems due to the short distance between the tool and the camera, which is mounted in the machine space to the side of the milling cutter, and due to chips and dirt on the inserts.

The indirect continuous measuring processes, which are able to determine the relevant disturbance, eg, tool wear, by measuring an auxiliary quantity and its changes, are generally less accurate than the direct methods. A valuable variable which can be measured for the purpose of indirect wear determination is the cutting temperature, which generally rises as the tool wear increases as a result of the increased friction and energy conversion. However, all the known measuring processes are pure laboratory methods for turning which are furthermore not feasible for milling and drilling, owing to the rotating tools. Continuous measurement of the electrical resistance between tool and workpiece is also not feasible for practical applications, on account of the required measures, such as insulation of the workpiece and tool, and to short circuits resulting from chips or cooling lubricant. Systems based on sound monitoring using microphones, for example, also have not yet reached industrial application owing to the problems caused by noise that is not generated by the machining process.

The philosophy of implementation of any sensing methodology for diagnostics or process monitoring can be divided into two simple approaches. In one approach, one uses a sensing technique for which the output bears some relationship to the characteristics of the process. After determining the sensor output and behavior for 'normal' machine operation or processing, one observes the behavior of the signal until it deviates from the normal, thus indicating a problem. In the other approach, one attempts to determine a model linking the sensor output to the process mechanics and then, with sensor information, uses the model to predict the behavior of the process. Both methods are useful in differing circumstances. The first is, perhaps, the most straightforward but liable to misinterpretation if some change in the process occurs that was not foreseen (that is, 'normal' is no longer normal). Thus some signal processing strategy is required.

The signal that is delivered by the sensor must be processed to detect disturbances. The simplest method is the use of a rigid threshold. If the threshold is crossed by the signal owing to some process change affecting the signal, collision or tool breakage can be detected. Since this method only works when all restrictions (depth of cut, workpiece material, etc.) remain constant, the use of a dynamic threshold is more appropriate in most cases. The monitoring system calculates an upper threshold from the original signal. The upper threshold time-lags the original signal. Slow changes of the signal can occur without violating the threshold. At the instant of breakage, however, the upper threshold is crossed and, following a plausibility check (the signal must remain above the upper threshold for a certain time duration), a breakage is confirmed and signaled. Because of the high bandwidth of the acoustic emission signal, fast response time to a breakage is insured. Of course, process changes not due to tool breakage (eg, some interrupted cuts) that affect the signal similarly to tool breakage will cause a false reading.

Another method is based upon the comparison of the actual signal with a stored signal. The monitoring system calculates the upper and lower threshold values from the stored signal. In the case of tool breakage, the upper threshold is violated. When the workpiece is missing, the lower threshold is consequently crossed. The disadvantage of this type of monitoring strategy is that a 'teach-in' cycle is necessary. Furthermore, the fact that the signals must be stored means that more system memory must be allocated. These methods have found applicability to both force and AE signal-based monitoring strategies.

These strategies work well for discrete events such as tool breakage but are often more difficult to employ for continuous process changes such as tool wear. The continuous variation of material properties, cutting conditions, etc., can mask wear-related signal features or, at least, limit the range of applicability or require extensive system training. A more successful technique is based on the tracking of parameters that are extracted from signal features that have been filtered to remove process-related variables (eg, cutting speed), eg, using parameters of an auto-regressive model (filter) of the AE signal to track continuous wear. The strategy works over a range of machining conditions.

The combination of different, inexpensive sensors today is ever increasing to overcome shortages of single sensor devices. There are two possible ways to achieve a multi-sensor approach. Either one sensor is used that allows the measurement of different variables or different sensors are attached to the machine tool to gain different variables. The challenge in this is both electronic integration of the sensor and integration of the information and decision making.

1.2.2
Basic Sensor Classification

We now review a basic classification of sensors based upon the principle of operation. Several excellent texts exist that offer detailed descriptions of a range of sensors and these have been summarized in the material below [1–3]. We distinguish here between *a transducer* and a *sensor* even though the terms are often used interchangeably.

A transducer is generally defined as a device that transmits energy from one system to another, often with a change in form of the energy. A good example is a piezoelectric crystal which will output a current or charge when mechanically actuated. A sensor, on the other hand, is a device which is 'sensitive' to (meaning responsive to or otherwise affected by) a physical stimulus (eg, light) and then transmits a resulting impulse for interpretation or control [4]. Clearly there is some overlap as in the case of a piezoelectric actuator (responding to a charge and outputting a motion or force) and a piezoelectric sensor (outputting a charge for a given force or motion input). In one case, the former, the piezo device acts as a transducer and in the other, the latter, as a sensor. The terms can often be used interchangeably without problem in most cases.

A sensor, according to Webster's Dictionary is 'a device that responds to a physical (or chemical) stimulus (such as heat, light, sound, pressure, magnetism, or a particular motion) and transmits a resulting impulse (as for measurement or operating control)'. Sensors are in this way devices which first perceive an input signal and then convert that input signal or energy to another output signal or energy for further use. We generally classify signal outputs into six types:

- mechanical;
- thermal (ie, kinetic energy of atoms and molecules);
- electrical;
- magnetic;
- radiant (including electromagnetic radio waves, micro waves, etc.); and
- chemical.

Sensors now exist, and are in common use, that can be classified as either 'sensors' *on* silicon as well as 'sensors *in* silicon' [1]. We shall discuss the basic characteristics of both types of silicon 'micro-sensors' but introduce some of the unique features of the latter which are becoming more and more utilized in manufacturing. The small size, multi-signal capability, and ease of integration into signal processing and control systems make them extremely practical. In addition, as a result of their relatively low cost, these are expected to be the 'sensors of choice' in the future.

The six types of signal outputs listed above reflect the 10 basic forms of energy that sensors convert from one form to another. These are listed in Table 1.2-1 [3, 5, 6]. In practice, these 10 forms of energy are condensed into the six signal types listed as we can consider atomic and molecular energy as part of chemical energy, gravitational and mechanical as one, mechanical, and we can ignore nuclear and mass energy. The six signal types (hence basic sensor types for our discussion) represent 'measurands' extracted from manufacturing processes that give us insight into the operation of the process. These measurands represent measurable elements of the process and, further, derive from the basic information conversion technique of the sensor. That is, depending on the sensor, we will probably have differing measurands from the process. However, the range of measurands available is obviously closely linked to the type of (operating principle) of the sensor employed. Table 1.2-2, adapted from [7], defines the relevant measurands from a range of sensing technologies. The 'mapping' of these measurand/sensing pairs on to a manufacturing process is the basis of developing a sensing strategy for a process or system. The measurands give us important information on the:

- process (the electrical stability of the process, in electrical discharge machining, for example),
- effects of outputs of the process (surface finish, dimension, for example), and
- state of associated consumables (cutting fluid contamination, lubricants, tooling, for example).

Tab. 1.2-1 Forms of energy converted by sensors

Energy form	Definition
Atomic	Related to the force between nuclei and electrons
Electrical	Electric fields, current, voltage, etc.
Gravitational	Related to the gravitation attraction between a mass and the Earth
Magnetic	Magnetic fields and related effects
Mass	Following relativity theory ($E=mc^2$)
Mechanical	Pertaining to motion, displacement/velocity, force, etc.
Molecular	Binding energy in molecules
Nuclear	Binding energy in electrons
Radiant	Related to electromagnetic radiowaves, microwaves, infrared, visible light, ultraviolet, x-rays and γ-rays
Thermal	Related to the kinetic energy of atoms and molecules

Tab. 1.2-2 Process measurands associated with sensor signal types (after [7])

Signal output type	Associated process measurands
Mechanical (includes acoustic)	Position (linear, angular) Velocity Acceleration Force Stress, pressure Strain Mass, density Moment, torque Flow velocity, rate of transport Shape, roughness, orientation Stiffness, compliance Viscosity Crystallinity, structural integrity Wave amplitude, phase, polarization, spectrum Wave velocity
Electrical	Charge, current Potential, potential difference Electric field (amplitude, phase, polarization, spectrum) Conductivity Permittivity
Magnetic	Magnetic field (amplitude, phase, polarization, spectrum) Magnetic flux Permeability
Chemical (includes biological)	Components (identities, concentrations, states) Biomass (identities, concentrations, states)
Radiation	Type Energy Intensity Emissivity Reflectivity Transmissivity Wave amplitude, phase, polarization, spectrum Wave velocity
Thermal	Temperature Flux Specific heat Thermal conductivity

Finally, there are a number of technical specifications of sensors that must be addressed in assessing the ability of a particular sensor/output combination to measure robustly the state of the process. These specifications relate to the operating characteristics of the sensors and are usually the basis for selecting a particular sensor from a specific vendor, eg [7]:

- ambient operating conditions;
- full-scale output;
- hysteresis;
- linearity;
- measuring range;
- offset;
- operating life;
- output format;
- overload characteristics;
- repeatability;
- resolution;
- selectivity;
- sensitivity;
- response speed (time constant);
- stability/drift.

It is impossible to detail the associated specifications for the six sensing technologies under discussion here. A number of references have done this for specific sensors for manufacturing applications, eg, Shiraishi [8–10] and Allocca and Stuart [2]. Others are referenced elsewhere in this volume or reviewed in [11].

1.2.3
Basic Sensor Types

1.2.3.1 Mechanical Sensors

Mechanical sensors are perhaps the largest and most diverse type of sensors because, as seen in Table 1.2-2, they have the largest set of potential measurands. Force, motion, vibration, torque, flow, pressure, etc., are basic elements of most manufacturing processes and of great interest to measure as an indication of process state or for control. Force is a push or pull on a body that results in motion/displacement or deformation. Force transducers, a basic mechanical sensor, are designed to measure the applied force relative to another part of the machine structure, tooling, or workpiece as a result of the behavior of the process. A number of mechanisms convert this applied force (or torque) into a signal, including piezoelectric crystals, strain gages, and potentiometers (as a linear variable differential transformer (LVDT)). Displacement, as in the motion of an axis of a machine, is measurable by mechanical sensors (again the LVDT or potentiometer) as well as by a host of other sensor types to be discussed. Accelerometer outputs, differentiated twice, can yield a measure of displacement of a mechanism. Shiraishi [9] relies on a number of mechanical sensing elements to measure the dimensions of a workpiece. Flow is commonly measured by 'flow meters', mechanical devices with rotameters (mechanical drag on a float in the fluid stream) as well as venturi meters (relying on differential pressure measurement, using another mechanical sensor) to determine the flow of fluids. An excellent review of other mechanical sensing (and transducing) devices is given in [2].

Mechanical sensors have seen the most advances owing to the developments in semiconductor fabrication technology. Piezo-resistive and capacitance-based devices, basic building blocks of silicon micro-sensors, are now routinely applied to pressure, acceleration, and flow measurements in machinery. Figure 1.2-2a shows the schematics of a capacitive sensor with applications in pressure sensing (the silicon diaphragm deflects under the pressure of the gas/fluid and modifies the capacitance between the diaphragm and another electrode in the device). Using a beam with a mass on the end as one plate of the capacitor and a second electrode (Figure 1.2-2b), an accelerometer is constructed and the oscillation of the mass/beam alters the capacitance in a measurable pattern allowing the determination of the acceleration. Figure 1.2-3 shows a TRW NovaSensor®, a miniature, piezoresistive chip batch fabricated and diced from silicon wafers. These sensor chips can be provided as basic original equipment manufacturer (OEM) sensor elements or can be integrated into a next-level packaging scheme. These devices are con-

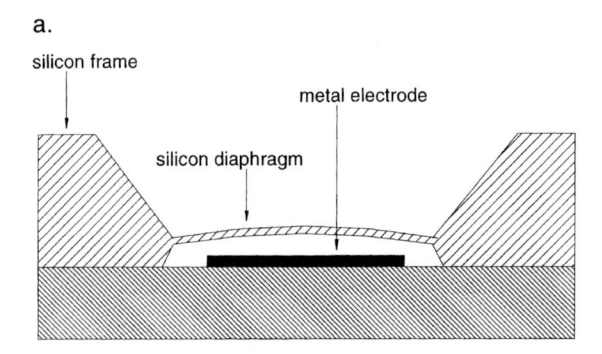

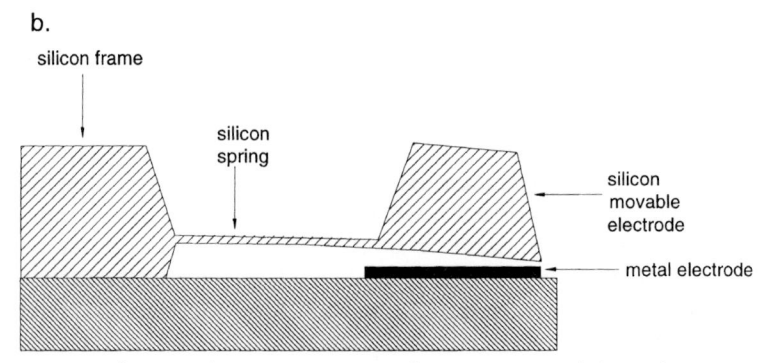

Fig. 1.2-2 Schematic of a capacitance sensor for (a) pressure and (b) acceleration

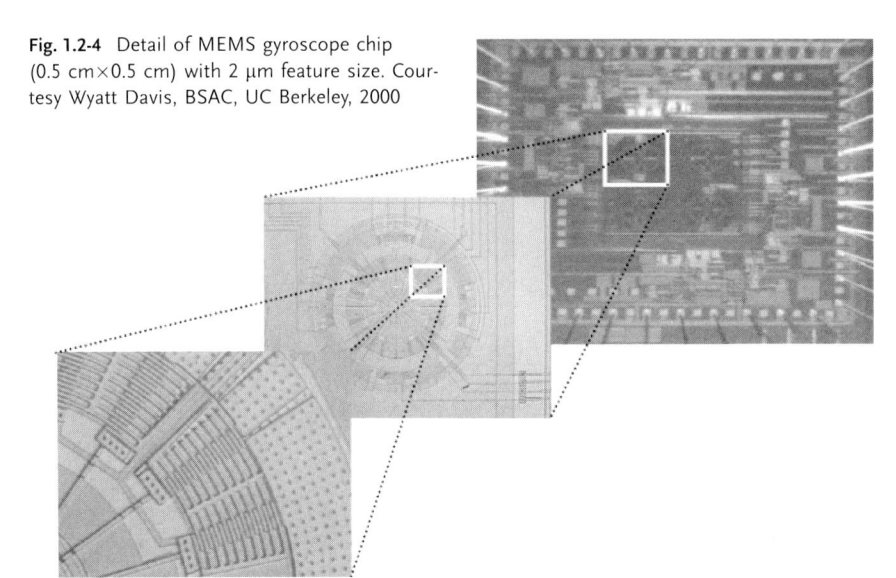

Fig. 1.2-3 Piezoresistive micromachined pressure die. Courtesy of Lucas NovaSensor, 2000

Fig. 1.2-4 Detail of MEMS gyroscope chip (0.5 cm×0.5 cm) with 2 µm feature size. Courtesy Wyatt Davis, BSAC, UC Berkeley, 2000

structed using conventional semiconductor fabrication technologies based on the semiconducting materials and miniaturization of very large scale integrated (VLSI) patterning techniques (see, for example, Sze [1] as an excellent reference on semiconductor sensors). The development of microelectromechanical sensing systems (so-called MEMS) techniques has opened a wide field of design and application of special micro-sensors (mechanical and others) for sophisticated sensing tasks. Figure 1.2-4 shows a MEMS gyroscope fabricated at UC Berkeley BSAC for use in positioning control of shop-floor robotic devices. In fact, most of the six basic sensor types can be accommodated by this technology. Accelerometers are built on these chips as already discussed. Whatever affects the frequency of oscillation of the silicon beam of the sensor can be considered a measurand. Coating the accelerometer beam with a material that absorbs certain chemical elements, hence changing the mass of the beam and its resonant frequency, changes this into a chemical sensor. Similar modifications yield other sensor types.

One particularly interesting type of micro-sensor for pressure applications, not based on the capacitance principles discussed above, is silicon-on-sapphire (SOS). This is specially applicable to pressure-sensing technology. Manufacturing an SOS transducer begins with a sapphire wafer on which silicon is epitaxially grown on the smooth, hard, glass-like surface of the sapphire. Since the crystal structure of the silicon film is similar to sapphire's, the SOS structure appears to be one crystal with a strong molecular bond between the two materials. The silicon is then etched into a Wheatstone bridge pattern using conventional photolithography techniques. Owing to its excellent chemical resistance and mechanical properties, the sapphire wafer itself may be used as the sensing diaphragm. An appropriate diaphragm profile is generated in the wafer to create the desired flexure of the diaphragm and to convey the proper levels of strain to the silicon Wheatstone bridge. The diaphragm may be epoxied or brazed to a sensor package. A more reliable method of utilizing the SOS technology involves placing an SOS wafer on a machined titanium diaphragm. In this configuration titanium becomes the primary load-bearing element and a thin (thickness under 0.01 in) SOS wafer is used as the sensing element. The SOS wafer is bonded to titanium using a process similar to brazing, performed under high mechanical pressure and temperature conditions in vacuum to ensure a solid, stable bond between the SOS wafer and the titanium diaphragm. The superb corrosion resistance of titanium allows compatibility with a wide range of chemicals that may attack epoxies, elastomers, and even certain stainless steels. The titanium diaphragm is machined using conventional machining techniques and the SOS wafer is produced using conventional semiconductor processing techniques. SOS-based pressure sensors with operating pressures ranging from 104 kPa to over 414 MPa are available.

Acoustic sensors have benefited from the developments in micro-sensor technology. Semiconductor acoustic sensors employ elastic waves at frequencies in the range from megahertz to low gigahertz to measure physical and chemical (including biological) quantities. There are a number of basic types of these sensors based upon the mode of flexure of an elastic membrane or bulk material in the sensor is employed. Early sensors of this type used vibrating piezoelectric crystal plates referred to as a quartz crystal microbalance (QCM). It is also called a thickness shear-mode sensor (TSM) after the mode of particle motion employed. Other modes of acoustic wave motion employed in these devices (with appropriate design) include surface acoustic wave (SAW) for waves travelling on the surface of a solid, and elastic flexural plate wave (FPW) for waves travelling in a thin membrane. The cantilever devices described earlier are also in this class.

1.2.3.2 Thermal Sensors

Thermal sensors generally function by transforming thermal energy (or the effects of thermal energy) into a corresponding electrical quantity that can be further processed or transmitted. Other techniques for sensing thermal energy (in the infrared range) are discussed under radiant sensors below. Typically, a non-thermal signal is first transduced into a heat flow, the heat flow is converted into a change in temperature/temperature difference, and, finally, this temperature difference is converted into an electrical signal using a temperature sensor. Microsensors employ thin membranes (floating membrane cantilever beam, for example). There is a large thermal resistance between the tip of the beam and the base of the beam where it is attached to the device rim. Heat dissipated at the tip of the beam will induce a temperature difference in the beam. Thermocouples (based on the thermoelectric Seebeck effect whereby a temperature difference at the junction of two metals creates an electrical voltage) or transistors are employed to sense the temperature difference in the device outputting an electrical signal proportional to the difference. Recent advances in thermal sensor application to the 'near surface zone' of materials for assessing structural damage (referred to as photo-thermal inspection) were reported by Goch et al. [12]. This review also covers other measurement techniques such as micromagnetic.

Thermal sensors are also employed in flow measurement following the well-known principle of cooling of hot objects by the flow of a fluid (boundary layer flow measurement anemometers). They can also be applied in thermal tracing and heat capacity measurements in fluids. All three application areas are suitable for silicon micro-sensor integration.

Thermal sensors have also found applicability traditionally in 'true-rms converters'. Root mean square (rms) converters are used to convert the effective value of an alternating current (AC) voltage or current to its equivalent direct current (DC) value. This is accomplished simply by converting the electrical signal into heat with the assistance of a resistor and measuring the temperature generated.

1.2.3.3 Electrical Sensors

Electrical sensors are intended to determine charge, current, potential, potential difference, electric field (amplitude, phase, polarization, spectrum), conductivity and permittivity and, as such, have some overlap with magnetic sensors. Power measurement, an important measure of the behavior of many manufacturing processes, is also included here. An example of the application of thermal sensors for true rms power measurement was included with the discussion on thermal sensors. The use of current sensors (perhaps employing principles of magnetic sensing technology) is commonplace in machine tool monitoring [11]. Electrical resistance measurement has also been widely employed in tool wear monitoring applications [8]. Most of the discussion on magnetic sensors below is applicable here in consideration of the mechanisms of operation of electrical sensors.

1.2.3.4 Magnetic Sensors

A magnetic sensor converts a magnetic field into an electrical signal. Magnetic sensors are applied directly as magnetometers (measuring magnetic fields) and data reading (as in heads for magnetic data storage devices). They are applied indirectly as a means for detecting nonmagnetic signals (eg, in contactless linear/angular motion or velocity measurement) or as proximity sensors. Most magnetic sensors utilize the Lorenz force producing a current component perpendicular to the magnetic induction vector and original current direction (or a variation in the current proportional to a variation in these elements). There are also Hall effect sensors. The Hall effect is a voltage induced in a semiconductor material as it passes through a magnetic field. Magnetic sensors are useful in nondestructive inspection applications where they can be employed to detect cracks or other flaws in magnetic materials due to the perturbation of the magnetic flux lines by the anomaly. Semiconductor-based magnetic sensors include thin-film magnetic sensors (relying on the magnetoresistance of NiFe thin films), semiconductor magnetic sensors (Hall effect), optoelectronic magnetic sensors which use light as an intermediate signal carrier (based on Faraday rotation of the polarization plane of linearly polarized light due to the Lorenz force on bound electrons in insulators [1]) and superconductor magnetic sensors (a special class).

1.2.3.5 Radiant Sensors

Radiation sensors convert the incident radiant signal energy (measurand) into electrical output signals. The radiant signals are either electromagnetic, neutrons, fast neutrons, fast electrons, or heavy-charge particles [1]. The range of electromagnetic frequencies is immense, spanning from cosmic rays on the high end with frequencies in the 10^{23} Hz range to radio waves in the low tens of thousands of Hz. In manufacturing applications we are most familiar with infrared radiation (10^{11}–10^{14} Hz) as a basis for temperature measurement or flaw/problem detection. Silicon-on-insulator photodiodes and phototransistors based on transistor action are typical micro-sensor radiant devices [1] for use in these ranges.

1.2.3.6 Chemical Sensors

These sensors are becoming particularly more important in manufacturing process monitoring and control. It is important to measure the identities of gases and liquids, concentrations, and states, chemical sensors for worker safety (to insure no exposure to hazardous materials or gases), process control (to monitor, for example, the quality of fluids or gases used in production; this is especially critical in the semiconductor industry which relies on complex process 'recipes' for successful production), and process state (presence or absence of a material, eg, gas or fluid). Chemical sensors have been successfully produced as micro-sensors using semiconductor technologies primarily for the detection of gaseous species. Most of these devices rely on the interaction of chemical species at semiconductor surfaces (adsorption on a layer of material, for example) and then the change caused by the additional mass affecting the performance of the device. This was discussed under mechanical sensors where the change in mass altered the frequency of vibration of a silicon cantilever beam providing a means for measuring the presence or absence of the chemical and some indication of the concentration. Other chemical effects are also employed such as resistance change caused by the chemical presence, the semiconducting oxide powder- pressed pellet (so called Taguchi sensors) and the use of field effect transistors (FETs) as sensitive detectors for some gases and ions. Sze [1] gives a comprehensive review of chemical micro-sensors and the reader is referred to this for details of this complex sensing technology.

1.2.4
New Trends – Signal Processing and Decision Making

1.2.4.1 Background
Human monitoring of manufacturing processes can attribute its success to the ability of the human to distinguish, by nature of the physical senses and experience, the 'significant' information in what is observed from the meaningless. In general, humans are very capable as process monitors because of the high degree of development of sensory abilities, essentially noise-free data (unique memory triggers), parallel processing of information, and the knowledge acquired through training and experience. Limitations are seen when one of the basic human sensor specifications is violated; something happening too fast to see or out of range of hearing or visual sensitivity owing to frequency content. These limitations have always served as some of the justification for the use of sensors. Sensors, of course, are also limited in their ability to yield an output sensitive to an important input. Hence we need to consider the use of signal processing and along with that feature extraction. In most cases the utilization of any signal processing methodology has as its goal one or more of the following: the determination of a suitable 'process model from which the influence of certain process variables can be discerned; the generation of features from sensor data that can be used to determine process state; or the generation of data features so that the change in the performance of the process can be 'tracked. Figure 1.2-5 shows the path from process (and the source of the measurants) through the sensor, extraction of a control signal, and application to process control for both heuristic and quantitative methodologies.

An overview of signal processing and feature extraction is summarized in Rangwala [13] (Figure 1.2-6). The measurement vector extracted from the signal representation from the sensor (basic signal conditioning) is the 'feedstock' for the feature selection process (local conditioning) resulting in a feature vector. The characteristics of the feature vector include signal elements that are sensitive to the parameters of interest in the process. The 'decision-making' process follows. Based on a suitable 'learning' scheme which maps a teaching pattern (ie, process characteristics that we desire to recognize) on to the feature vector, a pattern association is generated. The 'pattern association' contains a matrix of associations between

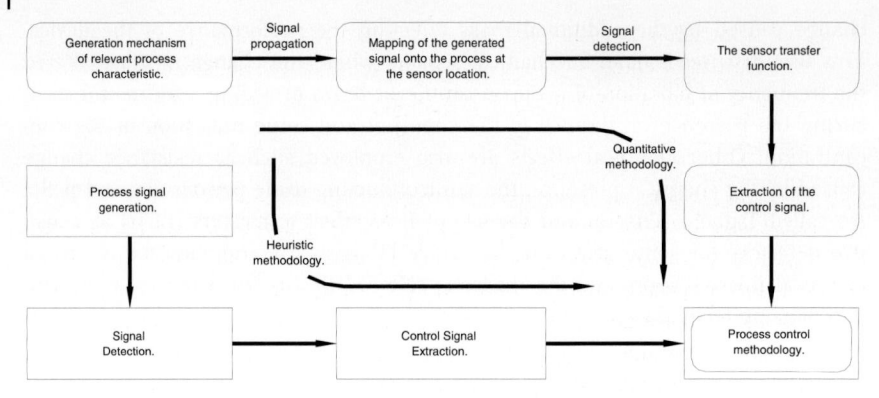

1 Fundamentals

Fig. 1.2-5 Quantitative and heuristic paths for the development of in-process monitoring and control methodologies

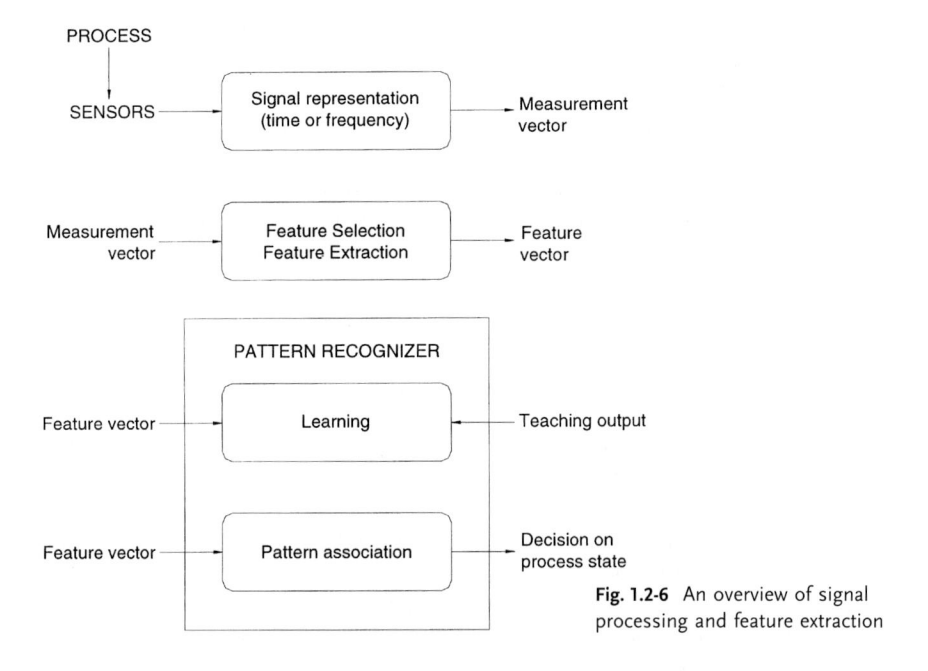

Fig. 1.2-6 An overview of signal processing and feature extraction

the desired characteristics and features of the sensor information. In application, the pattern association matrix operates on the feature vector and extracts correlation between features and characteristics – these are taken to be 'decisions' on the state of the process if the process characteristics are suitably structured (eg, tool worn, weld penetration incomplete, material flawed, etc.). In Figure 1.2-6, the measurement vector is the signal in the upper left corner. The feature vector in this case consists of the mean value shown in the upper right corner. Decision making, based on experience or 'training', sets the threshold at a level corresponding to excessive tool wear. When the feature element 'mean value' crosses the threshold a 'decision' is made that the tool is worn. The success of this strategy depends upon the degree to which the mean value of the sensor output actually represents the state (and progress) of tool wear.

1.2.4.2 Sensor Fusion

With a specific focus for the monitoring in mind, researchers have developed over the years a wide variety of sensors and sensing strategies, each attempting to predict or detect a specific phenomenon during the operation of the process and in the presence of noise and other environmental contaminants. A good number of these sensing techniques applicable to manufacturing have been reviewed in the early part of this chapter. Although able to accomplish the task for a narrow set of conditions, these specific techniques have almost uniformly failed to be reliable enough to work over the range of operating conditions and environments commonly available in manufacturing facilities. Therefore, researchers have begun to look at ways to collect the maximum amount of information about the state of a process from a number of different sensors (each of which is able to provide an output related to the phenomenon of interest although at varying reliability). The strategy of integrating the information from a variety of sensors with the expectation that this will 'increase the accuracy and ... resolve ambiguities in the knowledge about the environment' (Chiu et al. [14]) is called sensor fusion.

Sensor fusion is able to provide data for the decision-making process that has a low uncertainty owing to the inherent randomness or noise in the sensor signals, includes significant features covering a broader range of operating conditions, and accommodates changes in the operating characteristics of the individual sensors (due to calibration, drift, etc.) because of redundancy. In fact, perhaps the most advantageous aspect of sensor fusion is the richness of information available to the signal processing/feature extraction and decision-making methodology employed as part of the sensor system. Sensor fusion is best defined in terms of the 'intelligent' sensor as introduced in [15] since that sensor system is structured to utilize many of the same elements needed for sensor fusion.

The objective of sensor fusion is to increase the reliability of the information so that a decision on the state of the process is reached. This tends to make fusion techniques closely coupled with feature extraction methodologies and pattern recognition techniques. The problem here is to establish the relationship between the measured parameter and the process parameter. There are two principal ways to encode this relationship (Rangwala [13]):

- theoretical – the relationship between a phenomenon and the measured parameters of the process (say tool wear and the process); and
- empirical – experimental data is used to tune parameters of a proposed model.

As mentioned earlier, reliable theoretical models relating sensor output and process characteristics are often difficult to develop because of the complexity and variability of the process and the problems associated with incorporating large numbers of variables in the model. As a result, empirical methods which can use sensor data to tune unknown parameters of a proposed relation are very attractive. These types of approaches can be implemented by either (a) proposing a relationship between a particular process characteristic and sensor outputs and then using experimental data to tune unknown parameters of a model, or (b) associating patterns of sensor data with an appropriate decision on the process state without consideration of any model relating sensor data to the state. The second approach is generally referred to as pattern recognition and involves three critical stages (Ahmed and Rao [16]):

- sampling of input signal to acquire the measurement vector;
- feature selection and extraction;
- classification in the feature space to permit a decision on the process state.

The pattern recognition approach provides a framework for machine learning and knowledge synthesis in a manufacturing environment by observation of sensor data and with minimal human intervention. More important, such an approach allows for integration of information from multiple sources (such as different sensors) which is our principal interest here.

Sata et al. [17, 18] were among the first researchers to propose the application of pattern recognition techniques to machine process monitoring. They attempted to recognize chip breakage, formation of built-up edge and the presence of chatter in a turning operation using the features of the spectrum of the cutting force in the 0–150 Hz range. Dornfeld and Pan [19] used the event rate of the rms energy of an acoustic emission signal along with feed rate and cutting velocity in order to provide a decision on the chip formation produced during a turning operation. Emel and Kannatey-Asibu [20] used spectral features of the acoustic emission signal in order to classify fresh and worn cutting tools. Balakrishnan et al. [21] use a linear discriminant function technique to combine cutting force and acoustic emission information for cutting tool monitoring.

The manufacturing process may be monitored by a variety of sensors and, typically, the sensor output is a digitized time-domain waveform. The signal can then be either processed in the time domain (eg, extract the time series parameters of the signal) or in the frequency domain (power spectrum representation). The effect of this is to convert the original time-domain record into a measurement vector. In most cases, this mapping does not preserve information in the original signal. Usually, the dimension of the measurement vector is very high and it becomes necessary to reduce this dimension due to computational considerations. There are two prevalent approaches at this stage: select only those components of the measurement vector which maximize the signal-to-noise ratio or map the measurement vector into a lower dimensional space through a suitable transformation (feature extraction). The outcome of the feature selection/extraction stage is a lower dimensional feature vector. These features are used in pattern recognition techniques and as inputs to sensor fusion methodologies. This was illustrated in Figure 1.2-6.

1.2.5
Summary

The subject of sensors for manufacturing processes is well covered in other chapters of this book. The material in this chapter serves to acquaint the reader with the classification of sensor systems and some of the measurands that are associated with these sensors. How these sensor types and measurands map on to the various manufacturing processes will be the subject of the rest of the text. One important factor in the implementation of sensors in manufacturing is clearly the rapid growth of silicon micro-sensors based on MEMS technology. This technology already allows the integration of traditional and novel new sensing methodologies on to miniaturized platforms, providing in hardware the reality of multi-sensor systems. Further, since these sensors are easily integrated with the electronics for signal processing and data handling, on the same chip, sophisticated signal analysis including feature extraction and intelligent processing will be straightforward (and inexpensive). This bodes well for the vision of the intelligent factory with rapid feedback of vital information to all levels of the operation from machine control to process planning.

1.2.6
References

1 SZE, S.M. (ed.) *Semiconductor Sensors*; New York: Wiley, 1994.
2 ALLOCCA, J.A., STUART, A., *Transducers: Theory and Applications*; Reston, VA: Reston Publishing, 1984.
3 BRAY, D.E., McBRIDE, D. (eds.) *Nondestructive Testing Techniques*; New York: Wiley, 1992.
4 *Webster's Third New International Dictionary*; Springfield, MA: G.C. Merriam, 1971.
5 USHER, M.J., *Sensors and Transducers*; New Hampshire: Macmillan, 1985.
6 MIDDLEHOEK, S., AUDET, S.A., *Silicon Sensors*; New York: Academic Press, 1989.
7 White, R.M., *IEEE Trans. Ultrason. Feroelect. Freq. Contr.* **UFFC-34** (1987) 124.
8 SHIRAISHI, M., *Precision Eng.* **10(4)** (1988) 179–189.
9 SHIRAISHI, M., *Precision Eng.* **11(1)** (1989) 27–37.
10 SHIRAISHI, M., *Precision Eng.* **11(1)** (1989) 39–47.
11 BYRNE, G., DORNFELD, D., INASAKI, I., KETTLER, G., KÖNIG, W., TETI, R., *Ann. CIRP* **44(2)** (1995) 541–567.
12 GOCH, G., SCHMITZ, B., KARPUSCHEWSKI, B., GEERKINS, J., REIGEL, M., SPRONGL, P., RITTER, R., *Precision Eng.* **23** (1999) 9–33.
13 RANGWALA, S., *PhD Thesis*; Department of Mechanical Engineering, University of California, Berkeley, CA, 1988.
14 CHIU, S.L., MORLEY, D.J., MARTIN, J.F., in: *Proceedings of 1987 IEEE International Conference on Robotics and Automation*; Raleigh, NC: IEEE, 1987, pp. 1629–1633.
15 DORNFELD, D.A., *Ann. CIRP* **39** (1990)
16 AHMED, N., RAO, K.K., *Orthogonal Transforms for Digital Signal Processing*; New York: Springer, 1975.
17 SATA, T., MATSUSHIMA, K., NAGAKURA, T., KONO, E., *Ann. CIRP* **22(1)** (1973) 41–42.
18 MATSUSHIMA, K., SATA, T., *J. Fac. Eng. Univ. Tokyo (B)* **35(3)** (1980) 395–405.
19 DORNFELD, D.A., PAN, C.S., in: *Proceedings of 13th North American Manufacturing Research Conference*, University of California, Berkeley, CA: SME, 1985, pp. 285–303.
20 EMEL, E., KANNATEY-ASIBU, E., JR., in: *Proceedings of 14th North American Manufacturing Research Conference*, University of Minnesota, MN: SME, 1986, pp. 266–272.
21 BALAKRISHNAN, P., TRABELSI, H., KANNATEY-ASIBU, JR., E., EMEL, E., in: *Proceedings of 15th NSF Conference on Production Research and Technology*, University of California, Berkeley, CA: SME, 1989, pp. 101–108.

1.3
Sensors in Mechanical Manufacturing –
Requirements, Demands, Boundary Conditions, Signal Processing, Communication Techniques, and Man-Machine Interfaces

T. MORIWAKI, Kobe University, Kobe, Japan

1.3.1
Introduction

The role of sensor systems for mechanical manufacturing is generally composed of sensing, transformation/conversion, signal processing, and decision making, as shown in Figure 1.3-1. The output of the sensor system is either given to the operator via a human-machine interface or directly utilized to control the machine. Objectives, requirements, demands, boundary conditions, signal processing, communication techniques, and the human-machine interface of the sensor system are described in this section.

1.3.2
Role of Sensors and Objectives of Sensing

An automated manufacturing system, in particular a machining system, such as a cutting or grinding system, is basically composed of controller, machine tool and machining process, as illustrated schematically in Figure 1.3-2. The machining command is transformed into the control command of the actuators by the CNC

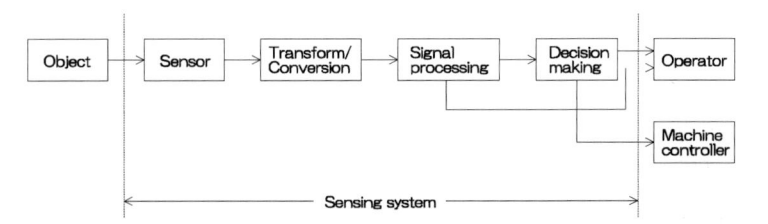

Fig. 1.3-1 Basic composition of sensor system for mechanical manufacturing

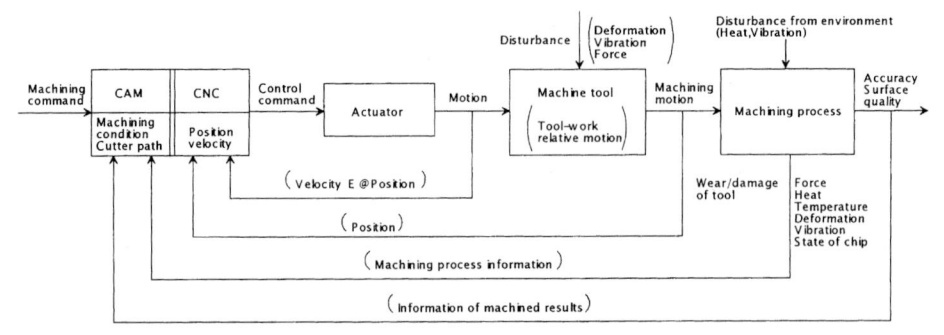

Fig. 1.3-2 Role of sensors in automated machining system

controller, which controls the motion of the actuators and generates the actual machining motion of the machine tool. The motion of the actuator, or the machining motion of the machine tool, is fed back to the controller so as to ensure that the relative motion between the tool and the work follows exactly the predetermined command motion. Motion sensors, such as an encoder, tacho-generator or linear scale, are generally employed for this purpose.

The machining process is generally carried out beyond this loop, where finished surfaces of the work are actually generated. Most conventional CNC machine tools currently available on the market are operated under the assumption that the machining process normally takes place once the tool work-relative motion is correctly given. Some advanced machine tools equipped with an AC (adaptive control) function utilize the feedback information of the machining process, such as the cutting force, to optimize the machining conditions or to stop the machine tool in case of an abnormal state such as tool breakage.

The machining process normally takes place under extreme conditions, such as high stress, high strain rate, and high temperature. Further, the machining process and the machine tool itself are exposed to various kinds of external disturbances including heat, vibration, and deformation. In order to keep the machining process normal and to guarantee the accuracy and quality of the work, it is necessary to monitor the machining process and control the machine tool based on the sensed information.

The objectives and the items to be sensed and monitored for general mechanical manufacturing are summarized in Table 1.3-1 together with the direct purposes of sensing and monitoring. Some items can be directly sensed with proper sensors, but they can be utilized to estimate other properties at the same time. For instance, the cutting force is sensed with a tool dynamometer to monitor the cutting state, but its information can be utilized to estimate the wear of the cutting tool simultaneously.

Almost all kinds of machining processes require sensing and monitoring to maintain high reliability of machining and to avoid abnormal states. Table 1.3-2 gives a summary of the answers to a questionnaire to machine tool users asking about the machining processes which require monitoring [1]. It is understood that monitoring is imperative especially when weak tools are used, such as in tapping, drilling, and end milling.

Tab. 1.3-1 Objects, items, and purposes of sensing

Object of sensing and monitoring	Items to be sensed	Purpose of sensing and monitoring
Work	State of work clamping Geometrical and dimensional accuracy Surface roughness Surface quality	Maintain high quality Avoid damage and loss of work
Machining process	Force (torque, thrust) Heat generation Temperature Vibration Noise and sound State of chip	Maintain normal machining process Predict and avoid abnormal state
Tool	Tool edge position Wear Damage including chipping, breakage, and others	Manage tool changing time, including dressing Avoid damage or deterioration of work
Machine tool, and auxiliary facility	Malfunction Vibration Deformation (elastic, thermal)	Maintain normal condition of machine tool and assure high accuracy
Environment	Ambient temperature change External vibration Condition of cutting fluid	Minimize environmental effect

Tab. 1.3-2 Machining processes which require sensing

Kind of machining	Number of answers	Percentage
Tapping	67	19.8
Drilling	66	19.2
End milling	55	16.8
Internal turning	51	15.1
External turning	30	8.9
Face milling	25	7.4
Parting	17	5.0
Thread cutting	13	3.9
Others*	15	4.4
Total	338	100

* Grinding, reaming, deep hole boring, etc.

1.3.3
Requirements for Sensors and Sensing Systems

The most important and basic part of the sensor is the transducer, which transforms the physical or sometimes chemical properties of the object into another physical quantity such as electric voltage that is easily processed. The properties of the object to be sensed are either one-dimensional, such as force and temperature, or multi-dimensional, such as image and distribution of the physical properties. The multi-dimensional properties are treated either as plural signals or a time series of signals after scanning.

The basic requirements for the transducers and sensor systems for mechanical manufacturing are summarized in Table 1.3-3. Figure 1.3-3 shows a schematic illustration of the characteristics of a typical transducer, such as a force transducer.

Tab. 1.3-3 Basic requirements for transducers and sensing systems

Performance/ accuracy	Reliability	Adaptability	Economy
Sensitivity	Low drift	Compact in size	Low cost
Resolution	Thermal stability	Light in weight	Easy to manufacture
Exactness	Stability against	Easy operation	Easy to purchase
Precision	environment, such as	Easy to be installed	Low power requirement
Linearity	cutting, fluid and heat	Low effect of ma-	Easy to calibrate
Hysteresis	Low deterioration	chining process	Easy maintenance
Repeatability	Long life	and machine tool	
Signal-to-noise ratio	Fail safe	Safety	
Dynamic range	Low emission of noise	Good connectivity to	
Dynamic response		other equipment	
Frequency response			
Cross talk			

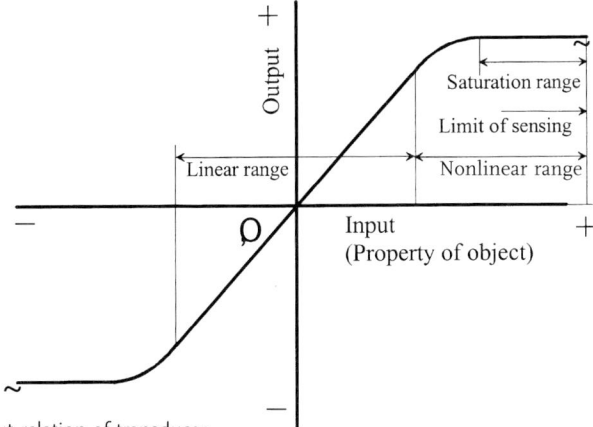

Fig. 1.3-3 Typical input-output relation of transducer

The figure represents the relation between the change in a property of the object, or the input and the output of the transducer. It is desirable that the transducer output represents the property of the object as exactly and precisely as possible. It is also essential for a transducer to output the same value at any time when the same amount of input is given. This characteristic is called repeatability. In most cases, the output increases or decreases in proportion to the input in the linear range, and then gradually saturates and becomes almost constant. When the amount of input exceeds the limit of sensing, the transducer becomes normally malfunctioning. The measurable range of the input is called the dynamic range of the sensor.

The ratio of output to input is called the sensitivity, and it is desirable that the sensitivity is high and the linear range of sensing is wide. The input-output relation is sometimes nonlinear depending on the principle of the transducer, as in the case of capacitive type proximeter (see Figure 1.3-4). Only a small range of linear input-output relation can be used in such a case when the accuracy requirement of sensing is high. When the nonlinear input-output relation is known exactly by calibration or by other methods in advance, the nonlinearity can be compensated afterwards by calculation. The nonlinear characteristics of thermocouples are well known, and the compensation circuits are installed in most thermometers for different types of thermocouples.

The input-output relation sometimes differs when the amount of input is increased and decreased, as shown in Figure 1.3-5. Such a characteristic is called hysteresis, and is sometimes encountered when a strain gage sensor is employed to measure the strain or the force. It is almost impossible to compensate for the hysteresis of the transducer, hence it is recommended to select transducers with small hysteresis.

The property of the object to be sensed in mechanical manufacturing is generally time varying or dynamic. The measurable dynamic range of the transducer is generally limited by the maximum velocity and acceleration of the output signal

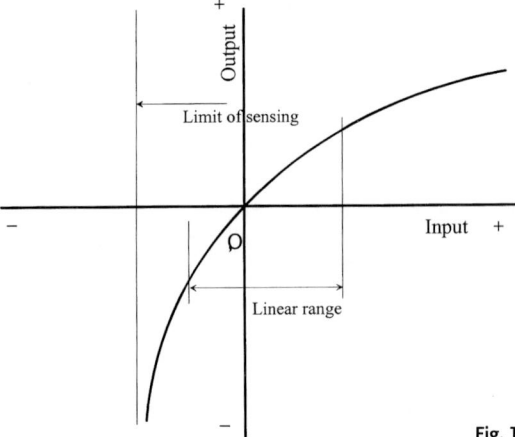

Fig. 1.3-4 Nonlinear input-output relation

Fig. 1.3-5 Hysteresis in input-output relation

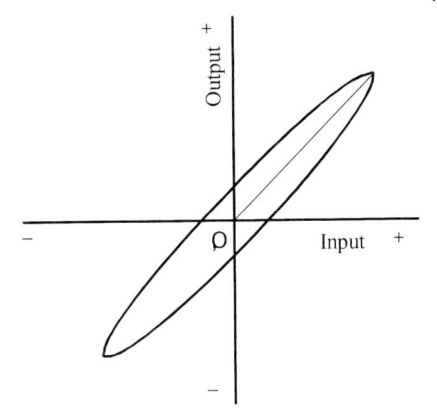

Fig. 1.3-6 Frequency response of typical transducers

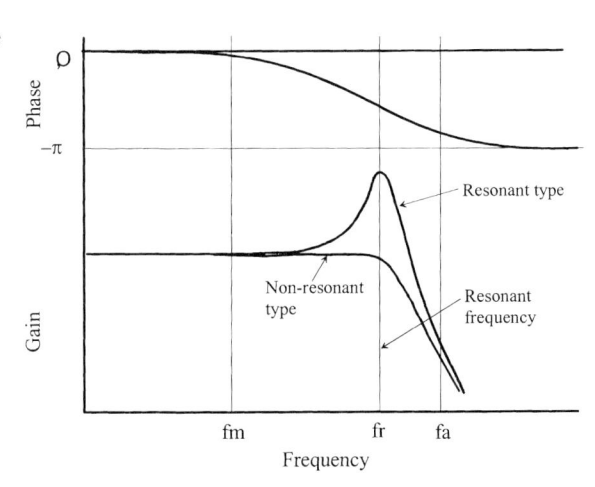

and also by the maximum frequency to which the change in the input property can be exactly transformed to the output. Figure 1.3-6 shows typical frequency characteristics of the transducers in terms of the frequency response. The vertical axis shows the gain or the ratio of the magnitudes of the output to the input, and also the phase or the delay of the output signal to the input.

Some transducers show resonance characteristics, and the gain in terms of output/input becomes relatively larger at the resonant frequency. It should be noted that the phase is shifted for about $\lambda/2$ at the resonant frequency. The phase shift in the output signal cannot be avoided generally even with well-damped type or non-resonant type transducers, as shown in the figure.

The sinusoidal wave forms of the input and the output at some typical frequencies are shown in Figure 1.3-7 to illustrate the changes in the gain and the phase. When the phase information is essential to identify the state of the object, it is important to select a transducer with resonant frequency high enough compared with the frequency range of the phenomenon to be sensed.

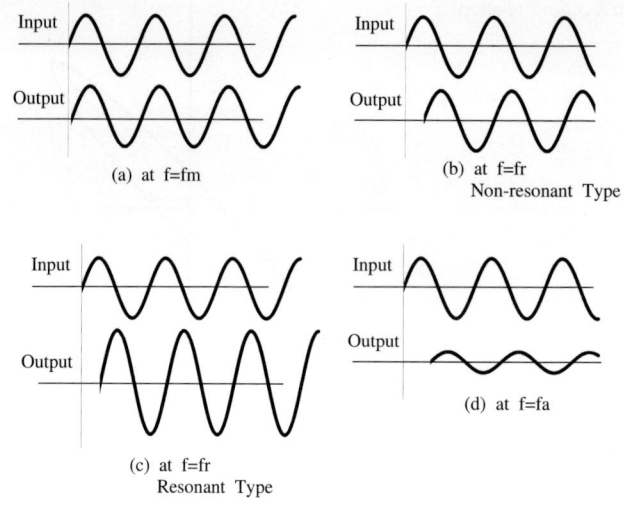

Fig. 1.3-7 Relation of input and output at some typical frequencies

(a) at f=fm
(b) at f=fr Non-resonant Type
(c) at f=fr Resonant Type
(d) at f=fa

As was mentioned before, the machining process normally takes place under high-stress, high-strain rate and high-temperature conditions with various kinds of external disturbances including the cutting and grinding fluids. It is therefore understood that high reliability and stability against various kinds of disturbances are the most important requirements for the sensors in addition to the basic performance and accuracy of the transducers. According to the answers given by industry engineers to the questionnaire concerning tool condition monitoring [2], the importance of technical criteria in selecting the sensors is in the order (1) reliability against malfunctioning, (2) reliability in signal transmission, (3) ease of installation, (4) life of the sensor, and (5) wear resistance of the sensor.

The importance of items in evaluating the monitoring system is also given in the order (1) reliability against malfunctions, (2) performance to cost ratio, (3) information obtained by the sensor, (4) speed of diagnosis, (5) adaptability to changes of process, (6) usable period, (7) ease of maintenance and repair, (8) level of automation, (9) ease of installation, (10) standard interface, (11) standardized user interface, (12) completeness of manuals, and (13) possibility of additional functions.

Table 1.3-4 summarizes items to be considered generally in selecting transducers and the sensors. It is basically desirable to implement on-line, in-process, continuous, non-contact, and direct sensing, but it is generally difficult to satisfy all of these requirements. The property of the object is directly sensed in the case of direct sensing, whereas in the case of indirect sensing it is estimated indirectly from other properties which can be easily measured and are related to the property to be measured. It should be noted that the property of object to be estimated indirectly must have a good correlation with the property to be measured. Indirect sensing is useful and is widely adopted when direct sensing is difficult.

Tab. 1.3-4 Items to be considered in selecting sensors

In-process sensing; between-process sensing; post-process sensing
On-line sensing; on-machine sensing; off-line sensing
Continuous sensing; intermittent sensing
Direct sensing; indirect sensing
Active sensing; passive sensing
Non-contact sensing; contact sensing
Proximity sensing; remote sensing
Single sensor; multi-sensor
Multi-functional sensor; single-purpose sensor

A typical indirect sensing is to estimate the wear and damage of a tool by sensing the cutting and grinding forces, the cutting temperature, the vibration, or the sound emitted. The wear and damage of the tool have a good correlation with those properties mentioned above, but they are also dependent on other conditions, such as the cutting and grinding conditions including the speed, the depth of cut and the feed, the cutting and grinding modes, the tool materials, etc. It is therefore necessary to have a good understanding of the correlation among the properties and the influencing factors.

1.3.4
Boundary Conditions

Sensing of the state of the machining process, the tool, the work, and the machine tool is not easy and it is restricted by many factors, as was mentioned earlier. Difficulties encountered in sensing, which are boundary and restrictive conditions for sensing, and their typical examples are summarized in Table 1.3-5. The most important requirements for sensing are to obtain the necessary information as accurately as possible under unfavorable conditions without disturbing the machining process, which normally takes place under high stress, high strain rate and high temperature.

It is always desirable to sense the properties of the object directly in-process and on-line, which is not generally easy to realize. When the cutting/grinding temperature and the acoustic emission (AE) signal are sensed, the sensors are normally attached apart from the cutting/grinding region, and hence the quality of necessary information deteriorates while the heat and the ultrasonic vibration are transmitted. It is more difficult to sense such signals when the transmission path is discontinuous, such as in the case of a rotating spindle or moving table. Fluid coupling is employed in the case of ultrasonic vibration.

The signal transmission is still difficult when the transducers are located on the rotating spindle or the moving table, even after the signals to be transmitted are converted to an electric signal by the transducers. The slip ring, wireless transmission with use of radio waves and the optical methods are commonly employed in such cases.

Tab. 1.3-5 Difficulties in sensing and examples

Items of difficulty	Example
In-process/on-line sensing is difficult	Geometrical and dimensional accuracy of work Surface roughness and quality of work Wear and damage of tool Thermal deformation of machine
Direct sensing is difficult	Tool wear and damage in continuous cutting Thermal deformation of machine
Distance between object and sensing position is large	Cutting/grinding point versus position where sensors can be placed
Installation of sensor should not affect machining process and rigidity of machining system	Reduction of rigidity of tool or machine elements to measure force by strain
Environment is not clean	Existence of cutting fluid Electrical noise due to power circuit
Signal is to be transmitted via rotating or moving element	Signal transmission from rotating spindle or fast-moving table Signal transmission via rotatable tool turret
Complicated correlation exists among many factors	Property of object to be sensed are affected by machining conditions, tool material, work material, etc.
Variety of machining method is large	Sensors are required to be effective for different machining methods, such as tapping, drilling, end milling, face milling, etc. on one machine

Another difficulty is that the sensors and the sensing systems are generally required to sense the properties of objects even though the combinations of the cutting/grinding methods, the machining conditions, the tool material, the work material, and even the machine itself are altered. In this sense, versatility is important for the sensors and the sensing systems.

1.3.5
Signal Processing and Conversion

1.3.5.1 Analog Signal Processing

The property of the object to be sensed is transformed into voltage, current, electrical charge, or other signal by the transducer. The signals other than the voltage signal are generally further transformed into a voltage signal which is easier to handle. The analog voltage signal is generally filtered to eliminate unnecessary frequency components and amplified prior to the digitization in order to be processed by computer.

There are basically two types of analog filters, the low-pass filter and the high-pass filter. The low-pass filter passes the signal containing the frequency components below the predetermined frequency, named the cut-off frequency, and prohibits the signal containing the frequency components above the cut-off frequency. The low-pass filter is commonly used when the high-frequency noise components, especially the electric noise components, are to be eliminated.

The high-pass filter passes the signal containing the frequency components above the cut-off frequency and prohibits the signal containing the frequency components below the cut-off frequency. The high-pass filter is commonly used when the AC (alternating current) components of the signal are utilized and the DC (direct current) components and the low-frequency components are eliminated. In other words, it is used when the dynamic components of the signal are utilized and the static or the low-frequency components are eliminated.

The combination of the low-pass and the high-pass filters constitutes the band-pass filter and the band-reject filter. The band-pass filter passes only the signal containing the frequency components within the specified frequency range, whereas the band-reject filter prohibits the signal containing the frequency components of that frequency range.

The band-pass filter is commonly used when the signal components of a particular frequency range are utilized, such as in the case when the signal components synchronizing to the rotational frequency of the spindle or the engagement of the milling cutter are to be monitored. The band-reject filter is used when the signal components of a particular frequency range are to be omitted.

The frequency characteristics of the filters are shown schematically in Figure 1.3-8 in terms of the output/input ratio. It should also be noted that the phase information is distorted when the signal is passed through the filters as shown in Figure 1.3-6.

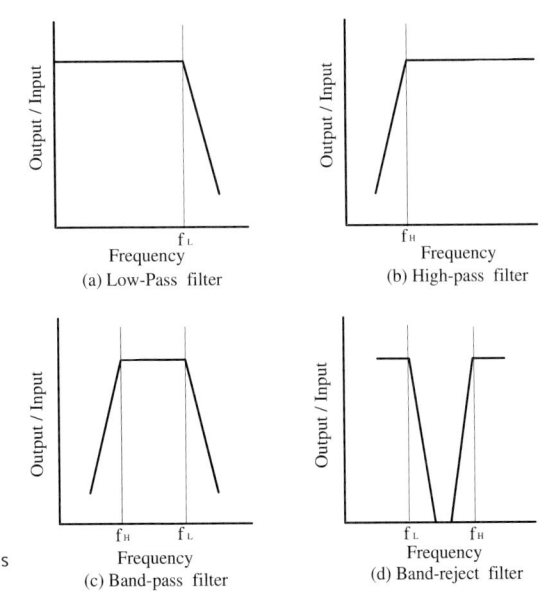

Fig. 1.3-8 Frequency characteristics of filters

Tab. 1.3-6 Typical processing and transformation of analog signal

---
Filtering (low-pass, high-pass, band-pass, band-reject)
Amplification
Differentiation
Integration
Logarithmic transformation

Tab. 1.3-7 Important parameters in AD conversion

Range of analog signal input
Number of digit (or resolution)
Sampling time Δt
Total number of sampled data M
Maximum frequency $f_{max}=1/2\Delta t$
Frequency resolution $\Delta f=1/M\Delta t$

The other transformation and processing of analog signals include the differentiation, integration, and logarithmic transformation, which are summarized in Table 1.3-6. The displacement signal can be transformed to a velocity signal by differentiation, and further to an acceleration signal, and vice versa. These signal transformations are often carried out after the signal is converted to a digital signal, which is explained below.

1.3.5.2 AD Conversion

The analog time series of electric signals is generally converted into digital values by the AD (analog-to-digital) converter prior to processing by computer. The important parameters of the AD converter are the input range, the number of digits of conversion, the sampling time, and the total number of sampled data (Table 1.3-7). The AD converter equally divides the voltage of the input range into the given digits and gives the corresponding number to the input voltage at a given sampling interval Δt. Comparison of the original analog signal and digitized samples is illustrated schematically in Figure 1.3-9.

When the input range of an 8-bit AD converter is ±1 V, the signal from +1 V to −1 V is converted to digital numbers from +127 to −127. This means that the electric signal is digitized with a resolution of 7.9 mV, or 1/127 V. The signal of 0.1 V is converted to 13, 0.5 V to 64, and so forth. The commonly used digits other than 8 bits are 10 bits (±511), 12 bits (±2047) and 16 bits (±8191). The AD conversion is always associated with the digitization error, but it can be ignored in practice if the number of digits is chosen to be high enough.

It is easily understood that the resolution of AD conversion is better if the number of digits is larger. However, it is useless to increase the resolution beyond the noise level of the original analog signal. The input signal is to be properly amplified prior to the AD conversion in such a way that the maximum voltage expected matches the input range of the AD converter.

Fig. 1.3-9 Schematic illustration of AD conversion

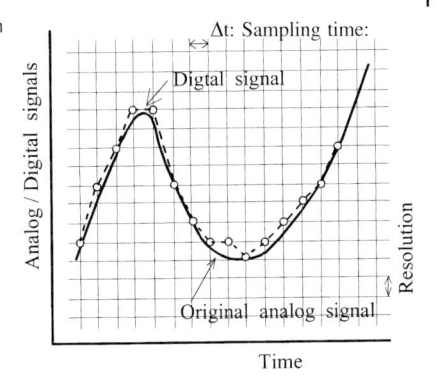

Fig. 1.3-10 Example of low sampling rate

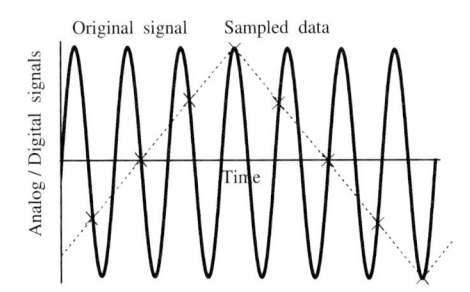

The sampling time Δt gives the time interval of successive AD conversion. A sampling time of 1 ms means that the signal is converted at a sampling rate of 1000 samples per second, or a sampling frequency of 1 kHz. If the sampling time is shorter or the sampling frequency is higher, the original signal can be better represented in a digital form, but the total number of digital data M for a given time period becomes larger and may require a longer processing time.

The sampling time Δt gives the upper limit frequency f_{max} of the digitized signal to be analyzed, or

$$f_{max} = 1/(2\,\Delta t) \tag{1.3-1}$$

This means that the frequency range of the digitized signal is limited below $1/(2\,\Delta t)$ Hz, and the frequency components of the original analog signal beyond this frequency are included in the frequency components of the digitized signal which is lower than f_{max}. This is called Shannon's sampling theorem.

An example of the case of a low sampling rate as compared with the frequency component of the original analog signal is depicted in Figure 1.3-10. It is understood that an original sinusoidal analog signal sampled at a sampling frequency lower than its frequency is represented as a low-frequency signal in digitized form. The signal components with frequencies beyond f_{max} are thus represented as components at lower frequencies in digital form. This phenomenon is called aliasing or folding.

In order to avoid such problems, an analog low-pass filter is generally employed prior to AD conversion, the cut-off frequency of which is matched to the sampling time. Another method is to employ digital filtering, which is a digital calculation equivalent to analog filtering. The original analog signal is sampled at a sampling frequency high enough to avoid folding, processed by the digital processor to eliminate the high-frequency components and then sampled again at a predetermined sampling frequency which is much lower than the original sampling frequency.

When two or more analog signals are to be digitized simultaneously, it is important that the signal of each channel must be sampled at the same time without any delay. This is realized either by employing several AD converters operated in synchronization, or employing the sample and hold circuits, which practically freezes the levels of the analog signals while the single AD converter scans all the analog signals and converts them into digital data.

1.3.5.3 Digital Signal Processing

Once the sensor signal has been converted into digital data, the latter are processed in many ways to extract the features and to give the basis for the identification and the decision making in the following process. Most of the signal data coming from the sensor are time series data, and they are primarily processed in the time domain or in the frequency domain after Fourier transformation. The multi-dimensional data, such as the image data, are treated as they are, or some distinctive features extracted from the image are utilized. Some typical methods of signal processing are summarized in Table 1.3-8. The wavelet transform is a

Tab. 1.3-8 Typical signal processing methods and distinctive values

Domain of signal processing	Method of signal processing	Distinctive value
Time domain	Selection of distinctive feature	Peak value
	Time series analysis	Rms value
	Correlation analysis	Differentiated value
		Integrated value
		Duration
		Filtered value
		Moving average
		Frequency
		Accumulated frequency
		Auto-correlation
		Cross-correlation
		Difference in arrival time
Frequency domain	DFT (digital Fourier transform)	Band power
		Power spectrum
		Cross spectrum
		Cepstrum
		Phase (difference)
Others	Wavelet transform	Wavelet
	Image processing	Pattern (image data)

new method which deals with the changes in the frequency characteristics of the signal. Some typical signal processing methods are explained below.

Let the digitized time series data of analog signal $x(t)$ be represented as $x(i)$, where i is an integer and

$$t = i\Delta t \tag{1.3-2}$$

The moving average $MA(i)$ of $x(i)$ is given by

$$MA(i) = \frac{1}{K}\sum_{j=0}^{K-1} a(j)x(i-j) \tag{1.3-3}$$

where $a(j)$ are coefficients normally chosen to be 1. The range of integration is sometimes chosen to be from $j=-K$ to $j=K$.

The algorithm of digital filtering mentioned above is practically the same as Equation (1.3-3). The function of the filter can be low-pass or high-pass depending on the coefficients of $a(j)$.

For a given set of time series data of $x(i)$ ($i=0, 1, 2, \ldots, M-1$), the auto-correlation function of $x(i)$ is given by

$$C_{xx}(k) = \frac{1}{M}\sum_{i=0}^{M-k-1} x(k+i)x(i) \quad (k = 0, 1, \ldots, h) \tag{1.3-4}$$

The cross-correlation function between $x(i)$ and $y(i)$ is given in the same way by

$$C_{xy}(k) = \frac{1}{M}\sum_{i=0}^{M-k-1} x(k+i)y(i) \quad (k = 0, 1, \ldots, h) \tag{1.3-5}$$

$$C_{xy}(k) = \frac{1}{M}\sum_{i=-k}^{M-1} x(k+i)y(i) \quad (k = -1, \ldots, -h) \tag{1.3-6}$$

The Fourier transform of $x(i)$ is given by

$$X(j2\pi k/M\Delta t) = \sum_{i=0}^{M-1} x(i)\exp(-j2\pi ki/M) \tag{1.3-7}$$

where $k = 0, 1, 2, \ldots, M/2$.

The discrete spectrum $X(j2\pi k/M\Delta t)$ is given at discrete frequencies $f=k/M\Delta t$. This means that the frequency resolution is given by dividing the maximum frequency f_{max} by $M/2$, as was shown in Equation (1.3-1) the maximum frequency is determined by the sampling time Δt and is given by $1/2\Delta t$. The frequency resolution Δf of the digitized data is then given by

$$\Delta f = 1/M\Delta t = 1/T \tag{1.3-8}$$

where T is the observation period of the signal.

In order to improve the frequency resolution and make Δf small, it is necessary to increase the number M or the observation period of the signal or to increase the sampling time Δt. The selection of sampling time Δt is restricted by the upper limit frequency or the maximum frequency, as explained before.

The Fourier spectrum $X(j2\pi k/M\Delta t)$ is a complex number, and it is divided into the real and the imaginary parts as

$$\mathrm{Re}(X) = A_k = \sum_{i=0}^{M-1} x(i) \cos(2\pi ki/M) \quad (k = 0, 1, \ldots, M/2) \quad (1.3\text{-}9)$$

$$\mathrm{Im}(X) = B_k = \sum_{i=0}^{M-1} x(i) \sin(2\pi ki/M) \quad (k = 1, 2, \ldots, M/2) \quad (1.3\text{-}10)$$

The relation between the original time series and the Fourier transform is shown schematically in Figure 1.3-11. The power spectrum P_k at a frequency $f=k/M\Delta t$ is given by

$$P_k = (A_k^2 + B_k^2)^{1/2} \quad (1.3\text{-}11)$$

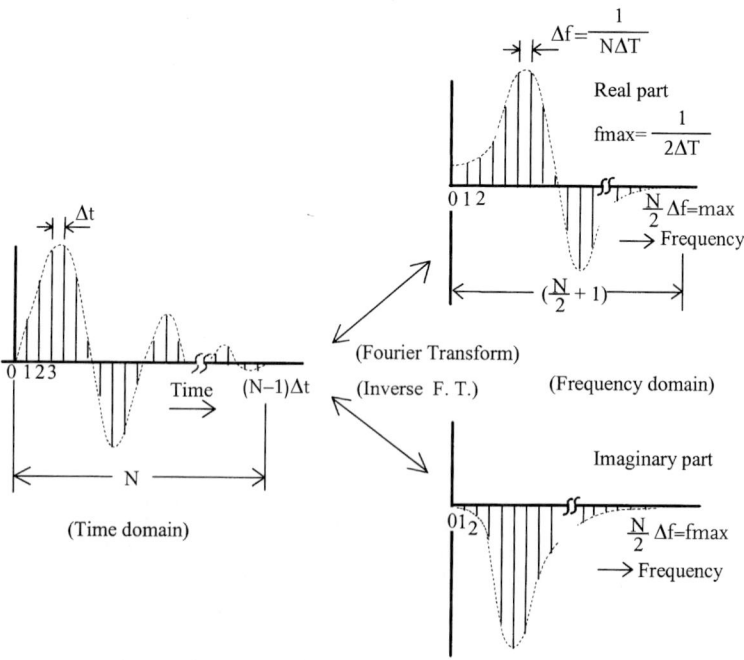

Fig. 1.3-11 Relation of time series data and its Fourier spectra

1.3.6
Identification and Decision Making

1.3.6.1 Strategy of Identification and Decision Making

The digitized sensor signals are used to extract their features, identify the state of the machining process and the conditions of the tool, the work, the machine, etc., and then make decisions to take necessary actions when it is necessary.

Figure 1.3-12 shows typical input-output relations between the input sensor signal and the output which is the status identified. In most cases, a single input signal is utilized to identify the specific state of the system, such as the condition of the tool as shown in case (a). Some sensor signals, such as the vibration signal or the force signal, contain information of various kinds of status, such as the tool wear, the chatter vibration, etc., and hence are utilized to identify those conditions as in case (b).

In order to increase the reliability of identification under varying conditions or to avoid the uncertainty in the identified results, it is useful to use several input signals instead of using a single input signal as in cases (c) and (d). Various kinds of algorithms or rules can be applied to the input signals. Such fusion of the input signals is becoming more popular to increase the quality of the identification.

The distinctive values of the processed signals, the extracted features or the identified parameters are mostly compared with the predetermined or given thresholds to identify the status by referring to these threshold values. In order to guarantee high accuracy of the identification, a reliable database must be prepared in advance based on the actual tests, etc. However, it is not easy to do so, as there are many combinations of the machining conditions, the tool, and the work, and this makes the identification difficult.

Another approach to identification is so-called model-based identification. Various kinds of analytical models or empirical models are employed which utilize

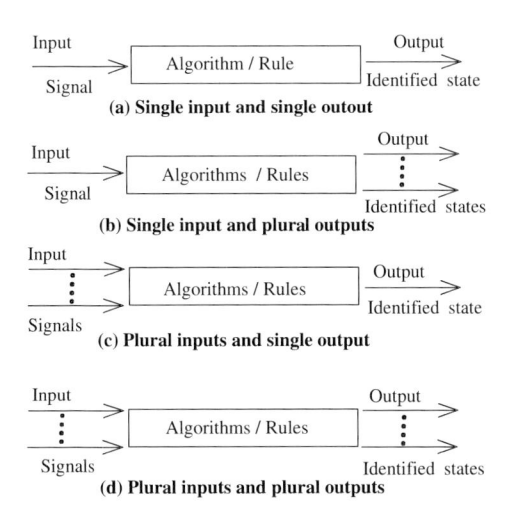

Fig. 1.3-12 Input-output relation of identification

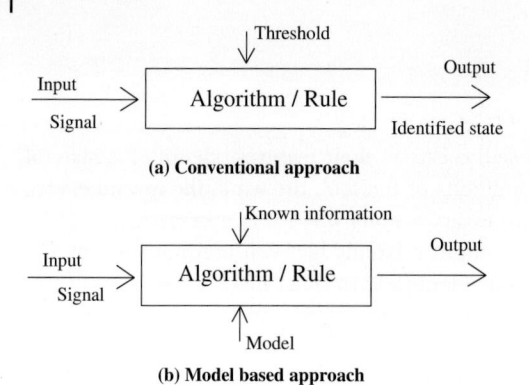

Fig. 1.3-13 Two approaches of identification

(a) Conventional approach

(b) Model based approach

Tab. 1.3-9 Typical decisions made and actions to be taken

Emergency stop or feed stop, and	Continue operation but change
• change tool • dress grinding wheel • change conditions (including NC program) to avoid chatter vibration, other damage, etc. • notify the operator	• spindle speed • feed speed • cutter path to compensate tool wear, thermal deformation or other error source

the known information, such as the cutting conditions. For instance, the generalized model parameters are extracted from the input signals and are compared with the database, or the hypothetical output of the system is calculated which is to be compared with the actual signal data. It is expected that both the reliability and the versatility of identification will be increased by introducing the model-based approach. The differences between the above two approaches of identification are shown schematically in Figure 1.3-13.

The final decision is made based on the results of the identification. Typical decisions made or actions to be taken in the case of machining are summarized in Table 1.3-9. When the abnormal state is identified, the machine is either to be stopped or continues to operate depending on the nature of the abnormal state and the control capability of the machine.

Various kinds of AI (artificial intelligence) technologies are applied to the identification and the decision making, which are briefly explained below.

1.3.6.2 Pattern Recognition

The pattern recognition method has been widely applied to identify the state of the machining process and the cutting tool, etc. [3–5].

It is based on the similarity between a sample to be identified and the patterns or classes that describe the target statuses. From a geometrical point of view, the monitoring indices, or the selected distinctive feature values extracted from the

Fig. 1.3-14 Separation of clusters by coordinate transformation

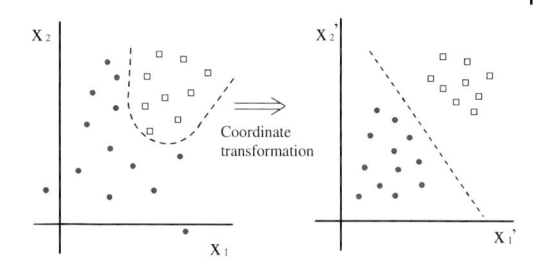

sensed signals, $x=(x_1, x_2, \ldots, x_m)$ span an m-dimensional space. In the span, each target status, h_j, is characterized by a pattern vector $p_j=(p_{j1}, p_{j2}, \ldots, p_{jm})$. The similarity between the sample with the feature values and a pattern is measured by the distance between the two vectors. The minimum distance is then used as the criterion for classifying the sample.

The clustering of the sample points, which belong to the particular patterns, is accomplished by a proper coordinate transformation in such a way that the mean square of the above mentioned distance becomes minimum. The transformed signal x' is given by

$$x' = [w]x \qquad (1.3\text{-}12)$$

where $[w]$ is the transformation matrix.

Figure 1.3-14 shows schematically how the original sample points are classified into distinctive classes by a proper transformation in a two-dimensional space. The most appropriate coordinate transformation is obtained by learning with given sample data.

1.3.6.3 Neural Networks

The neural network is basically an imitation of the neural system of animals, and it has been applied to identify the state of the cutting tool [6], the machining process [7, 8], and also the thermal deformation of the machine tool [9], etc. The advantages of neural networks over pattern recognition are that it can easily constitute optimum nonlinear multi-input functions for pattern recognition and that the accuracy of pattern recognition is easily improved by learning.

A neural network may consist of several layers and each layer has a number of neurons as shown in Figure 1.3-15. The output O_j^L of the jth unit in the Lth layer to its input X_j^L is generally given by

$$O_j^L = h(X_j^L - \theta_j^L) \qquad (1.3\text{-}13)$$

where θ_j^L is the threshold value. The well-known sigmoid monotonic input-output relation is generally adopted, which is given by

$$h(X_j^L - \theta_j^L) = \frac{1}{1 + 1/\exp(X_j^L - \theta_j^L)} \qquad (1.3\text{-}14)$$

Fig. 1.3-15 Basic structure of neural network

Input layer   Hidden layer   Output layer

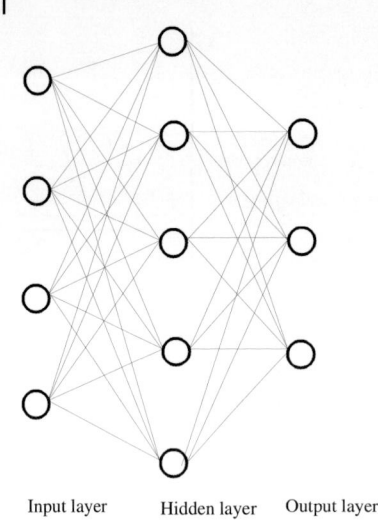

The input X_j^L of the jth unit in the Lth layer, except the input layer, is given by the weighted sum of the outputs from the units in the previous layer, or

$$X_j^L = \sum_{i=1}^{m} W_{ji}^{L-1} O_i^{L-1} \qquad (1.3\text{-}15)$$

where W_{ji}^{L-1} represents the weight which is given by the path from ith unit in the $(L{-}1)$th layer to the jth unit in the Lth layer, and m is the number of nodes in the $(L{-}1)$th layer.

The outputs of the network O_k are calculated based on the inputs following the paths of the network and the procedures mentioned above. The thresholds θ and the weights W are so determined that the sum of squares of the differences between the ideal outputs R_k and the calculated outputs O_k is minimized, or

$$X_j^L = \sum_{k=1}^{m} (R_k - O_k)^2 \qquad (1.3\text{-}16)$$

is minimized. The thresholds and the weights are further modified through learning as the additional data are given to the network.

1.3.6.4 Fuzzy Reasoning

Fuzzy reasoning was first introduced by Zadeh [10] and has been applied to state identification and decision making when there exists fuzziness in the process, such as the grinding process [11].

Fuzzy reasoning is a reasoning method based on the fuzzy production rules. The fuzzy production rules are given in such a way as

Fig. 1.3-16 Typical examples of membership functions

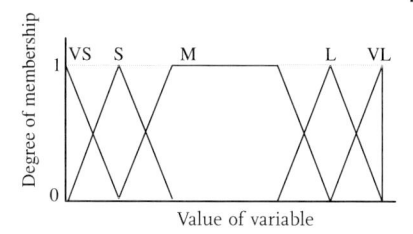

IF   (x_1 is very small) and (x_2 is medium)

THEN (x_k is small)

In the fuzzy approach, uncertain events are described by means of a fuzzy degree or a membership function. If A is an uncertain event as a function of x, A can be described by

$$A = \{x|\mu_A(x)\} \qquad (1.3\text{-}17)$$

where $\mu_A(x)$ the membership function. The membership function is a monotonous function $0 \leq \mu_A(x) \leq 1$, while '0' means certainly no and '1' means certainly yes. Some typical examples of the membership functions are shown in Figure 1.3-16, which represent the linguistic variables, such as VS (very small), S (small), M (medium), L (large) and VL (very large).

When a set of the input variables are given, the degrees of applicability of the rules are calculated according to the membership functions and they are applied to the production rules to give the quantified outputs. The detailed procedures of the fuzzy reasoning and examples of applications are given in Ref. [12].

Other AI technologies, such as expert systems, are employed for state identification, diagnosis, and decision making, but they are not explained in detail here.

1.3.7
Communication and Transmission Techniques

Communication and transmission of the signal within the sensing system are generally processed in digital form after digitization of the analog input signal. The analog transmission of the sensed signal prior to digitization requires special care, as the quality of the signal transmission directly influences the quality of sensing. The analog signal is easily deteriorated by the noise signal surrounding the transducers/sensors and the signal transmission cables. The high-frequency noise signals coming from the power circuits including the motors, the digital devices, etc., as well as those coming from the power supply can be major sources of noise signals.

The signal transmission requires special techniques when the signal is to be transmitted via relatively moving interfaces without contact. The slip ring, wireless transmission with use of radio waves and optical methods are generally employed in such cases.

The communication and transmission of digital signals and data can be easily conducted with the aid of current computer technology. A large amount of digital data can be transmitted between the I/O (input/output) devices and computers via an RS232C or RS422 serial interface at high speeds. Most computers and controllers are connected via the ether-net with the TCP/IP protocol, and the messages and the data can be easily transmitted with use of appropriate communication programs.

The internet services are available to transmit messages and data all over the world via a dedicated line or a commercial telephone line.

1.3.8
Human-Machine Interfaces

The outputs of the sensing system, which are the processed sensor signals, the identified states of the process or the system, or the decisions made, are transmitted to the machine controller and to the operator. At the same time, the operator has to input various kinds of commands to the sensing system. In this sense the human-machine interface plays an important role in the sensing system.

Typical I/O devices or media between the sensing system and the operators are listed in Table 1.3-10. The operators can input commands via dedicated switches or a keyboard, which is more versatile. A touch panel is widely adopted on the actual production floor, which is used to input commands by pressing the specified location on the screen displaying the various functions. The information from the pressed position on the screen is input into the computer via the touch sensor and transformed to a command input. Voice commands are not widely used in noisy environments.

Alarms are the most popular output to the operator when some malfunctions are identified in the system. The visual output, either a graphical presentation or a document, via the display, helps the operator to understand the situation. Oral output with use of a synthetic voice is also helpful.

Tab. 1.3-10 Typical input/output devices or media

Input devices/media	Output devices/media
Switch	Alarm (sound, light, etc.)
Keyboard	Voice (synthetic voice)
Touch panel	Display
Voice command	Printout

1.3.9
References

1 TECHNICAL COMMITTEE ON INTEGRATED MANUFACTURING SYSTEMS, *Questionnaire on Unmanned Operation and Cutting State Monitoring;* JSPE Technical Committee on IMS, 1980 (in Japanese).
2 MORIWAKI, T., *Result of Questionnaire on Tool Condition Monitoring. Activity Report of Technical Committee on IMS;* JSPE, 1994, pp. 58–68.
3 MONOSTORI, L., *Comput. Ind.* **7** (1986) 53–64.
4 MORIWAKI, T., TOBITO, M., *Trans. ASME J. Engl. Ind.* **112** (1990) 214–218.
5 DU, R. et al., *Trans. ASME J. Engl. Ind.* **117** (1995) 121–132.
6 DORNFELD, D., *Ann. CIRP* **39(1)** (1990) 101–105.
7 MORIWAKI, T., MORI, Y., in: *Mechatronics and Manufacturing Systems;* Amsterdam: North-Holland, 1993, pp. 497–502.
8 DU, R. et al., *Trans. ASME J. Eng. Ind.* **117** (1995) 133–141.
9 MORIWAKI, T., ZHAO, C., in: *Proceedings of IFIP TC5/WG5.3, 8th International PRO-LAMAT Conference;* 1992, pp. 685–697.
10 ZADEH, L. A., *Trans. IEEE* **SMC-3** (1973) 28.
11 SAKAKURA, M., INASAKI, I., *Ann. CIRP* **42(1)** (1993) 379–382.
12 MAMDANI, E. H., GAINES, B. R., *Fuzzy Reasoning and Its Applications;* New York: Academic Press, 1981.

2
Sensors for Machine Tools and Robots

H. K. TÖNSHOFF, *Universität Hannover, Hannover, Germany*

Sensors in production systems such as machine tools or robots may be classified into four categories (Figure 2-1). They are activated either during operation or in the set-up phase. Three types of sensors may be applied during the operation: those which measure kinematic values such as position, velocity, orientation or angular velocity, sensors which are applied to control the process in adaptive control systems, and sensors which are used to monitor the production systems and to provide diagnostic functions to assure a high availability of the systems. Sensors for process control functions are shown in Chapters 3 and 4 and will not be mentioned here. Sensors which are applied in the set-up phase are used after assembly to adjust or test the accuracy of the systems or to calibrate the interacting members of the kinematic chain.

2.1
Position Measurement

In this section sensors are described which measure linear or rotational movements mainly during operations of machine tools and robots. These sensors became of es-

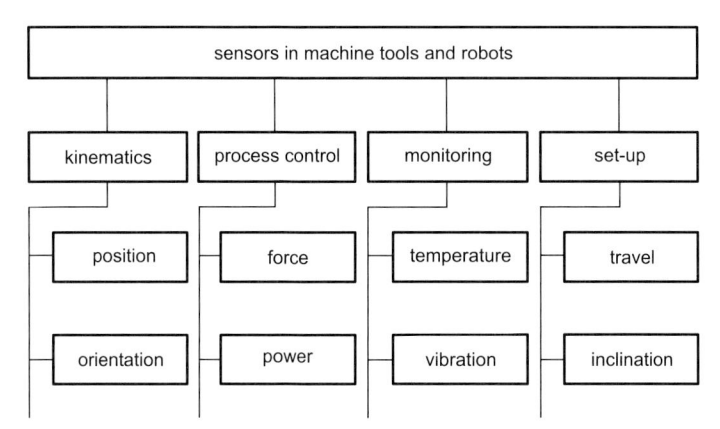

Fig. 2-1 Sensors in production systems

sential importance with the introduction of position control loops for numerical controlled machines (Figure 2-2) [1]. The applied electric and in a few cases hydraulic or pneumatic drives are not able to be positioned by themselves in a control chain but need a feedback of a position signal. An exception to these control loop-based drives are the digital controllable stepping motors. Their application is limited because of accuracy, dynamics, and power. Thus every numerical controlled axis of a machine, tool, or robot that means every feed movement needs at least one position sensor. Since the feed movement in a machine tool, eg, the tool movement of a turning machine, determines the accuracy of the machine, properties of the sensors namely their resolution, their repeatability, their drift velocity, and others (see Chapter 1.3) are of fundamental importance for the accuracy of a production system.

A sensor can be set either directly or indirectly. Indirect means that the travel or the position of a moved machine part such as a slide or a table of a machine tool or the arm of a robot are not directly measured but by means of a movement- transforming device (Figure 2-3).

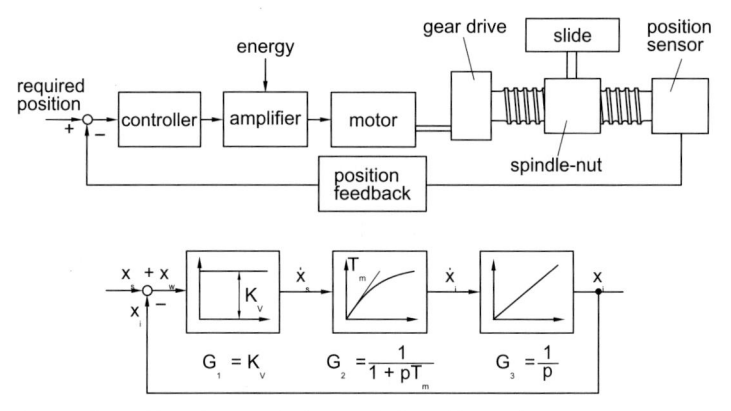

Fig. 2-2 Control loop in numerical controlled machine tools

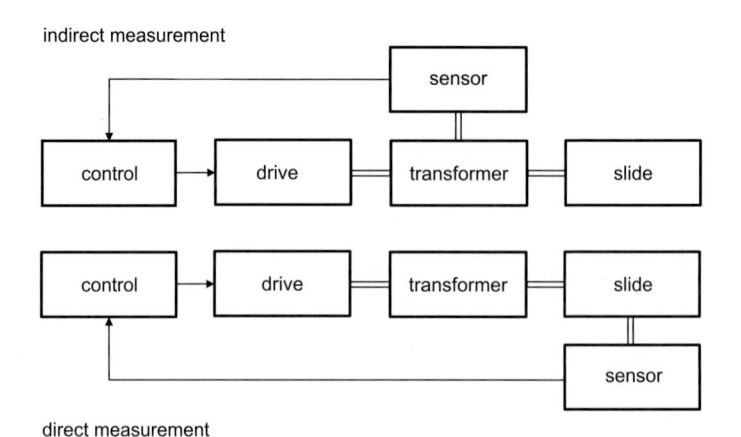

Fig. 2-3 Placement of position sensors

This is normally a transformer from rotational to translatory movements for linear moving slides such as a ball screw (see Figure 2-2), a pinion-rack, a screw-rack, a roll-band, or a wheel-track device. Strongly speed-reducing transmissions such as harmonic drives, worm drives, and others are used for rotational moving tables or arms. Because of the high speed reduction, the demands on the accuracy of a rotational sensor are much higher if it is placed behind the transformer than before it. Indirect sensing in general has the advantage that there is no need for cost-effective measuring devices so that simple and reliable seals can be used. It has the disadvantage that errors of the transmission system are introduced in the measured quantity. These can be, for instance, thermal or elastic deformations of the ball screw or geometric and kinematical aberrations of the transmission system in robots (Figure 2-4).

Therefore, the direct measuring principle should be used if high accuracy and small aberrations are required, eg, for radial positioning in grinding or turning machines. On the other hand, it has to be considered that indirect measurement very often gives a better chance to follow Abbe's principle in machine tools (Figure 2-5). This principle demands that the probe, in this case the travel of the ma-

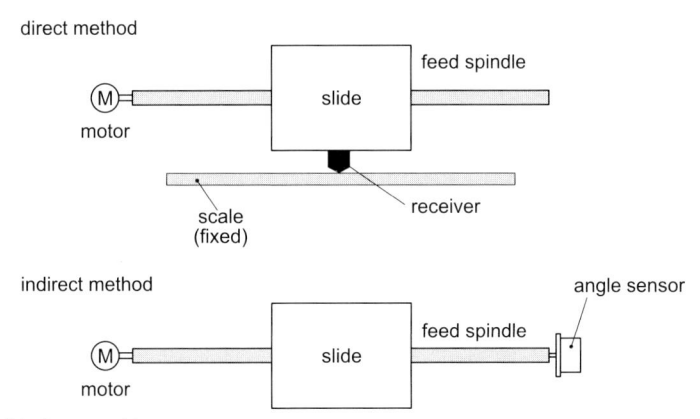

Fig. 2-4 Direct and indirect position measurement

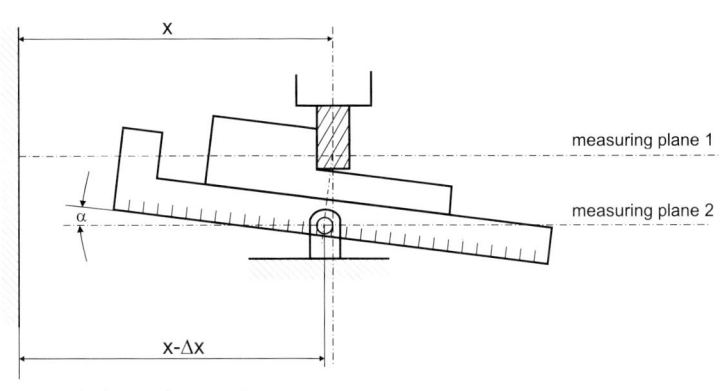

Fig. 2-5 Abbe's principle for machine tools

chine component, and the scale for measuring the travel be in alignment, otherwise errors can occur by non-orthogonality and by tilting effects. It is possible to reach the necessary alignment for indirect measuring systems approximately whereas the direct measuring system is usually placed parallel to the slide and is thus sensitive against tilting errors.

Sensors can be separated in accordance with the kind of signal into analog and digital position-measuring devices. An example of an analog system is the voltage divider (Figure 2-6). The sensor is applied for limited relative resolutions. It can be based on a resistance wire or a vapor-deposited layer of carbon. The resistance of this element is

$$R_b = \rho \cdot \frac{l}{A} \tag{2-1}$$

where the specific resistance ρ and the cross section A of the conducting wire or layer may be erroneous. In addition, the voltage U_x has to be measured within the tolerance of the resolution. For example, the technical limit can be assumed to be 0.1 mV of a 10 V maximal voltage. That means a relative resolution of 10^{-5} or a positional resolution of 1 µm limits the maximum travel to 100 mm.

The digital sensor determines the position either by counting increments (digital-incremental measurement) or by reading coded numbers (digital-absolute measurement) (Figure 2-7). The resolution is given by the width of the incremental unit. The length of measurement l and the incremental width τ are connected by

$$l = 2^n \tau \tag{2-2}$$

where n is the number of bits necessary to describe the maximum length with a resolution of τ. For the example of $\tau = 1$ µm and $l = 250$ mm,

$$n = \frac{\log\left(\frac{l}{\tau}\right)}{\log 2} = 17.9 \tag{2-3}$$

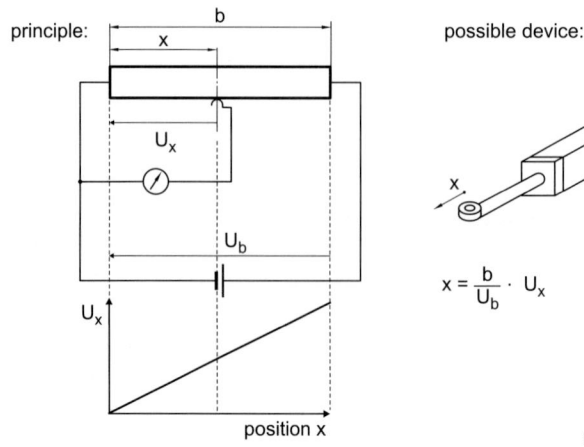

Fig. 2-6 Potentiometric sensor

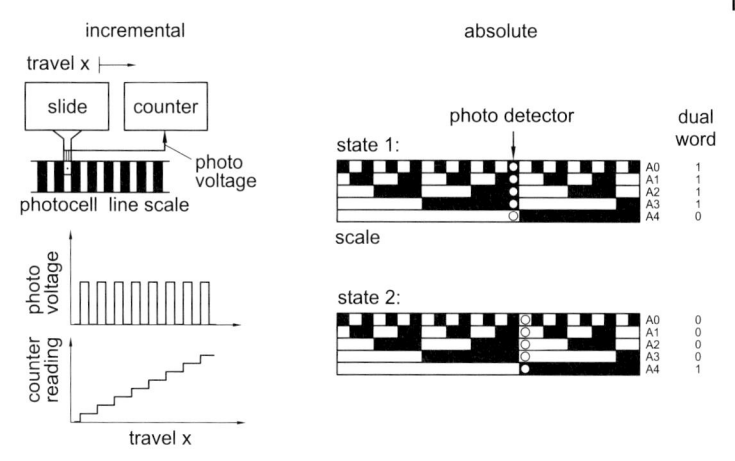

Fig. 2-7 Digital measuring principle

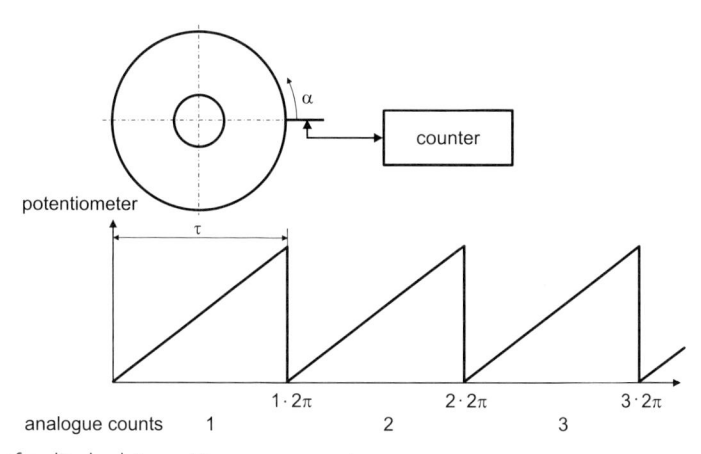

Fig. 2-8 Principle of cyclic-absolute position measurement

This means that 18 bits have to be implemented in the counter of an incremental system or 18 channels have to be incorporated in an absolute system. The incremental measurement has the disadvantage that the position can only be determined relatively. The digital absolute sensor, on the other hand, has the disadvantage that the system needs a large number of channels for a long measuring length and a high resolution and is therefore expensive. All three mentioned kinds of measurement, the analog, the absolute-digital and the incremental-digital principle, have their specific advantages. Therefore, the idea of combining favorable abilities is not unreasonable. This leads to the cyclic-absolute measuring principle. It takes advantage of the absolute character of the analog system applying it only on limited measuring lengths and of the incremental sensing principle with its simple structure and robustness. It is shown schematically in Figure 2-8.

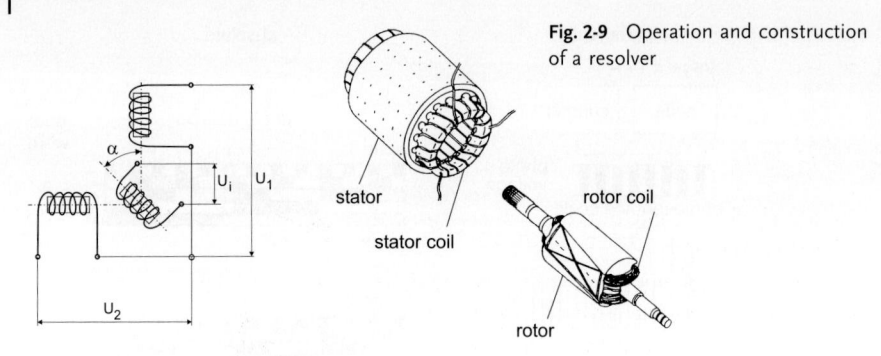

Fig. 2-9 Operation and construction of a resolver

The resolver is an absolute-cyclic measuring system. It is based on the inductive principle and measures angular or rotational movements. The resolver is often applied in an indirect measuring system. It is basically a transformer consisting of three windings (Figure 2-9). The stator has two windings whose active directions are exactly placed at 90°. The rotor carries a third coil, the secondary system. High-frequency voltages are acting at the two stator systems which are 90° electrically phase shifted:

$$U_1 = \hat{U}_0 \sin \omega t$$
$$U_2 = \hat{U}_0 \cos \omega t$$

They induce corresponding voltages in the secondary coil which is rotated against the stator by α:

$$U_{i1} = k\hat{U}_0 \sin \omega t \cos \alpha$$
$$U_{i2} = k\hat{U}_0 \cos \omega t \sin \alpha$$

where k is the coupling factor of the transformer. The two voltages U_{i1} and U_{i2} are added to give

$$U_i = k\hat{U}_0 \sin(\omega t + \alpha) \tag{2-4}$$

Comparing the phase between U_1 and $U_{i1,2}$, the searched for angle α can be directly determined. A phase comparison can be made by a phase-locked loop. In another kind of implementation the rotor system is supplied with $U_R = \hat{U} \sin \omega t$. The induced voltage in the stator coils is modulated by the angle of rotation α due to the spatial arrangement:

$$U_{s1} = k\hat{U}_0 \sin \omega t \sin \alpha$$
$$U_{s2} = k\hat{U}_0 \sin \omega t \cos \alpha$$

In this implementation the quotient of the stator- induced voltages is calculated by

$$\frac{U_{s1}}{U_{s2}} = \frac{\sin a}{\cos a} = \tan a \tag{2-5}$$

The searched for angle a is determined by an arctan algorithm. This is called the ratiometric method.

Resolvers are supplied with alternating current of high frequency, hence the space requirement can be minimized. An upper limit is given at 0.4–1 kHz because of the iron within the transmitter. The resolution might reach $1.5 \cdot 10^{-3}$ degrees, but the accuracy of the sensor is mainly determined by the manufacturing accuracy, which influences the costs substantially. The resolver is therefore mostly applied for less critical resolutions where it is fairly inexpensive.

The resolver principle is also used for linear sensors. The inductosyn® sensor is basically a resolver which is straightened in the plane. It is a very common applied cyclic absolute measuring device (Figure 2-10).

Similarly to the resolver, the scale or the reader can be supplied with a high-frequency alternating current (120 kHz). The signal processing methods are the same. If the alternating voltage is applied to the scale the following voltage is induced in the reader with coupling constant k:

$$U_{R1} = k\hat{U}_s \sin \omega t \cos \frac{2\pi}{\tau} x$$
$$U_{R2} = k\hat{U}_s \sin \omega t \sin \frac{2\pi}{\tau} x \tag{2-6}$$

because the reader includes a longitudinal phase shift of $\tau/4$. The position x can therefore be determined, for instance, by the ratiometric method. Compared with the resolver, the resolution can be 1000 times higher. The measurement is analog within the domain of the division of 2 mm or 0.1 in. The signal is repeated cyclically. The divisions have to be counted or resolvers have to be applied additionally

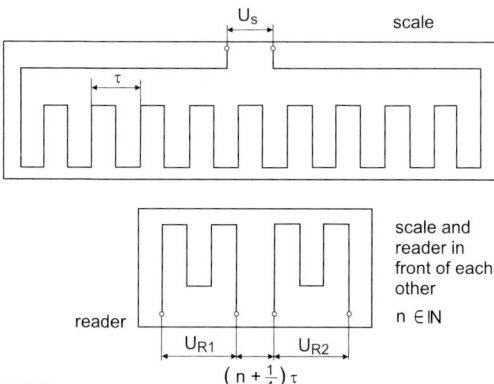

Fig. 2-10 Principle of the inductosyn® sensor

to provide the coarse measurement. Inductosyn® devices are available in modules of 250–1000 mm. They can be serially mounted. The assembly has to be very accurate to avoid errors at the joints. It is usually made by interferometric means.

A digital-incremental sensor is shown in Figure 2-11. The information is given only relatively. The impulses have to be counted. A reference point must be given, for instance, by driving to a micro limit switch when starting the measurement. These sensors often work by the optical principle. The divisions are applied on glass scales by vapor deposition. The sampling can work by direct or transmitted light. The principle is explained by the transmitted light method. Fine lines are applied on the glass scale. The transparent and black lines can be equal in width. Divisions of 10 µm are used in practice. The width is limited because of the wavelength of the applied light. A scanning reticle is moved along the scale with the table whose position or travel is to be measured. Using a scale with several divisions means an increase in the light energy which is received by the photo detector.

The width ratio of the transparent to the non-transparent slots can vary. Using a narrower sampling slot ($\delta \ll \tau/2$), the light intensity at the receiver is stronger. The receiver gives almost a rectangular signal (Figure 2-12; left).

The properties of this principle are:

- the received light energy is comparatively low;
- the signal is of digital nature;
- it is only appropriate for coarse divisions;
- the information is gained by counting the signal impulses.

According to another principle, the non-transparent and the transparent sections are equal in width. The photo receiver delivers a value-continuous signal (Figure 2-12 right). The properties are:

- the signal is of analog nature and the resolution is determined only by the signal-to-noise ratio;
- the received light energy is comparatively large;
- a further increase in resolution is possible.

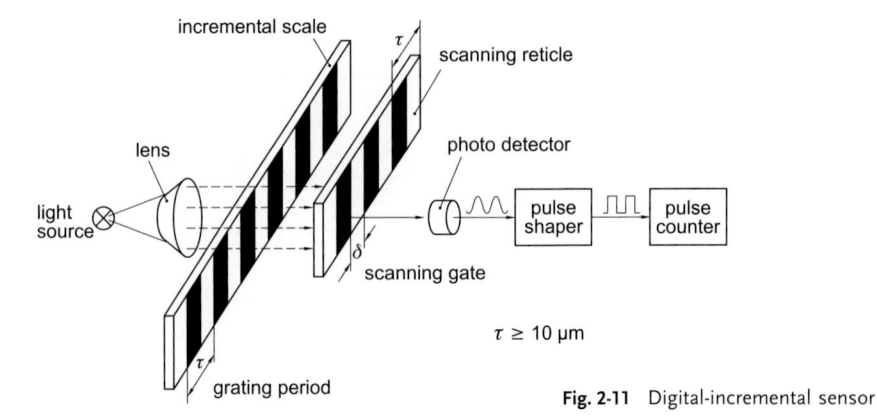

Fig. 2-11 Digital-incremental sensor

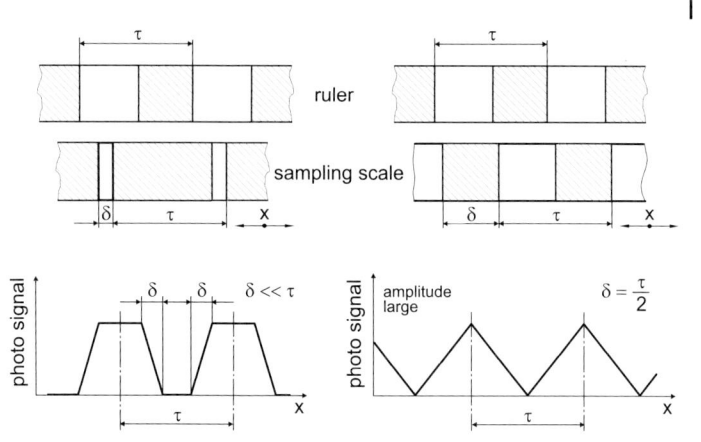

Fig. 2-12 Influence of sampling slot on photosignal

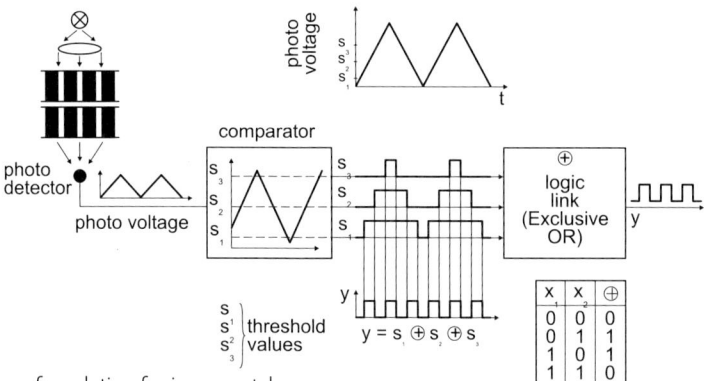

Fig. 2-13 Increase of resolution for incremental sensors

The value of the continuous signal can be used for further improvement of the resolution as can be seen from Figure 2-13. Instead of the direct shaping of one impulse, several impulses are gained from the triangular signal by a comparison with threshold values, a kind of interpolation.

The Moiré effect can be used for scanning glass scales (Figure 2-14). The scale and the reader are tilted against each other by a small angle. Moiré's stripes are generated using the through-light method. These stripes move with much higher speed than the scale. A photo receiver delivers a sine signal of the moving Moiré's stripes. A sine and a cosine signal are received if two photo receivers which are displaced by one quarter of the period of Moiré's stripes are applied. Information on the position can be obtained with high accuracy by an electronic interpolation unit. The two signals are amplified by defined factors and added according to the following equation:

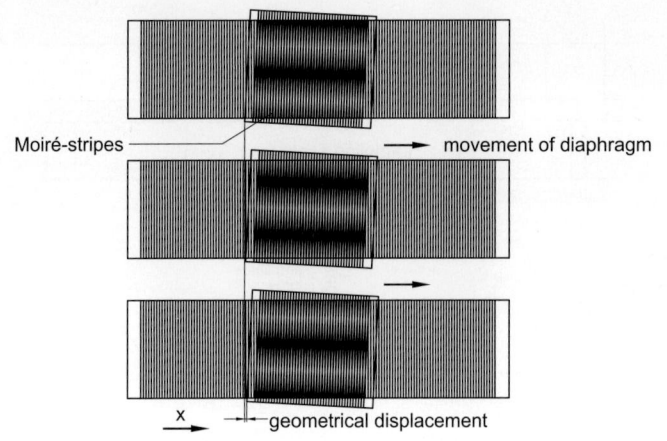

Fig. 2-14 Scanning with Moiré stripes

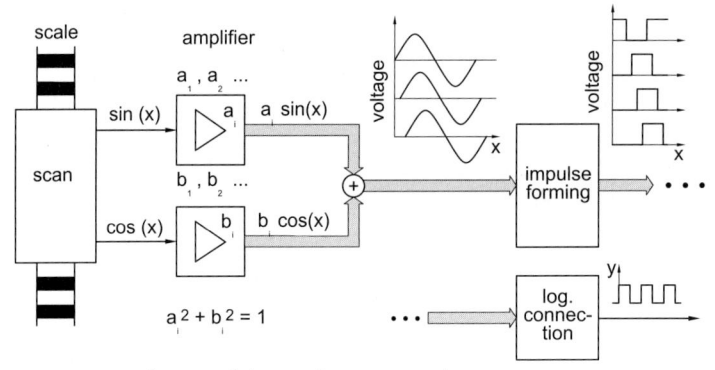

Fig. 2-15 Magnification of division for incremental sensors

$$a \sin x + b \cos x = \sqrt{a^2 + b^2} \sin\left(x + \arctan \frac{b}{a}\right) \qquad (2\text{-}7)$$

Several phase-shifted signals are achieved by this algorithm, which can generate a high resolution of the sensor after an impulse shaping operation (Figure 2-15). The measuring step can be 1/20 or 1/200 of the division period, which is an interpolation factor up to 1:100. A resolution of 50 µm can be reached by such incremental sensors.

The resolution is limited without further means by the scale division or the normal which is applied for digital measuring systems. The required accuracy for high-precision machine tools is <0.5 µm. It is difficult to produce scales with such divisions with the necessary accuracy. One measure to overcome this limitation is the application of interpolation methods. One increment can be divided electronically into an integral number of partitions.

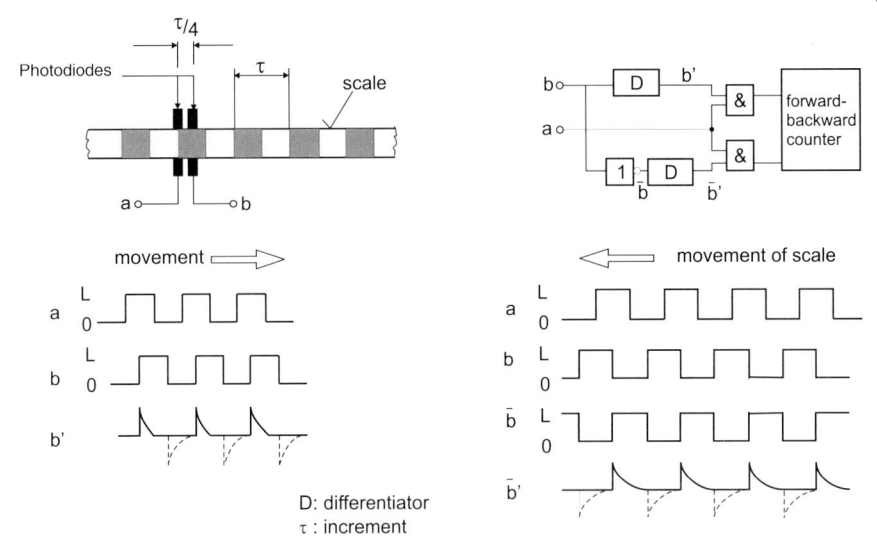

Fig. 2-16 Direction discriminator. Source: Herold, Massberg, Stute

The number of impulses is counted in incremental systems. A discrimination of the moving direction is necessary for this method, otherwise vibrations already existing between the scale and the reading device can falsely indicate a movement. Figure 2-16 shows an electronic discriminator. Impulses are taken from an incremental scale by a reading device. This is, for instance, a scanning reticle which consists of a lamp and two photo diodes (a, b). The signals are shifted against each other by $\tau/4$. Thus the information of direction can be deduced. A wiring diagram is given in Figure 2-16. The signal of diode b is differentiated, inverted and compared with signal a. If the original signal b' is equal to a, the counter goes forward; if b is equal to a it counts backward. Only the positive parts of the b' or b signals are relevant in the device.

The time deviation of travel or distance can be used to measure speeds or velocities indirectly. Direct measurements of speeds are possible using the induction principle, when moving a magnet against an electric coil. This is an electrodynamic speed sensor (Figure 2-17, left). The induced voltage U is

$$U = wlBv \tag{2-8}$$

with the number of windings w and their length l, the magnetic flux density B and the speed v determined. As can be seen, the speed is proportional to the measured voltage. The linearity is usually better than 1% of the full measuring range. Typical measurable speed domains are $0.5 \cdot 10^{-3}$–0.5 m/s. The maximum measuring range is 6 mm with commercially available devices.

Fig. 2-17 Measurement of speed and acceleration

electrodynamic velocitymeter

piezoelectric accelerometer

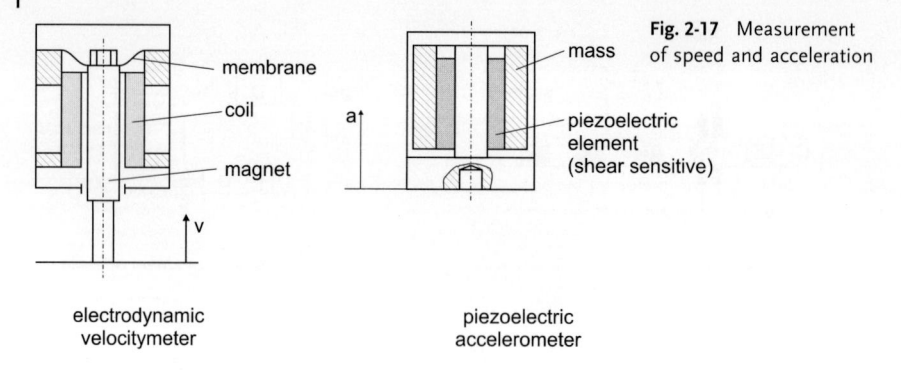

The time deviations of travel or speed may be used to measure accelerations indirectly. The linear acceleration a can also be determined by the reaction force F of a known mass m according to

$$F = ma \tag{2-9}$$

The piezoelectric principle is commonly used for this purpose. Figure 2-17 (right) shows an accelerometer which contains a piezoelectric element sensitive to shear. Other sensors use the piezo effect for normal stresses, ie, in a cantilever or a rod. Such sensors can be built to small sizes and thus have high limit frequencies up to 100 kHz. The measuring domain may range from 10^{-2} to 10^6 m/s^2. The sensors can be designed in a triaxial manner to determine acceleration components in the space.

2.2
Sensors for Orientation

The orientation of a machine tool or robot component is given by the angle of the relevant direction to a reference plane. To determine orientation means measuring the inclination angle.

The autocollimator is an optical sensor which is used for measuring small angles of inclination. It works like a telescope (Figure 2-18) using the collimator lens twice. From a light source a beam is focused in the plane of an ocular scale. The beam is made parallel in the collimator lens (telescope adapted to infinity). It is reflected by a measuring mirror which is perpendicular to the plane to be measured. If the mirror is inclined by a small angle a this generates a shift x of the ocular scale image. The measured angle a and the shift x are independent of the distance between the collimator lens and mirror because of the parallel light between them. The shift x can be transformed to an electrical signal by a linear CCD camera. High-precision autocollimator sensors work with an accuracy of 10^{-7} (0.1 µm over 1 m). They are used to measure the straightness of machine tool guideways or planeness of tables.

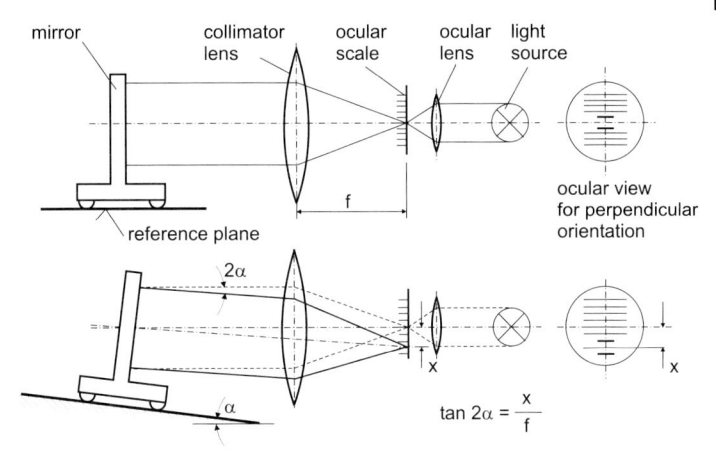

Fig. 2-18 Principle of autocollimator

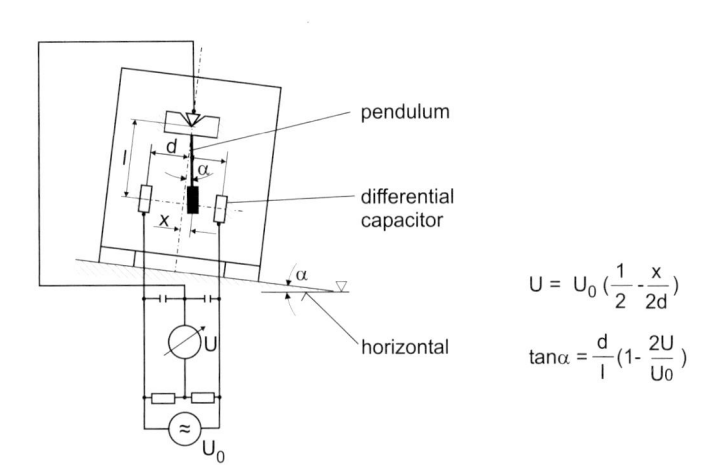

$$\tan 2\alpha = \frac{x}{f}$$

$$U = U_0 \left(\frac{1}{2} - \frac{x}{2d}\right)$$

$$\tan\alpha = \frac{d}{l}\left(1 - \frac{2U}{U_0}\right)$$

Fig. 2-19 Electronic level, capacity principle

Inclinations can also be measured by the capacity principle using a gravity pendulum in an electronic level (Figure 2-19). The end of the pendulum is one part of a differential capacitor. By this arrangement the electric signal U is proportional to the inclination angle with a good linearity better than 1%. The pendulum is strongly damped by oil. If the pendulum is built in a mechanical or optical indexing head, any direction can be measured.

Furthermore, angles can be measured by the resolver principle as shown in Figure 2-9. According to the principle shown in Figure 2-20 there are incremental sensors for rotational movements provided that the rulers are not designed linearly but as an incrementally or absolute digitally divided disk.

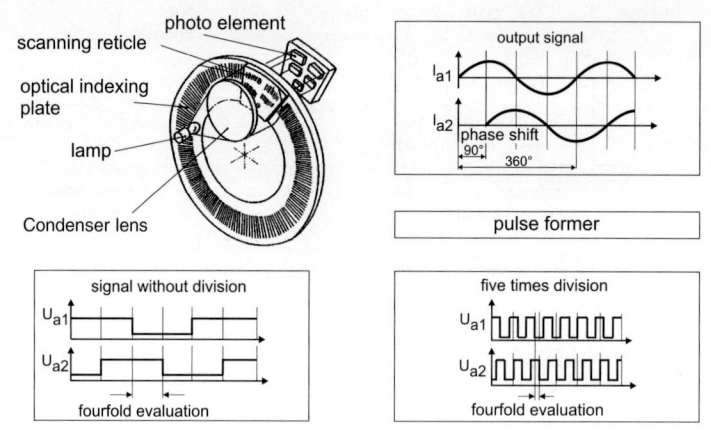

Fig. 2-20 Incremental angle sensor. Source: Heidenhain, Traunreut

2.3
Calibration of Machine Tools and Robots

Calibration means in general determine the connection between a measured quantity and its true value. Calibration as a qualification step for machine tools and robots in the sense as it is used here is the sensorial determination of geometric deviations and of the feed motion errors of the tool center point. The calibration can to a certain extent be the basis for corrective measures mainly introduced by the control of the machine tool or robot. In this sense geometric and kinematic deviations can be defined in the six degrees of a rigid body, in three rotational and three linear deviations. Therefore, the already mentioned sensors for positions and orientations (Sections 2.1 and 2.2) can be applied for such calibration measurements. Usually the values measured by the built-in measuring devices or sensors are compared with independent measured quantities. The difference of the vector of the measured values and of the required quantities is equal to the compensation vector which can be applied in all six components – or in a simplified manner – to correct the required quantities. The independent measurements can be based on the known mechanical, optical or electronic principles for position and orientation determination. According to a general rule of precision measurements, the measuring accuracy should be at least 10 times better than the deviations to be determined. In some cases this is not achievable and then statistical methods may be applied.

A typical and easily implementable sensor is explained here, the double ball bar (DBB) device (Figure 2-21). The DBB method is an integrative approach which takes the machine tool or robot kinematics into consideration as well as the dynamics of the feed drives and the control. In the past, machining tests were used to evaluate the geometric and kinematic quality of a machine tool. Such procedures require great efforts with respect to time and costs and the different influ-

Fig. 2-21 Structure of DBB instrument.
Source: Kakino

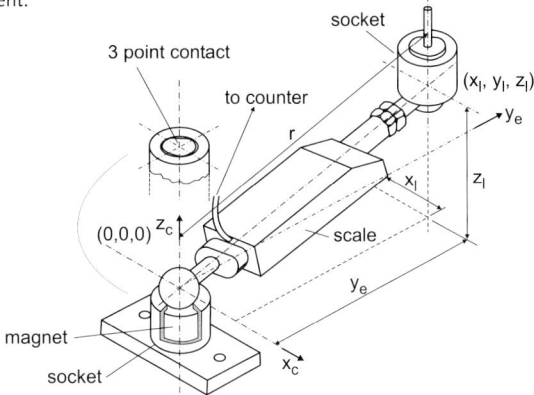

ences depending on the process itself and the machine cannot be exactly separated.

The DBB method was introduced in order to avoid the named disadvantages. It is actually a circular form test. Two or more axes of a machine tool or robot are numerically controlled in a such way that the tool center point moves on a circle. The radial deviation, ie, the deviation of the measured path from the required circle, is determined [2]. The radius of a required circle is denoted r and its deviation Δr. The coordinates of its center point in the space are x_c, y_c, and z_c and their deviations are u_c, v_c, and w_c. The coordinates of the circle line are x_l, y_l, and z_l and their respective deviations are u_l, v_l, and w_l. Then the circle equation in the space can be written as

$$(r+\Delta r)^2 = [(x_c+u_c)-(x_l+u_l)]^2 + [(y_c+v_c)-(y_l+v_l)]^2 + [(z_c+w_c)-(z_l+w_l)]^2 \quad (2\text{-}10)$$

This equation can be simplified by neglecting small quantities of second order to give

$$\Delta r = [(x_c-x_l)(u_c-u_l) + (y_c-y_l)(v_c-v_l) + (z_c-z_l)(w_c-w_l)]/r \quad (2\text{-}11)$$

Deterministic errors can be recognized by algorithms based on the above mentioned equations. Thus the influence of different kinematic or control errors can be evaluated. These are:

- deviations depending on the position such as errors of orthogonality, of straightness, and of backlash of guideways;
- deviations depending on the feed motion such as errors of mismatch of control loop gains, of stick slip, and of servo drive response.

Figure 2-22 shows an example for a DBB test of a machining center of medium size.

① error of quadrant change over

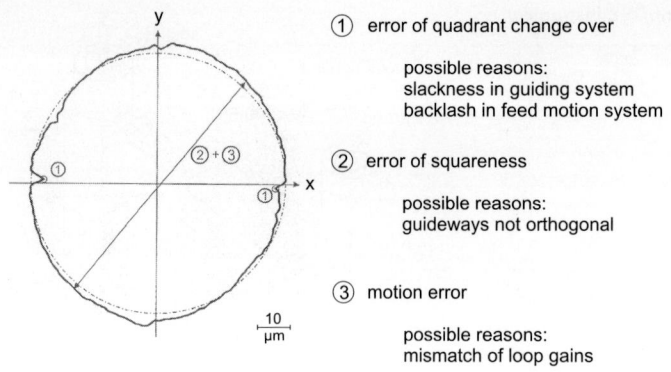

possible reasons:
slackness in guiding system
backlash in feed motion system

② error of squareness

possible reasons:
guideways not orthogonal

③ motion error

possible reasons:
mismatch of loop gains

Fig. 2-22 DBB test, deviation from a circular path

2.4
Collision Detection

Collisions can occur between parts of the machine tool or of the robot, between workpieces and workpiece clamping devices, and tools. Collisions generate major disturbances in production and can damage tools, workpieces and parts of the machine. The adjustment of relevant machine parts such as spindles, spindle housings, tables, and slides can be destroyed. The time needed for repair can be hours, days, or even weeks. It is therefore of great importance to avoid collisions and to detect critical situations which can lead to such accidents. Measures can be derived from the causes which may lead to collisions [3]. Three phases are distinguished:

- the programming phase;
- the set-up and try-out phase; and
- the operation phase.

In each phase there are different causes, as shown in Table 2-1.

According to an investigation which was made in the early 1980s [4], human causes are the main reason for collisions, constituting 72% of all collisions (Figure 2-23).

The programming phase constitutes with 13% on average but depends on the complexity of the parts to be machined, as can be seen from Figure 2-24 [5]. An investigation of 1848 new NC programm showed a significant correlation with the number of numerical control (NC) axes.

The measures to avoid such failures are based on geometric or kinematic test procedures or on simulation techniques. Geometric and kinematic tests which run in the background of the central cycle or are actuated at suitable incidents or times can be applied in the set-up and operation phase. The computation time for the necessary geometric calculation is a limiting factor. Even if the movements of the machine components are simulated in advance and are checked on collisions, collision monitoring by software cannot be established exclusively in the back-

2.4 Collision Detection

Tab. 2-1 Causes of collisions and measures to avoid them

Phase	Functions leading to failure	Measure to avoid collisions
Programming	Of tool path Of reference planes Of tools Of clamping device	Geometric test procedure
Set-up try-out	At manual test By clamping system By raw material At choice of tool	Definition of collision space Collision sensors
Operation	By control system By measuring systems By tool changing By wrong part	Definition of collision space Collision sensors

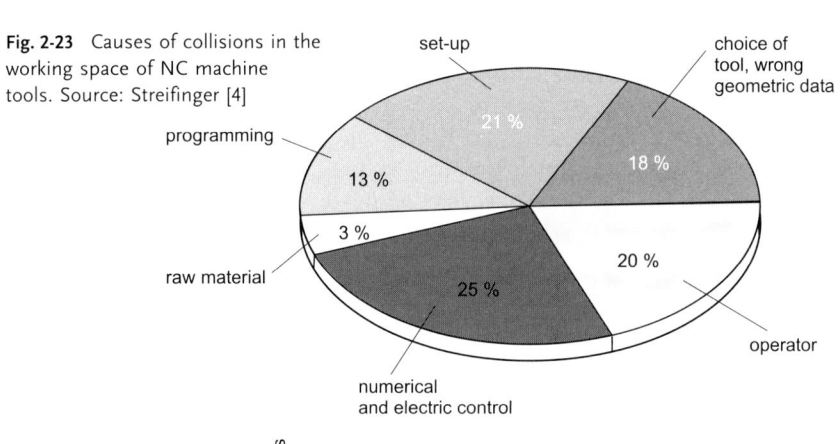

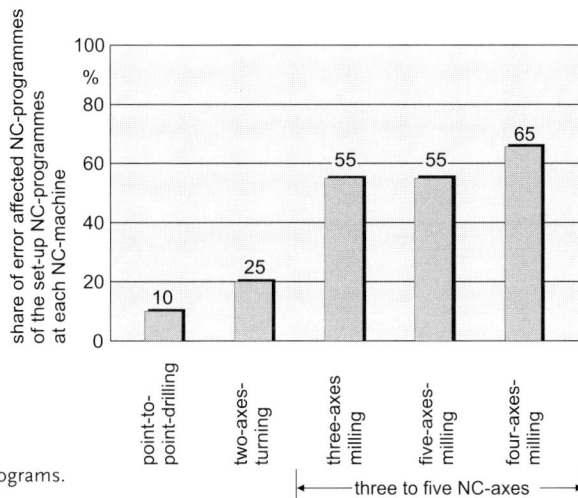

Fig. 2-23 Causes of collisions in the working space of NC machine tools. Source: Streifinger [4]

- set-up: 21 %
- choice of tool, wrong geometric data: 18 %
- operator: 20 %
- numerical and electric control: 25 %
- raw material: 3 %
- programming: 13 %

Fig. 2-24 Error rate of NC programs. After Flavell [5]

- point-to-point-drilling: 10
- two-axes-turning: 25
- three-axes-milling: 55
- five-axes-milling: 55
- four-axes-milling: 65

(three to five NC-axes)

share of error affected NC-programmes of the set-up NC-programmes at each NC-machine

ground of the NC control without affecting the process. Especially for high speeds and short sets of the parts program the machine may have to wait for the release of the next program set. Therefore, sensorial approaches have been sought.

Sensors to avoid collisions have so far been based mainly on force monitoring [5]. These devices have the disadvantage that they cannot really prevent the contact between tool, workpiece, clamping device, and machine or robot components, but can only avoid major damage. Such sensors send alarm signals to the control or interrupt autonomously the force flow in the feed drives. An example is given in Figure 2-25. The force sensor gives a continuous signal to a control unit which requires a fast processor. The force signal is reached during the set-up phase. A signal corridor takes care of minor alterations of the measured force and this corridor is adjusted to a systematic force development with respect to tool wear. If the corridor is exceeded by an actual signal an interrupt is given to the feed drive of the machine. Such signals can occur either by tool breakage or by collision. Both incidents need a fast reaction of the feed drive. In addition to these functions the unit can serve as a wear monitoring system by setting fixed upper limits for the force. The described system is operating in practice. It is applicable only for medium-or long-running batches.

A force sensor for a machining center is shown in Figure 2-26. It is applied to interrupt the flow of forces which are in this case mainly generated by mass inertia. The device is placed in the feed drive system of a machining center with a moving column. The kinetic energy stored in the drive chain is

$$E_{kin} = \frac{1}{2} m v_f^2 + \frac{1}{2} \sum_{i=1}^{n} \Theta_i \cdot \omega_i^2 \tag{2-12}$$

where m is the mass of the translatorially moved components such as the column including the head stock and spindle and v_f is the feed speed. Θ_i are the mo-

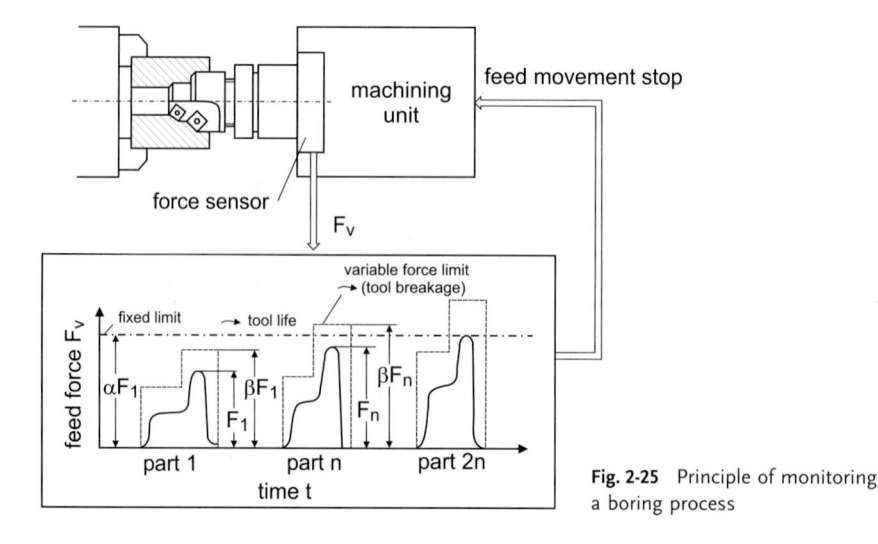

Fig. 2-25 Principle of monitoring a boring process

Fig. 2-26 Feed force collision sensor. Source: Bohle

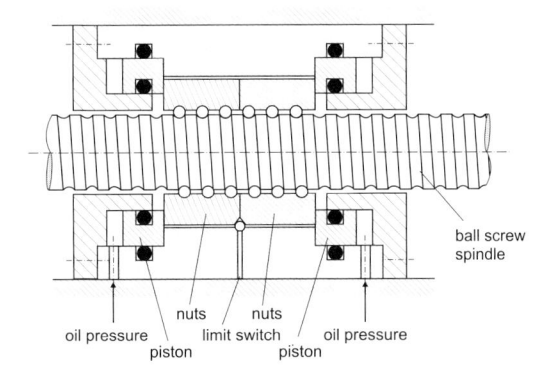

ments of inertia of the different components i in the system rotating with angular speed ω_1. If the rotating components such as motor, gear drive, and ball screw are decoupled, the collision energy is decreased by one to two orders of magnitude because of the fast-rotating parts and the square order influence of the speed. The figure shows a ball screw drive whose two nuts are fixed by axially acting hydraulic cylinders. If the acting feed force for instance by collision exceeds the hydraulic force, the respective piston with the nuts moves axially and actuates a limit switch which gives a signal to the control and stops the feed movement. The generated collision force depends, of course, on the hydraulic force, on the time delay from actuating the limit switch to the reaction of the feed drive, and on the compliance in the force chain.

Optical or acoustic sensors use cameras or acoustic sender and receiver systems. These principles have not been introduced into practice yet because they are sensitive to chips, coolant, and particles and therefore have a high disturbance potential.

2.5
Machine Tool Monitoring and Diagnosis

The growing investment machine tools and production systems requires their maximum availability. The complexity of such systems, which are highly automated and consist of many modules which are linked and have to work together without failure, increases the risk of breakdowns. Systems containing n modules which are coupled without redundancy and which have an availability of $P_j < 1$ each have a total availability of only

$$P = \prod_{1}^{n} P_j \qquad (2\text{-}13)$$

This means the availability of the total system is always lower than that of the weakest module. The failure behavior of machine tools was investigated over a

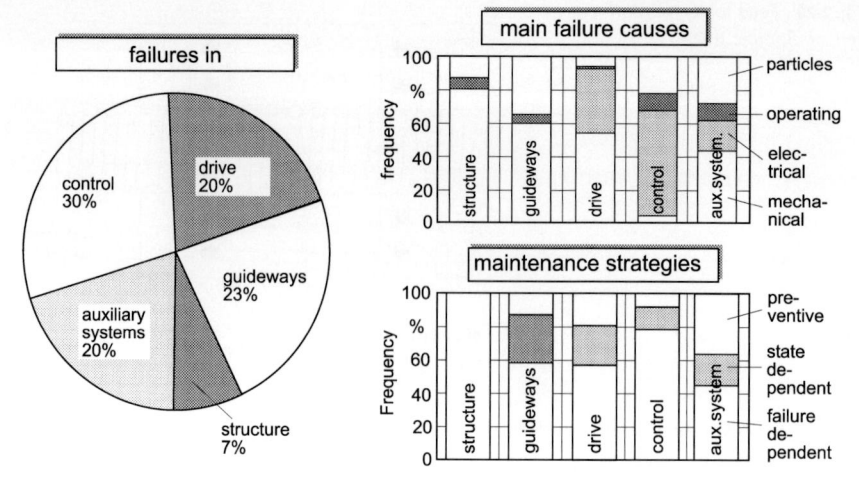

Fig. 2-27 Distribution of machine tool failures. Source: Seufzer [6]

large number of machine tools (540 lathes, 401 manufacturing systems and 151 milling machines) (Figure 2-27) [6]. It was seen that the electric and electronic systems and the auxiliary modules are of dominant interest as far as reliability is concerned. The repair strategies which were reported from the machine users are mainly failure or state related.

Monitoring and diagnostic devices are an interesting approach to increase the availability of a machine tool by decreasing the mean time to repair (MTTR). Monitoring is the automatic supervision of machine tool functions (or of processes, which is reviewed in Chapter 4). The monitoring system has to ensure that a machine works correctly without malfunctions. The result of its operation is a corresponding message about the machine state. This test can be performed according to a plan, periodically or continuously. A diagnosis system goes further, identifying the incorrect function and the reason for this malfunction. It gives an indication of the reasons and it is initiated when an incident occurs or upon demand.

Different methods can be applied for M & D (monitoring and diagnosis) [2]:

- signal-based M & D (heuristic);
- model-based M & D with signal prediction;
- model-based M & D on parameters;
- feature recognition or classification;
- knowledge-based M & D.

Heuristic signal-based systems use experienced states and the respective signals of machine components, eg, a temperature signal, and compare them with a threshold value of the signal which is set by former experience.

Model-based M & D systems need a description of the system to be monitored. This description or model is a parametric algorithm or another input-output corre-

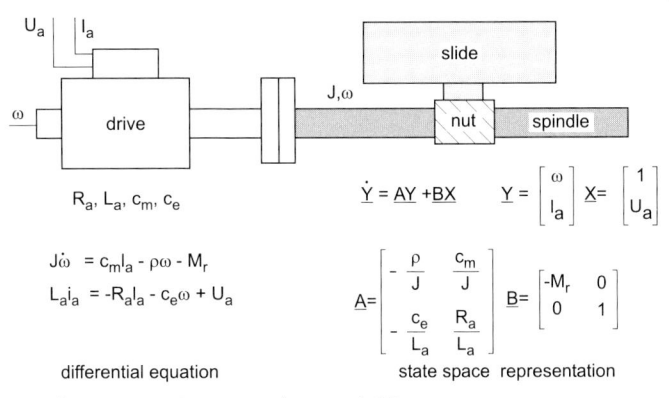

Fig. 2-28 Model structure of an NC axis. Source: Potthast et al. [7]

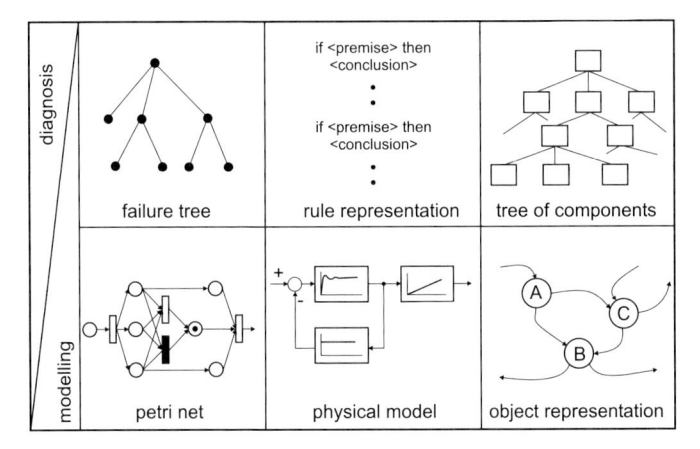

Fig. 2-29 Methods of knowledge representation. Source: Seidel [8]

lation of the system which is experimentally determined. The model is able to predict the relevant behavior of the machine. Figure 2-28 shows as an example the model structure of a DC drive of a machine tool table. The model is given by the system describing differential equations or the state variables [7]. Not a single signal is considered in feature recognition or classification systems but the constellation or cluster of different measured values or multi-dimensional information such as images is analyzed.

Knowledge-based methods store information about the system behavior and error-symptom relations. They are used to identify the causes of disturbances, failures, or interruptions.

The knowledge representation is made by different methods; Figure 2-29 shows examples. An often applied method is the failure tree representation. The failure tree is a graphical description of logic links between the failure entrances leading to a defect (Figure 2-30). The figure shows the malfunction of a valve. Cause

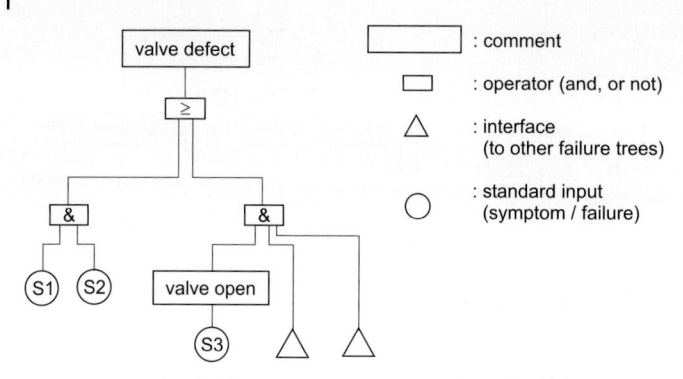

Fig. 2-30 Example of failure tree representation. Source: DIN 25424

chains for all sources or consequences of the failure are stored in failure trees. The knots in the tree may be assigned a probability for different errors. The tree entrances, ie, the elements in the bottom level, are classifications of observable process or machine states or extracted features.

The advantage of a knowledge representation by a failure tree is that it can be easily applied in practice. Further, there is no need for an engineer with special knowledge to generate a failure tree normally. The failure tree can be easily transformed into algorithms for computers. The main disadvantages are the statistical representation of the knowledge and the limited appropriation for dynamically transmitting failures. The time dependence cannot or only with difficulty be introduced in failure trees. Also, only known failure descriptions can be considered in those trees. This might lead to problems for complex systems.

The formulation of heuristic knowledge in rules, ie, 'if–then', representations is also widespread. It is often preferred because the experience and knowledge of service technicians and machine operators can be easily modeled. It is similar to the failure tree representation applied to post-failure analysis, which means assisting the operating or service personnel to find out the cause of failure in case of malfunction. Compared with failure trees, the rule-based representation is more flexible but it is not simple for more complex applications. Another disadvantage of the rule-based representation is that the run time depends on the number of rules and hence it might not be applicable to on-line diagnosis of larger systems. The speed of rule-based knowledge systems can be accelerated by dividing the knowledge basis and by import and export of rules, by efficient inference algorithms, and by using compilers for the knowledge basis.

M & D systems should be well integrated into the control system of a machine tool. Nevertheless, the machine should be able to operate even if the M & D system fails. This means that the M & D functions have to work independently and parallel to the machine functions but they have to be synchronized with them. Set data of the control and additionally status and sensor signals of the machine are used for the synchronization. The following criteria should be taken into consideration under the prerequisites of flexibility, modularity, and extensibility:

2.5 Machine Tool Monitoring and Diagnosis

- the data processing functions should be implemented in small independent and reusable modules;
- data and data processing functions should be clearly separated;
- the mechanism to control the data processing functions should be interchangeable to be able to apply different methods with variable flexibility;
- the architecture must be multiprocessor treatable and multicomputer applicable.

Figure 2-31 shows the internal structure of a system which follows these requirements. The complete system is built up as a network of separate, equally structured run-time modules which act autonomously.

The internal elements of a run-time system and the interfaces between them are shown in Figure 2-32. In addition to the communication interface the run control is the central kernel which determines the data processing functions accord-

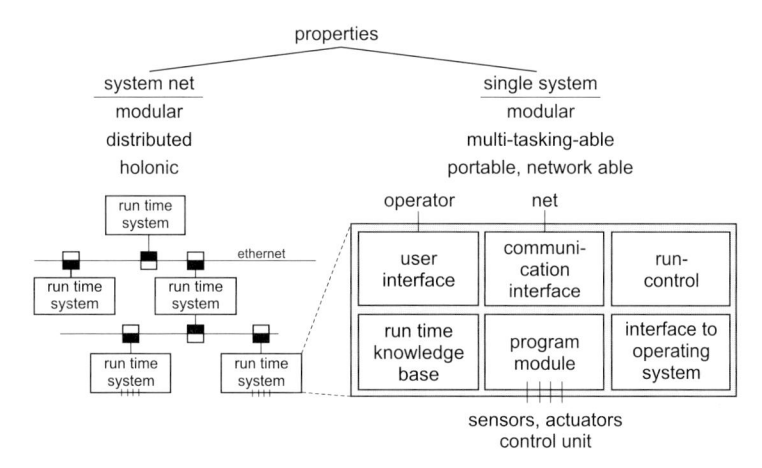

Fig. 2-31 Internal structure of the M & D kernel system. Source: Seidel [8]

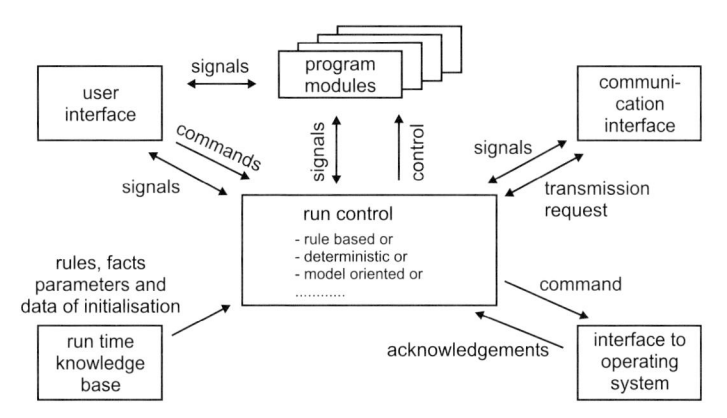

Fig. 2-32 Information exchange between elements of the run-time system. Source: Seidel [8]

ing to stored data of the knowledge base and the actual state of the machine. The data processing functions are implemented in separate program modules in which algorithms for signal processing tasks, communication functions, and functions to switch on sensors and actuators are integrated.

Demands on the availability of machine tools grow with the investment. Their complexity needs specific know-how to maintain or repair them. The consequence is that teleservice and telediagnostics are of growing interest. The economic benefits result from the fact that M & D systems may be expensive. It is interesting to use them only if they are necessary by telecommunication and not to place them locally. Several machine tool manufacturers already offer tele-M & D systems. Via an RS-232 interface the service personnel can communicate with the machine over long distances. The test can be run by a local machine operator in collaboration with the service technician at the manufacturer's office. It is also possible to run a part program to repeat critical states of the machine. Some machine control units are provided with ISDN interfaces. The service center can load specific diagnosis programs and evaluate the test results.

2.6
References

1 TÖNSHOFF, H.K., *Werkzeugmaschinen-Grundlagen*; Berlin: Springer, 1995.
2 TÖNSHOFF, H.K., WULFSBERG, J.P., KALS, H.J.J., KOENIG, W., VAN LUTTERVELT, C.A., *Ann. CIRP* 37 (1988) 611–622.
3 WECK, M., *Werkzeugmaschinen*, Vol. 2; Berlin: Springer, 1997.
4 STREIFINGER, E., *Dr.-Ing. Dissertation;* TU München, 1983.
5 FLAVELL, N.L., in: *Proceedings of 20th Annual Meeting and Technical Conference;* Cincinnati, OH: Numerical Control Society, 1983.
6 SEUFZER, A., *Dr.-Ing. Dissertation;* Universität Hannover, 1999.
7 POTTHAST, A., ZUGHAIBI, N., SUWALSKI, I., et al., *ZwF* 87 (1992) 269–273.
8 SEIDEL, D., *Dr.-Ing. Dissertation;* Universität Hannover, 1993.

3
Sensors for Workpieces

3.1
Macro-geometric Features
A. WECKENMANN, *Universität Erlangen-Nürnberg, Erlangen, Germany*

Measurement of macro-geometric characteristic variables involves the acquisition of features of geometric elements that are defined in design by dimensions and tolerances for dimensional, form, and positional deviations (Figure 3.1-1). The term 'dimension' refers both to the diameter of rotationally symmetrical workpieces and to distances and angles between planes and straight lines and to cone angles.

The sensors used for measurement can be classified according to the method used to acquire the measured value into mechanical, electrical, and optoelectronic sensors. A small proportion work by other methods, eg, pneumatic measuring methods.

The sensors mainly work with point-by-point, usually tactile measured value acquisition. Contactless and wide-area measurements of characteristic variables of the rough shape are possible with optical sensors.

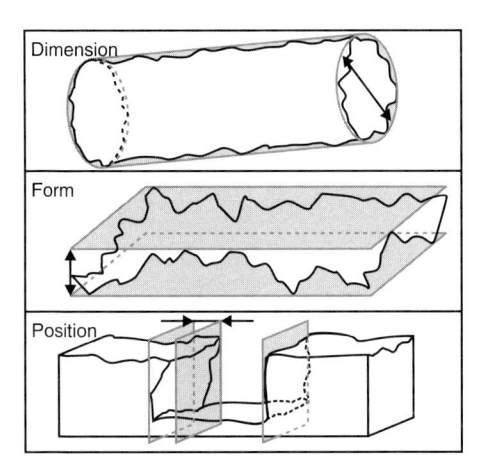

Fig. 3.1-1 Deviations of the macro shape of workpieces

3.1.1
Mechanical Measurement Methods

By far the greatest number of measuring systems used in dimensional metrology work with tactile probes and mechanical transmission of the measured value. For acquisition and indication of the measured value, a linear scale is usually used or the measured value is transmitted to deflection of a needle, say, by means of a rack and pinion. Indication is analog. Measuring instruments with a digital display usually use measuring systems with capacitive, inductive, or optoelectronic (Section 3.1.4) measured value acquisition.

3.1.1.1 Calipers

The various designs of calipers (DIN 862) are used for outside, inside, and depth measurements. The measured length is transmitted mechanically and a scale with millimeter divisions that can be read absolutely is used. Use of a Vernier scale provides an additional means of displaying 1/10, 1/20, or 1/50 mm graduations (Figure 3.1-2). The function, eg, of the 1/10 mm Vernier scale, is based on providing a length of 39 mm with 10 graduation marks at equal intervals. The point at which a graduation mark on the main scale is aligned with a graduation mark on the Vernier scale indicates the number of 1/10 mm on the measured length. Sometimes a division with 20 graduation marks or a rotary dial is used instead of the Vernier scale with 10 graduation marks.

Except for the depth gage, the scale of a caliper and the measuring object cannot be fully aligned. This violation of Abbe's comparator principle causes a sine deviation between the scale and the slider due to an angular deviation (Figure 3.1-3, Table 3.1-1). When expanding into a Taylor series, the angle of the tilt is included linearly in the result error. We therefore refer to it as a first-order error.

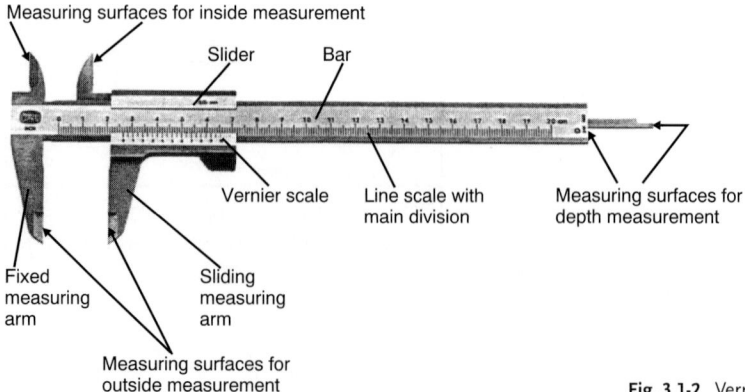

Fig. 3.1-2 Vernier caliper

Tab. 3.1-1 Sine deviation for a measuring arm length $l=40$ mm

	Angular deviation, φ			
	1′	5′	10′	1°
Sine deviation, f (μm)	11.6	58.2	116.4	698.1

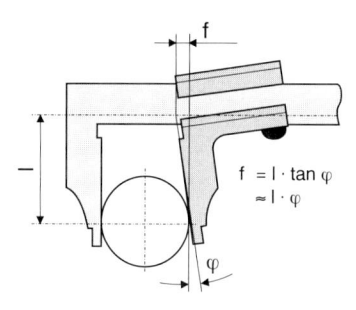

$$f = l \cdot \tan \varphi \approx l \cdot \varphi$$

Fig. 3.1-3 Violation of the comparator principle on a caliper

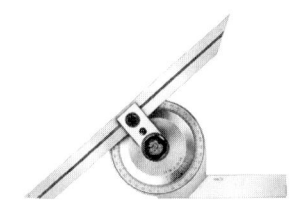

Fig. 3.1-4 Universal protractor (courtesy: Brown and Sharpe)

3.1.1.2 Protractors

A measuring instrument which works in an analogous way to the caliper is the universal protractor for measuring angles (Figure 3.1-4). The universal protractor also has an absolute angular scale and a Vernier scale, which allows the user to read off angular dimensions in steps of 5′. Models with a digital display are also available. Their smallest graduation is 1′.

3.1.1.3 Micrometer Gages

Some types of micrometer gages (DIN 863) can be used for the same tasks as calipers. Micrometer calipers (Figure 3.1-5) are used for outside measurements and inside measurements (measuring range usually about 25 mm) and depth micrometers for depth measurements. Drill-hole diameters can be measured using three-point inside micrometer gages.

A threaded spindle is used to transfer the measured value to the scale on the sleeve. The graduations on the sleeve indicate steps, each of which corresponds to one turn of the threaded spindle. A further, finer subdivision is also marked on a circumferential division on the scale thimble. The scale interval is usually 0.01 mm. A slip clutch ensures that the measuring force is limited. Insulation ensures

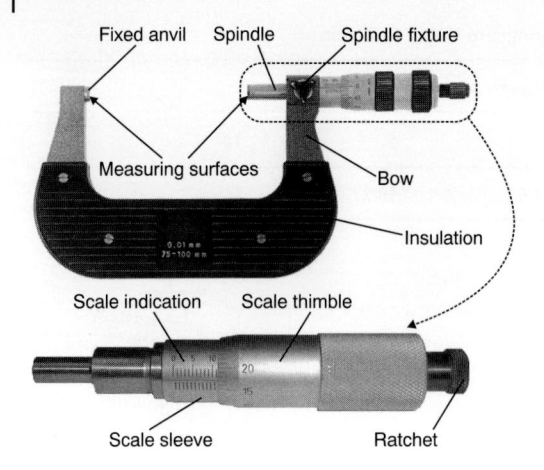

Fig. 3.1-5 Micrometer caliper with measuring head

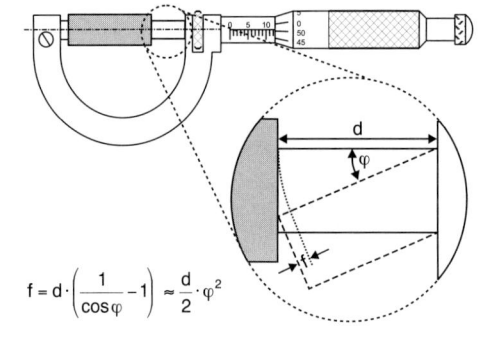

$$f = d \cdot \left(\frac{1}{\cos \varphi} - 1 \right) \approx \frac{d}{2} \cdot \varphi^2$$

Fig. 3.1-6 Cosine deviation in measurement using a micrometer caliper

that heat from the hands is not transferred to the measuring instruments, which could otherwise cause a thermally induced alteration in length.

Special inserts for the fixed anvil and the measuring surface of the spindle permit an extension of the application range. For example, if a notch and cone are used, it is possible to measure flank diameters on threads, and larger measuring contacts are used to measure tooth widths. Models with numerical or digital displays also exist.

Micrometer gages ensure that the measuring object and the scale are aligned. Since the comparator principle is not violated, no first-order measuring error can occur; only a second-order error remains (also called a cosine deviation, Figure 3.1-6), which is much less significant (Table 3.1-2). According to the measuring range, the maximum total discrepancy span is specified between 4 and 13 μm (DIN 863-1).

Tab. 3.1-2 Cosine deviation for spindle length $d = 20$ mm

	Angular deviation, φ			
	1′	5′	10′	1°
Cosine deviation, f (μm)	0.001	0.021	0.85	3.046

3.1.1.4 Dial Gages

With their comparatively short plunger travel (3 or 10 mm), dial gages (Figure 3.1-7a, DIN 878) are mostly used for differential measurements. Their applications are checking straightness, parallelism, or circularity. To determine an absolute dimension with a dial gage and stand, it is first necessary to set the required specified dimension with a material measure, say, a parallel gage block, and then to adjust the needle to a defined deflection (calibration).

The displacement of the measuring bolt is transmitted to a gear-wheel mechanism via a rack, converting the distance measured to needle deflection. The result is displayed on a circumferential scale with a scale interval of typically 0.01 mm. Since dial gages indicate a width of backlash, measurements should be performed only touching the measuring object in the same direction as when calibrating. Radial run-out measurements can therefore be afflicted with systematic errors. On dial gages, the needle can revolve around the scale several times over the entire plunger travel; a small pointer then counts the number of revolutions. Dial gages are also available in digital versions. The probe tip diameter is usually 3 mm, but numerous other probe styluses are available, eg, pointed, cutting edge, plane or ball measuring contacts, balls of other diameters, or measuring rollers. According to the measuring range, the maximum total discrepancy span is specified between 9 and 17 μm (DIN 878).

a) Dial gage   b) Comparator dial   c) Lever-type test indicator

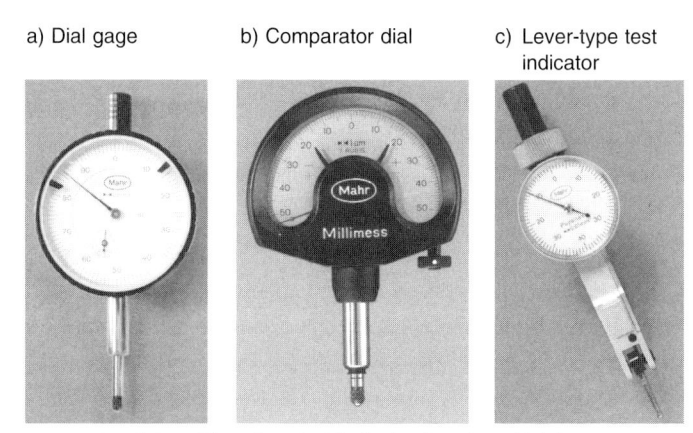

Fig. 3.1-7 Dial gage, comparator dial, and lever-type test indicator (courtesy: Mahr)

3.1.1.5 Dial Comparators

Dial comparators (Figure 3.1-7b) are also mainly used for differential measurements, but the measuring range is smaller than that of dial gages, usually under 1 mm, with a smaller scale interval, starting at 0.5 µm according to the standards (DIN 879-1, DIN 879-3). The needle deflection only extends over the angular range of the scale, and the motion of the measuring bolt is transmitted to the point via a lever mechanism or a torsion spring, indicating a negligible width of backlash. Comparator dials with contact limits are used, for example, to indicate violation of tolerance ranges with a special display unit. The maximum total discrepancy span is specified as 1.2 times the scale interval (DIN 879-1).

3.1.1.6 Lever-type Test Indicators

Lever-type test indicators (Figure 3.1-7c, DIN 2270) are similar to comparator dials in both form and function. The angular deflection of the stylus is also transmitted to the needle via a lever mechanism. A circumferential scale with a scale interval of 0.002 mm is used for display. The measuring range is smaller than 1 mm. Although lever-type test indicators use a circumferential scale, unlike on a dial gage, multiple revolutions of the needle around the scale are not recorded with an additional small needle. The admissible deviation is specified.

3.1.2
Electrical Measuring Methods

Electrical dimensional measurement has clear advantages over mechanical methods:

- low measuring forces;
- small dimensions of the measured value pickup;
- separation of the measured value pickup and the display unit;
- simple amplification and combination of measuring signals;
- possibility of electrical further processing of the measured length;
- easy connection to a computer and data processing.

This is offset by a greater handling effort.

It is possible to distinguish between three types of electrical dimensional measurement (Figure 3.1-8):

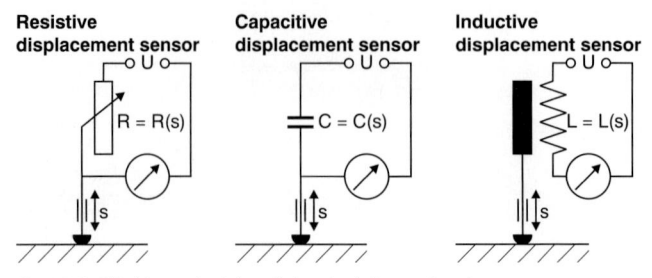

Fig. 3.1-8 Working principle of electrical dimensional measurements

- resistive displacement sensors;
- capacitive displacement sensors;
- inductive displacement sensors.

A length can be acquired either continuously and analog or incrementally. In incremental systems, numerous basic measuring elements (eg, magnets) are arranged consecutively at defined intervals on a scale and the number of zero crossings that the measuring bolt produces in the measured signal as it passes the measuring elements is counted. The measured value is therefore digitized. Common incremental methods of electrical dimensional measurement function magnetically, capacitively, or inductively. What all incremental measured value sensors have in common is a reference mark that they require to permit absolute measurements. The incrementally determined intervals then refer to this reference mark which is approached as soon as the instrument is switched on.

3.1.2.1 Resistive Displacement Sensors

Resistive displacement sensors in the form of potentiometers permit the measurement of lengths and angles. The resistance is varied in direct proportion to the linear or angular displacement via a sliding contact. The voltage, which depends on the resistance, is measured (Figure 3.1-9). Given a sufficiently high input resistance in the voltmeter, the following applies:

$$U_a = \frac{s}{s_0} \cdot U_0 \quad \text{or} \quad U_a = \frac{\varphi}{\varphi_0} \cdot U_0 \,. \tag{3.1-1}$$

Resistance displacement pickups are available with a wound resistance wire on an insulating main body, or with a continuous resistive layer applied to a material substrate. The disadvantage is the wear on the sliding contact.

3.1.2.2 Capacitive Displacement Sensors

Capacitive displacement measurement makes use of the effect that the capacitance of a plate capacitor depends on the distance between the capacitor plates.

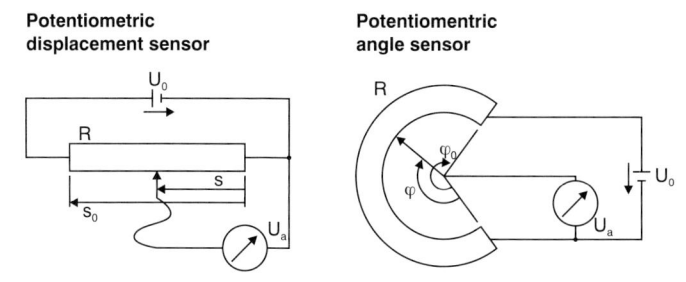

Fig. 3.1-9 Potentiometer length and angle measurement

On electrically conductive workpieces, contactless measurement is possible; the surface of the workpiece is then used as a moveable capacitor plate itself. The advantage lies in the almost inertialess measured value acquisition which, for example, permits circular or axial measurement on cylindrical parts rotating at high speed. One of its applications is therefore in-process monitoring of spindles in machine tools. On workpieces with insufficient electrical conductivity, the dimensional measurement has to be transmitted to a moving capacitor plate via a rigid measuring bolt.

If all the capacitor plates of a capacitive displacement sensor used in the differential method are identical, it is possible to measure voltage U_a depending on length s (Figure 3.1-10 shows a setup of a capacitive displacement sensor). The following applies:

$$U_a = \frac{s}{2s_0} \cdot U_0 . \tag{3.1-2}$$

In dimensional measurement, capacitive displacement sensors are actually used fairly rarely. They have become common as filling level meters and for the contactless measurement of material thicknesses.

3.1.2.3 Inductive Displacement Sensors

Most electrical dimensional measurement sensors function inductively, there being two different types of inductive displacement sensors: the plunger core sensor, in which the inductance of a coil varies as a function of the length measured, and the transformer sensor, in which the transformational coupling between two coils varies as a function of the length measured.

Inductive probes make use of the effect that in a coil carrying AC, an AC voltage is induced having the opposite polarity to the excitation voltage. The magnitude of the voltage depends on the inductance of the coil. This inductance can be varied by moving a magnetic core (plunger core) in the magnetic field of a coil. Because the inductance measurable via the induction voltage depends on the displacement of the magnetic core in a nonlinear way, the coils are connected in a differential circuit on inductive probes that produce an output signal that depends linearly on the displacement of the magnetic core after phase-dependent rectification. Two different types of probes are in common use: half-bridge probes on the plunger core sensor principle and LVDT probes on the transformer sensor principle (Figure 3.1-11).

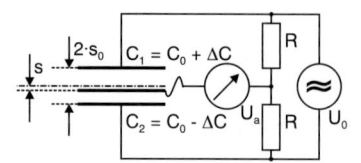

Fig. 3.1-10 Capacitive displacement sensors in the differential method

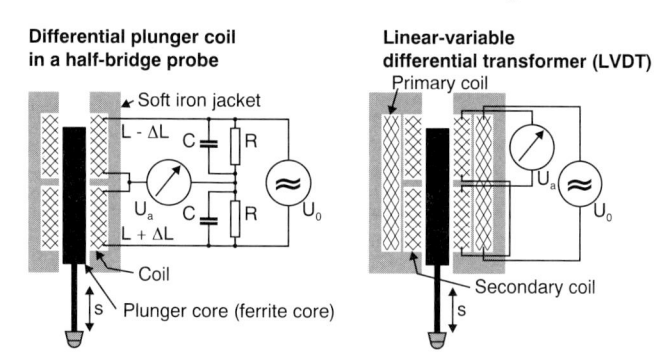

Fig. 3.1-11 Working principles of inductive probes

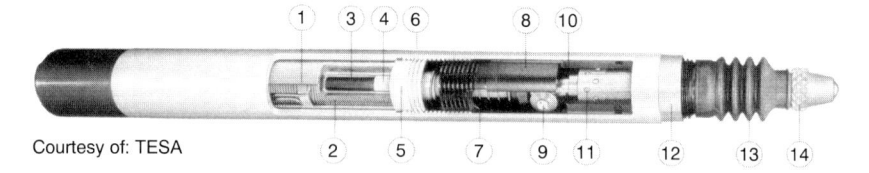

Courtesy of: TESA

1 Sleeve for magnetic core
2 Induction coils
3 Ferromagnetic core
4 Insulation for compensation of thermal expansions between the mechanics and the electronics
5 Stop for measuring force spring
6 Shaft
7 Compression spring
8 Guiding sleeve
9 Rotation protection guide
10 Measuring bolt
11 Ball cage
12 Travel limitation
13 Sealing bellows
14 Stylus

Fig. 3.1-12 Design of an inductive half-bridge probe (courtesy: TESA)

On half-bridge probes (Figure 3.1-12), both coils are directly fed an AC voltage of approximately 10 kHz. For the measurement signal, the ferrite core functions as a voltage divider. For the measured induction voltage U_a the following applies:

$$U_a = \frac{1}{2K} \cdot \frac{\Delta L}{L} \cdot U_0 \tag{3.1-3}$$

where ΔL is proportional to the displacement s and K is a constant. If the plunger core is precisely in the center between the two coils, the induction voltage is zero. The induction voltage increases if the plunger core is moved out of the central position toward one of the two coils. The maximum value is present if only one coil is completely covered by the plunger core. If it is moved further along the coil in the same direction, the induction voltage decreases again. The linearity

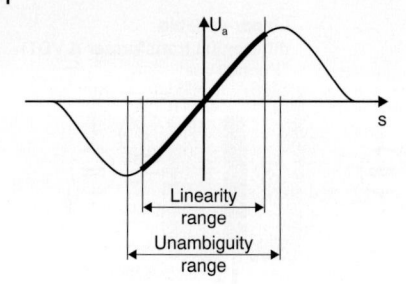

Fig. 3.1-13 Unambiguity and linearity of the measurement signal of an inductive displacement sensor

range in which the measurement signal is directly proportional to the displacement of the plunger core is smaller than and included in the unambiguity range (Figure 3.1-13).

LVDT probes, on the other hand, have one primary coil and two secondary coils that are arranged concentrically around the moveable plunger core. The primary coil receives an AC voltage of approximately 5 kHz that is transmitted to the secondary coils in phase opposition. The measurement signal U_a derived from the differential connection of the two secondary coils is directly proportional to the displacement s of the measuring bolt. The following applies:

$$U_a = K \cdot s \cdot U_0 . \tag{3.1-4}$$

Inductive displacement sensors can be operated with very small measuring forces (down to 0.02 N) on some types. Resolutions down to 0.01 µm and small linearity errors of below 1% permit high-precision dimensional measurements. They are also suitable for static and dynamic measurements. They are frequently used in multi-gaging measuring instruments and automatic measuring machines. When using inductive probes, the thermally induced zero point drift in µm/K, stating how the measured value indicated varies as a function of the temperature for a constant measured quantity, must be taken into account.

Eddy current measurement is a special case of inductive dimensional measurement, which is suitable for contactless distance measurement, if the workpiece material is electrically conductive. If a coil that forms a magnetic field is brought close to an electrically conductive body, eddy currents form within it which, in turn, form a magnetic flux with opposite polarity. This causes a reduction in inductance in the coil, which is electrically measurable. The change in inductance depends on the distance between the coil and the measuring object. For eddy current sensors in a differential circuit, a linear relationship is established between the distance and the change in inductance.

Fig. 3.1-14 Working principle of magnetic incremental linear measuring systems

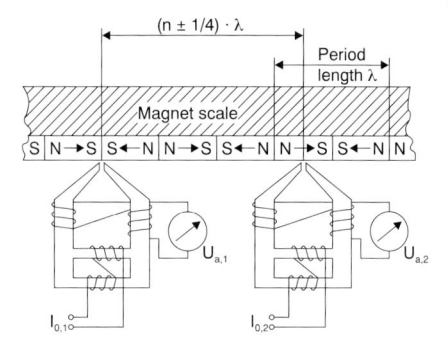

3.1.2.4 Magnetic Incremental Sensors

Incremental measuring systems based on magnets use a scale with permanent magnets as a material measure. The magnets are attached to the scale with alternating opposite polarity. The reading is obtained using ferromagnetic heads into which an excitation current with a defined frequency is injected.

Depending on the position of the magnetic poles of the reading head with respect to the permanent magnets in the scale, a different voltage will be generated at the output coils. If the poles of the magnet head are precisely symmetrical with respect to one pole of the scale, the magnetic fluxes produced by the excitation coil are shorted. In that case, no signal is present at the output coil. However, if the two pole shoes are precisely in the center between a north and a south pole of the scale, induction is caused in the output coil with double the frequency of the excitation current (Figure 3.1-14). Because of the different responses of the reading head to different positions with respect to the magnetic scale, it is possible to count how many period lengths of the individual permanent magnets are passed during motion along the scale. The direction of motion can be detected if two magnet heads are used for reading, which are arranged such that the measurement signal determined at the same time has a phase offset of one fourth of a period length. The direction of motion can be determined from the time sequence of the signal progressions between the heads.

3.1.2.5 Capacitive Incremental Sensors

Capacitive incremental sensors use scales on which a graduation of thin metal foil is attached. An identical metal foil is attached to the measuring element opposite the scale, so that the measuring element and scale together act as a capacitor. If the measuring element moves along the scale, the capacitance varies sinusoidally. It is possible to derive the number of graduation periods passed from the number of zero crossings. Additional interpolation of the sinusoidal signal permits resolutions up to 0.1 µm. Here, too, it is possible to detect the direction of movement by the fact that a second division is offset by a fourth of the graduation period. Capacitive incremental scales are used in calipers and micrometer calipers. The low energy requirement permits uninterrupted use of the measuring instrument with a battery for about 1 year.

3.1.2.6 Inductive Incremental Sensors

The best known example of an inductive incremental sensor is the inductosyn. Depending on the type, inductosyn measuring systems can be used for angular measurement (rotary inductosyn) or displacement measurement (linear inductosyn). A meandering conductor path is applied to a nonmagnetic, flat substrate material, eg, made of glass by means of an etching process. A scale with the required measurement length and a shorter, moveable cursor, on which two separate conductor paths offset by one fourth of a division period are the basic structure of an inductosyn (Figure 3.1-15).

An alternating voltage is applied to the conductor path of the scale. This induces a voltage in the conductor paths on the cursor according to the transformer principle. The amplitude of the induced voltage depends on the relative positions of the conductor paths on the scale and cursor. If the two paths coincide exactly, the amplitude of the induced voltage is at a maximum. If the conductor path of the cursor is precisely in the gap between the conductor paths of the scale, no induction occurs (Figure 3.1-16). The measurement signals are evaluated in an analogous way to the methods described above for magnetic and capacitive incremental sensors.

The inductosyn is frequently used for positioning in machine tools because it is largely insensitive to dirt, contrary to optoelectronic measurement methods de-

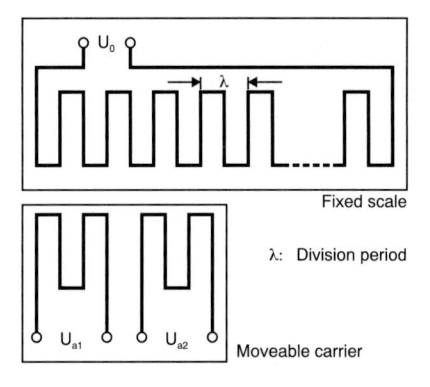

Fig. 3.1-15 Design of the linear inductosyn

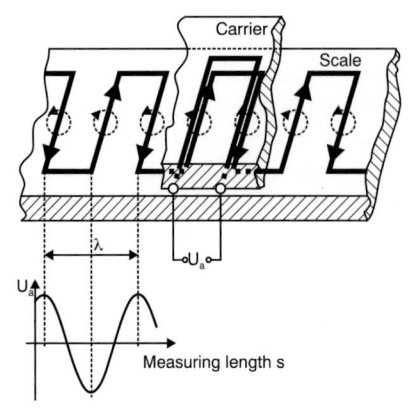

Fig. 3.1-16 How the measurement signal is obtained in an inductosyn

scribed below. On linear guideways it allows almost any measurement length by adjoining several scales. With a pole pitch of typically 1–2 mm, resolutions down to 1 µm can be achieved.

3.1.3
Electromechanical Measuring Methods

Each modern coordinate measuring machine (CMM) needs at least one sensing device for the workpiece data acquisition. Originally, sets of hard probes (spheres, cones, disk, and cylinders) were the only probes available for use with a CMM. Different probe tips are shown in Figure 3.1-17.

Today, electromechanical devices are used exclusively. The structure of such a sensing device system is crucial for the measuring accuracy of the CMM, which is appropriate within the range from a few µm to 0.1 µm. Today's touch trigger probe are always afflicted by their geometric structure with a first-order error (violation of the Abbé principle, triangle characteristics of the six-point support) (something similar applies also to measuring sensing devices). While probing the workpiece the touching strength results in a deflection and bending of the stylus shaft via stylus tip. The deflection is passed on to the measuring element (points of support, Figure 3.1-18) which are not aligned with the stylus shaft and affect the first-order error. These errors (first-order error, bending of stylus shaft) are compensated nowadays by time-consuming software calibrations. The result depends, however, on geometry of linkage, touching rate, touching force, and temperature.

Probes are made in a touch trigger or scanning mode.

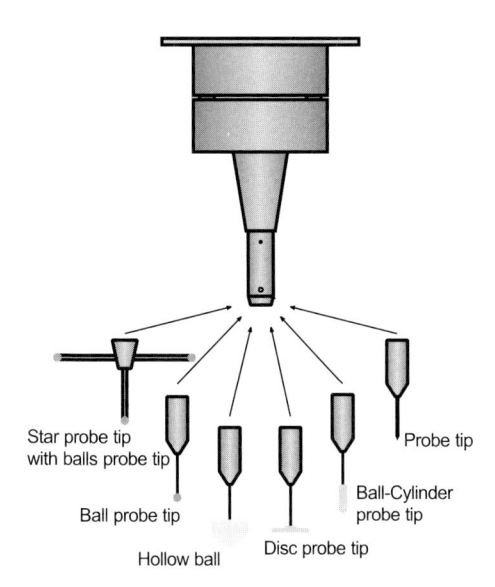

Fig. 3.1-17 Different probe tips

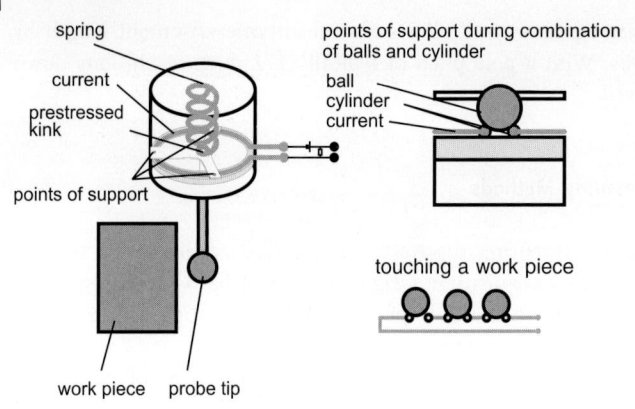

Fig. 3.1-18 Touch trigger probe

3.1.3.1 Touch Trigger Probe

The simplest sensor is the so-called touch trigger probe. It possesses a prestressed kink, which can be yielded in five or six coordinate directions and also in each intermediate direction. After sensing, the sensing device moves and returns in a spatially fixed resting position. This is uniquely made by three supporting points. A well-known principle is the combination of cylinder and balls.

The sensor signal is produced by opening and closing of one or more mechanical contacts (Figure 3.1-18) or establishing an external electrical contact between the probe and the workpiece. This signal leads to immediate freezing of length-measured values in all axes at their current value. This measuring procedure is called dynamic, since the measurement takes place during the movement of the probe relative to the workpiece.

For high accuracy, a rigid structure of the device is necessary, otherwise the acceleration forces necessary for movement cause deformations and thus inaccuracies. This principle is not completely free from accuracy losses, which are dependent on the preload and the measuring direction. These problems are overcome with piezo sensor touch trigger probes generating a trigger signal sensitive to tension and compression. Touch trigger probes are used at workpieces whenever individual points should be taken as fast as possible.

3.1.3.2 Continuous Measuring Probe System

The tactile three-dimensional precision measuring technique achieved in 1973 a new quality of three-axes probe measurement. By this means it is possible to read off the length-measured values with complete deadlock of the measuring axes of the coordinate measuring machines. In contrast to dynamic measurement with a switching probe, the measurement takes place statically (Figure 3.1-19). The result was a substantial increase in the measuring accuracy of coordinate measuring machines.

Fig. 3.1-19 Differences between dynamic and static measurement

Static taking of coordinates of point:
probe tip motionless, no deflection

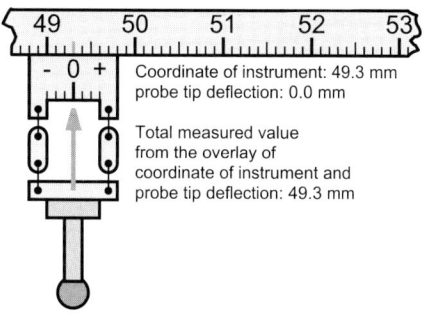

Coordinate of instrument: 49.3 mm
probe tip deflection: 0.0 mm

Total measured value from the overlay of coordinate of instrument and probe tip deflection: 49.3 mm

Dynamic taking of coordinates of point:
probe tip in motion, deflected

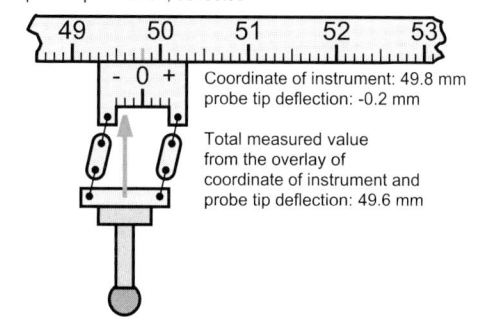

Coordinate of instrument: 49.8 mm
probe tip deflection: -0.2 mm

Total measured value from the overlay of coordinate of instrument and probe tip deflection: 49.6 mm

The base for all axes of the measuring probe is formed one by one by the use of three orthogonally arranged spring parallelograms. The deflection of the parallelograms is included in each axis by an inductive measuring system, where movement of a ferro magnetic core inside a coil produces distance-proportional analog signals. Each parallelogram can be actuated by a motor-operated locking mechanism in its center position (Figure 3.1-20).

Producing a defined measuring force for testing is complicated. For this purpose a moving coil is mechanically coupled on each parallelogram, moved itself in a toroidal magnet. The measuring force at a selectable level (between 0.1 and 1 N) is produced by a positive or negative current flow in the coil (according to the sensing direction).

The control automatically moves on clamping of the current sensing axes. The deflection after probing the workpiece with a stylus causes a combination of measuring force and a switching of the measuring carriage drive. This is position regulated by the signal of the inductive caliper. The drive is stopped by it at the zero point, ie, in the center position of the parallelogram. Also in the two other axes the probe, by clamping, is positioned at its spatial zero point. With probe heads which can simultaneously measure in all three axes, the force and direction vectors acting on the probe tip are determined from the measured total displacement. This simplifies the correction for bending in probe elements. The addition

Fig. 3.1-20 Continuous measuring probe system (courtesy: Zeiss)

of the probe values to the length-measured values at the CMM is applied in all three measuring axes; this is an important basic principle of the coordinate measuring technique with measuring sensors.

Commercially available probes of this type differ basically in the constructive arrangement of the three signal systems and the way in which they produce measuring forces.

3.1.4
Optoelectronic Measurement Methods

3.1.4.1 Incremental Methods
Optoelectronic dimensional measurement methods usually use scales (or angle disks) with an incremental division as the material measure. The applications range from encapsulated systems in probes with a measuring range up to approximately 50 mm to steel tapes with a measuring length of up to 30 m. The measured value is acquired by the front or rear illumination method. With the front illumination method (Figure 3.1-21), the indexing consists of two alternating types of line of the same width. One type of line (eg, made by coating with gold) reflects incident light directly, the other diffusely. The substrate material for the scale is usually steel. If the rear illumination method (Figure 3.1-22) is used, the division consisting of chrome lines is applied to a translucent glass substrate. To some extent glass ceramics are also used, which are insensitive to thermal expan-

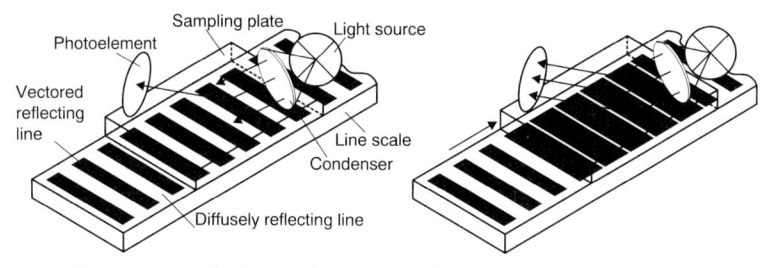

Fig. 3.1-21 Front illumination method (according to K. Tischler)

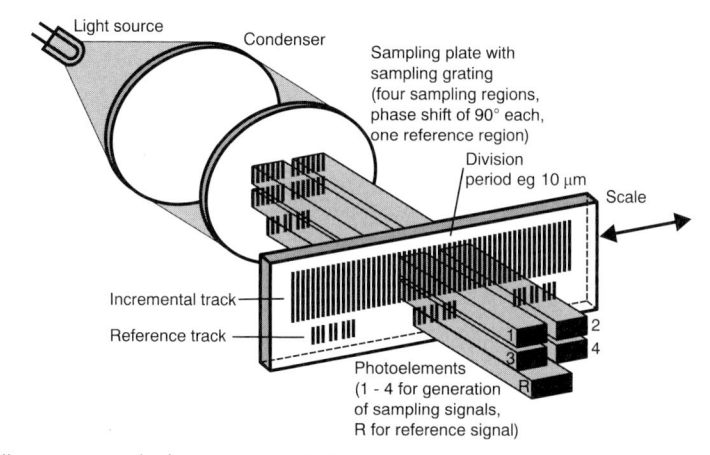

Fig. 3.1-22 Rear illumination method (courtesy: Heidenhain)

sion, eg, Zerodur. The chrome lines prevent light from passing through the glass scale at this point.

In both methods, measured value acquisition is based on the alternation between the maximum and minimum brightness with the relative motion of a sampling plate with respect to the scale, which can be detected by a light-sensitive sensor. The sampling plate consists of a translucent material with graduations of the same period as on the scale. High-precision scales in conjunction with interpolation algorithms for measured value acquisition permit resolutions down to the single-figure nanometer range.

Information about the direction of the relative motion between the scale and the sampling plate is obtained by using several sampling gratings each offset from the next by one fourth of a division period.

The absolute reference is established by passing a reference mark that is attached to a separate track of the scale. To avoid the nuisance of having to return to the reference mark on every power-on, especially on long scales, distance-coded reference marks on the reference track are often used. The reference marks have a defined different distance between them in the form of graduation marks on the main division so that the absolute position is known after no more than two distance-coded reference marks have been passed (Figure 3.1-23).

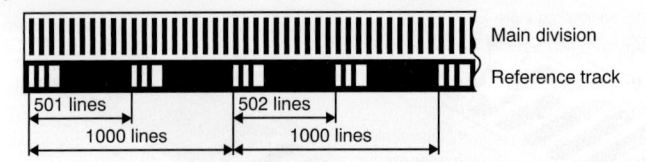

Fig. 3.1-23 Distance-coded reference marks (according to Heidenhain)

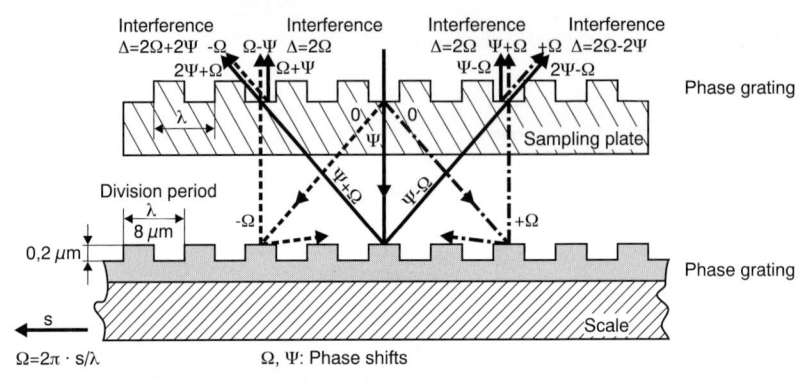

Fig. 3.1-24 Interferometric principle (courtesy: A. Ernst)

$\Omega = 2\pi \cdot s/\lambda$ Ω, Ψ: Phase shifts

The methods described with scales and sampling plates with alternating translucent and opaque zones use a change in the measured light amplitude for dimensional measurement.

A further incremental optoelectronic measuring method is based on the interferometric principle. A modulation of the light phase is used for dimensional measurement. The measurement method with front illumination uses a scale and a sampling plate with a phase grating. Steps made of gold are attached to the gold-coated, highly reflective scale with a division period λ of approximately 8 μm. The step increment is 0.2 μm, one fourth of the wavelength of the light to be measured. The sampling plate contains the same stepped structure. On its way through the air and sampling plate and on reflection on the scale, the light produced by a semiconductor light source is refracted three times. After its second passage through the sampling plate, beams having undergone different refractions and therefore having different phases interfere (Figure 3.1-24). Three parallel and interfering beams are directed on to photoelements through collecting lenses. These measure the light intensity which depends on the interference.

If a relative motion of length s is effected between the scale and the sampling plate, the first-order refraction of the light wave on the scale has phase offset $2\pi s/\lambda$. Displacement by one division period causes a positive displacement by one wavelength for the positive first order of refraction, and a displacement with an inverted sign for the negative first order of refraction. The relative displacement between the first two orders of refraction therefore corresponds to two wavelengths, which means that two signal periods per division period are measured.

3.1.4.2 Absolute Measurement Methods

The advantage of coded scales is that it is possible to determine the absolute position on the scale at any time without approaching the reference mark. They are used less frequently with front or rear illumination. They are much more complicated to handle and expensive to manufacture than incremental scales. Because of the underlying measuring principle, the distinction between light and dark, the coding on the scale is binary. Usually, binary code and Gray code are used (Figure 3.1-25).

Binary code has the property that more than one digit of the number representing the measured value can change value from one measuring increment to the next. Because of the limited precision of manufacturing of the division grating, optical detection of the edge between the graduation lines is only possible with a certain degree of fuzziness, so that it is not always possible to distinguish reliably between a logical 0 and 1. The probability of an erroneous value reading is different for different measured lengths. This behavior is countered by double sampling. Except for the track with the finest division, there are two read-off units on each track. Starting with the signal that is detected on the track with the finest division, the decision is made as to which of the two signals measured in the following track is to be used to form the measured value. This method is continued until the outer track is reached. If, for example, the signal 0 is applied to track n, the A signal is evaluated in track $n+1$ and the B signal if a 1 is applied.

Gray code is a single-step code: only one digit of the number representing the measured value changes value from one step to the next. In that case, one reading unit per track is sufficient.

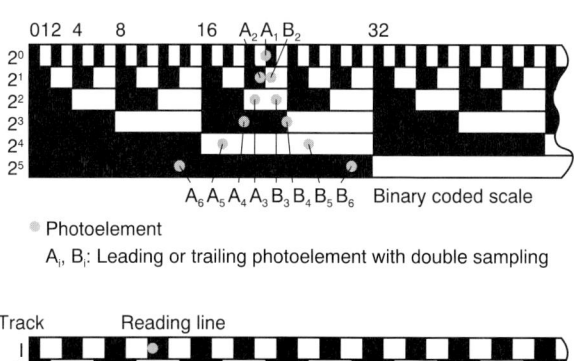

Binary Code	Gray Code	Decimal Code
00000	00000	0
00001	00001	1
00010	00011	2
00011	00010	3
00100	00110	4
00101	00111	5
00110	00101	6
00111	00100	7
01000	01100	8
01001	01101	9
01010	01111	10
01011	01110	11
01100	01010	12
01101	01011	13
01110	01001	14
01111	01000	15
10000	11000	16
...	...	...

Fig. 3.1-25 Absolute coded linear scales

3.1.5
Optical Measuring Methods

Optical measuring methods are being used for more and more applications in industrial quality control. Contactless access of the measuring object offers special advantages such as a high data transfer rate, ie, acquisition of a large number of sampling points within a short time. One disadvantage, however, is that the measurement results can depend on the surface properties of the measuring object, eg, its reflectivity, which is usually unrelated to its function. The comparability of the measurement results of optical and the established tactile measuring methods is not always given.

Depending on the measuring method, the surface of the measuring object is acquired point-, line-, or area-wise.

3.1.5.1 Camera Metrology

Camera measuring systems consist of an illumination facility, one or more charge-coupled device (CCD) cameras (line or matrix sensors) including imaging optics as well as image processing hardware and software. The scene to be observed can be illuminated by the front or rear illumination method. With the rear illumination method, the measuring object is located between the light source and the camera, so that its shadow is evaluated. With the front illumination method, the lighting and the camera are both on the same side of the measuring object, so that contrasting object details can also be detected. Moving objects can be measured by short exposure times (shutter or lighting with flash). Adaptation to different measuring fields is readily achieved by choosing a suitable focal length of the imaging optics.

Shape features are determined with computer support using special image processing algorithms, eg, for detecting edges, corners, or drill-holes in the gray-scale image acquired by the camera. It is then possible to calculate features such as distance, diameters, or angles.

A transversal magnification depending on the object distance may introduce systematic measuring deviations for absolute measurements. Telecentric lenses providing a constant magnification remedy this, as does calibration with defined test objects, eg, ruled gratings. The effective resolution can be increased by one order of magnitude over the physical resolution given by the number of active columns and rows of the CCD imager using subpixel methods. This might correspond up to approximately 0.005% of the measuring range depending on the number of camera pixels. Camera measuring systems are used for numerous and varied tasks. Depending on the complexity of the task, the evaluation times range from a few milliseconds to several seconds.

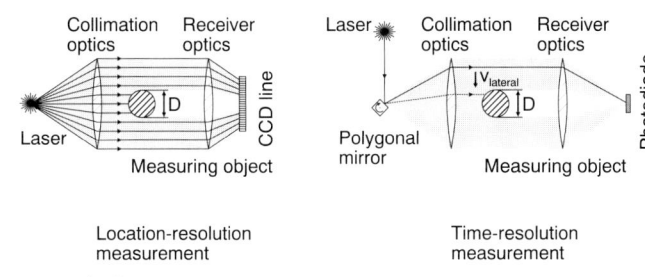

Fig. 3.1-26 Shadow casting methods

3.1.5.2 Shadow Casting Methods

In this measuring method, the measuring field is lit by collimated (parallel) light. The measuring object located in the measuring field then casts a shadow whose diameter D is measured (Figure 3.1-26). With position-resolution measurement, the shadow is imaged on a CCD line and, after calibration, the diameter D is determined directly from the position of the edges (bright-dark transitions) determined in the image. For smaller measuring fields, it is sometimes possible to dispense with the receiver optics.

For time-resolution measurement (laser scanner), a laser beam is directed at a polygonal mirror rotating with a constant angular velocity, which is located in the focal point of the collimation optics. By means of specially corrected $f\theta$ lenses, which provide a linear relation between the angle of incidence θ and the distance y, the beam is deflected parallel to the optical axis, and the beam is guided through the measuring field with constant lateral velocity. A photodiode acquires the time for which the laser beam is shaded and therefore the diameter D which is directly proportional to it.

With sampling rates of 10–100 MHz, it is possible to achieve 100–1000 measurements per second, so that it is also possible to measure moving objects. Depending on the size of the measuring field (typically 20–100 mm), resolutions down to 1 µm are possible.

3.1.5.3 Point Triangulation

With point triangulation, the light of a light-emitting or laser diode is focused on the surface of a measuring object. At a triangulation angle of typically 15–35°, the light point is imaged on to a linear detector (CCD line or position-sensitive photodiode). A change in the distance between the workpiece and the sensor causes a position change of the imaged light point on the detector which is usually inclined with respect to the optical axis of observation to increase the depth of focus according to the Scheimpflug condition:

$$\tan\theta \cdot \tan\beta = f/(d-f) \tag{3.1-5}$$

where θ denotes the angle of triangulation, β the Scheimpflug angle, f the focal length of the observing lens, and d the distance between the lens and the point of intersection of the optical axis of illumination and observation (Figure 3.1-27).

After calibration of the sensor, it is possible to calculate the distance between the workpiece and the sensor from the location of the light point imaged on the detector. A two- or three-dimensional measurement of the workpiece contour is possible by scanning, ie, by a defined change in the position of the workpiece relative to the sensor from one measurement to the next.

With working distances between 5 mm and 5 m and measuring ranges between 1 mm and 1 m with measuring times down to 0.1 ms, it is possible to achieve resolutions of down to 0.01% of the measuring range, with a lower limit of about 1 µm.

In addition to distance measurements and a completeness check, it is possible to implement thickness, flatness, or circularity measurements using several sensors and suitable handling equipment.

3.1.5.4 Light-section Method

The light-section method is an expansion of point triangulation. Using cylinder lenses or suitable prisms, or on high-quality systems using high-frequency oscillating or rotating mirrors, a line is projected on to the surface of the measuring object instead of a point. This line is monitored with a CCD matrix camera (Figure 3.1-27). The profile of the workpiece along the projected light line can be determined by its image on the CCD image sensor. The lateral offset of the line in each measuring point is evaluated, which depends on the distance from the sensor of the corresponding point on the surface of the measuring object.

Three-dimensional measurement of workpiece topography is possible if the projected line is guided over the workpiece surface, as in welded seam tracking in automated welding. With evaluation times of 0.1–1 s for one profile, the working distance, measuring range, and resolution are approximately the same as for point triangulation sensors.

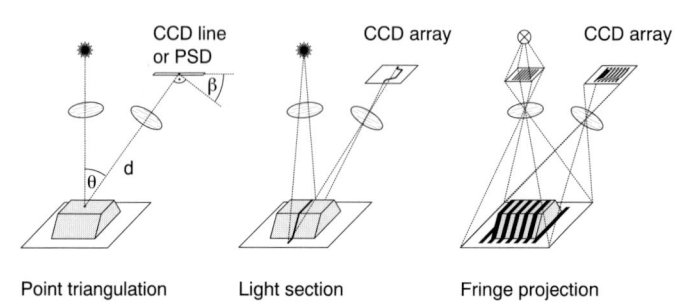

Fig. 3.1-27 Triangulation methods with active (structured) illumination of the measuring object

3.1.5.5 Fringe Projection

This term covers a number of associated techniques which are an expansion of the light section method that permit an area-wise acquisition of the surface of the measuring object. A white light source, eg, a halogen lamp, illuminates a mask (negative), an LCD (liquid crystal display), or a DMD (digital mirror device) array. Subsequently, a fringe pattern is projected on to the measuring object in accordance with the transparency of the mask or activation of the individual pixels of the LCD or DMD array, and acquired by a CCD camera. It is possible to calculate the topography of the measuring object from the deformation of the fringe pattern (Figure 3.1-27). Because unique assignment of the fringes observed is generally not possible, especially on objects with height steps, a sequence of fringes is projected and evaluated for a stationary object.

These might consist, for example, of a number of equidistant binary fringes, in which each individual fringe is projected in a unique sequence of light or dark (coded light approach, eg, Gray code). In that way, one can identify all pixels in the gray-scale image that are lit by a certain fringe. Evaluation of the lateral offset of each fringe is then performed as for the light section method. With this technique, a resolution of down to 1% of the longitudinal measuring range is possible for measuring and evaluation times in a range of seconds and minutes, respectively.

A more precise measurement is achieved by projecting a sequence of fringe patterns of the same period with a sinusoidal progression of intensity for a phase angle varying step by step (phase shift method). It is then possible to determine the phase in each measuring point from the detected intensities and to calculate a height value from it. Because the phase can only be uniquely determined cyclically, subsequent post-processing of the measured values is required to eliminate discontinuities (phase unwrapping). Unique determination of the phase is possible if, in addition, a Gray code sequence or further sinusoidal fringe patterns with another period are projected.

3.1.5.6 Theodolite Measuring Systems

Theodolites such as those used in surveying are used to measure workpieces of large dimensions, eg, in marine or aviation engineering. With two theodolites, whose spatial position with respect to each other is known, readily locatable points on the measuring object, eg, corners or attached gage marks, are aimed at. This is done by directing the telescopes on the theodolites at that point. The vertical and horizontal angular coordinates of the telescopes are then acquired by the evaluation computer. From these it is possible to calculate the three-dimensional coordinates of the point aimed at by triangulation.

3.1.5.7 Photogrammetry

Photogrammetry is also a technique that originates from surveying. The application is therefore similar to that used by theodolite measuring systems to measure large objects. A further common feature is that for calculation of three-dimensional information, well-defined features such as gage marks attached to the workpiece surface have to be acquired. One or more camera systems are used, imaging the measuring object from at least three different positions. If common identification of the marks acquired is possible in the camera images, ie, one mark in one image can be uniquely assigned to one mark in another image, it is possible to calculate the spatial position of the mark by means of triangulation. To facilitate the necessary assignment of point marks, they can contain, for example, a unique coding.

3.1.5.8 Interferometric Distance Measurement

Interferometric distance measuring systems are based on the principle of a Michelson interferometer. The light from a laser source is divided by means of a beamsplitter into a measuring and a reference beam. The reference beam is reflected at a stationary triple mirror (cube corner) and the measuring beam at a mobile mirror, which can be moved along the measuring path. Compared with plane mirrors, they possess the advantage of reflecting a beam back into the direction of inclination independent from a slight tilt or lateral shift of the mirror. When shifting the mobile triple mirror along the measuring path, the optical path of the measuring beam changes, so that both beams, after a second passage through the beamsplitter, alternatively interfere constructionally or destructively. The corresponding light/dark changes of the interference signal are detected with a photodiode (Figure 3.1-28).

While moving the triple mirror, the distance between two detected intensity maxima corresponds to half the wavelength of the laser, which represents the solid measure of this measuring method. For high-precision measurements, there-

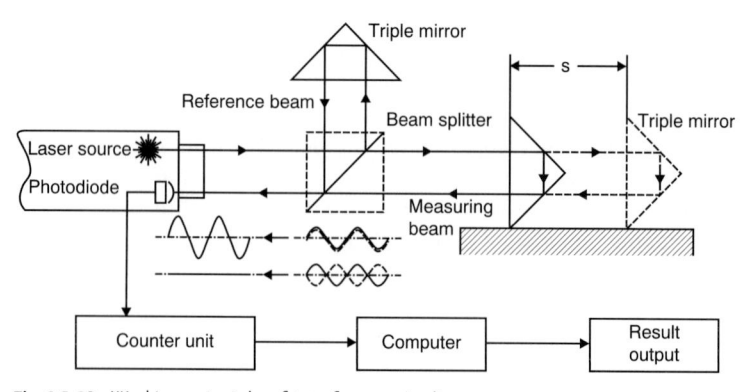

Fig. 3.1-28 Working principle of interferometric distance measurement

fore, the dependence of the light wavelength λ on the refractive index n of air is problematic:

$$\lambda = \lambda_0/n. \tag{3.1-6}$$

Whereas the vacuum wavelength λ_0 can be kept almost constant by suitable measures (eg, frequency stabilization of the laser), the refractive index depends on chemical composition, temperature, humidity, and atmospheric pressure of the air in the environment of the measuring system. In order to compensate for its influence on the measuring result, different procedures are common. With the parameter procedure temperature, humidity and atmospheric pressure are measured and the refractive index is estimated according to the Edlén equation to calculate the wavelength λ. It is unfavorable, however, that the composition of air is not involved here and so, as a cause of systematic measuring deviations, is not compensated. A more exact but also more complex alternative consists in a direct measurement of the refractive index by means of a refractometer. With this procedure relative measuring uncertainties smaller than $5 \cdot 10^{-7}$ can be realized.

Apart from the linear distance measurement with a resolution up to 5 nm with measuring lengths into the range of meters, with a suitably modified set-up measurements of straightness or angular deviations are also possible. Because of sensitivity to environmental conditions, which cannot always be controlled sufficiently, laser interferometers are used less in the production process than as a reference measuring system within the acceptance procedure of machine tools or coordinate measuring machines.

3.1.5.9 Interferometric Form Testing

Apart from distance measurements, a multiplicity of interferometric measuring methods for access to the area of the workpiece surface exist, which are continuously spreading within the area of form testing of workpieces with high-precision machined surfaces, eg, optical components.

Their application in machine engineering is limited by boundary conditions inherent to the measuring method, eg, sensitivity to vibrations or the comparatively greater roughness of technical surfaces, which cause speckle when lit with coherent light and so disturb the interference signal.

Depending on the geometry of the workpiece under test, different types of interferometers are suitable, eg, Michelson interferometers for plane or Twyman-Green interferometers for spherical geometry.

Additional measuring methods are based on the application of diffractive optical components (holographic interferometry) or are applicable at optically rough surfaces (speckle interferometry, white light interferometry).

3.1.5.10 Autofocus method

The autofocus method permits a high-precision pointwise distance measurement. To achieve this, the light from a laser diode is first collimated and then projected on to the surface of the measuring object with a movable lens. With position control, the lens position is corrected such that the focus in the illumination beam path lies on the surface of the measuring object. Deviations from this are detected using the light reflected back on to a focus detector via a lens, collimator, and beamsplitter, eg, in accordance with Foucault's method consisting of an aperture stop and a differential photodiode (Figure 3.1-29).

If the measuring object is in focus, it is imaged sharply between the two segments of the differential photodiode. Otherwise, the aperture stop causes the fuzzy image of the projected light spot to light only one of the two segments and the position of the lens can be corrected. A position sensor on the guide of the lens acquires its position and provides the measuring signal. A two- or three-dimensional measurement of the contour is achieved by scanning the measuring object.

The advantage of this method is its very small lateral expansion of the measuring point (approximately 1 µm) with a longitudinal resolution of approximately 10 nm; the disadvantages are the small working distance (maximum 15 mm) and measuring range (maximum 300 µm), and the relatively long measuring times due to the sequential measuring value acquisition.

Autofocus systems are used to measure the topology of small components and in roughness metrology (only in the case of waviness).

3.1.6
Pneumatic Measuring Systems

Pneumatic dimensional measurement methods (DIN 2271) are based on the principle by which the narrowest cross section of a flow duct is influenced by the

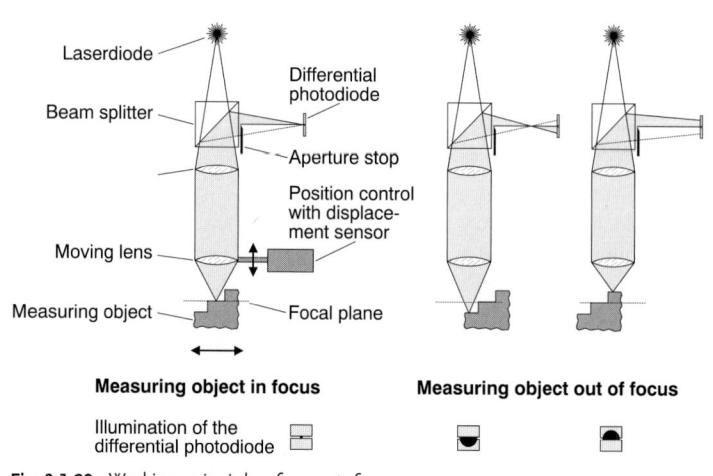

Fig. 3.1-29 Working principle of an autofocus sensor

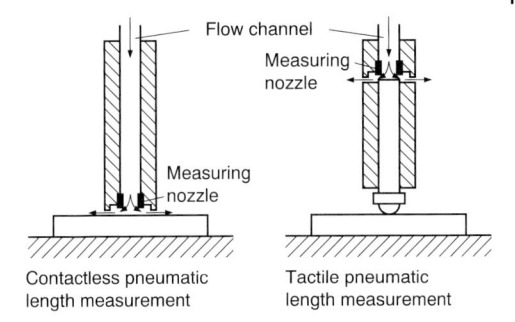

Fig. 3.1-30 Pneumatic dimensional measurement systems

length to be measured, which provides measurable variations in the volume of air flowing through. If the narrowest cross section is the outlet opening of the flow duct, it is immediately possible to determine contactlessly the distance between the outlet opening and the workpiece by the volume flow-rate, because within a certain linearity range this is directly proportional to distance s. In addition to contactless sensors, tactile pneumatic probes are also available with which the narrowest cross section of the flow duct is influenced by displacement of a measuring bolt (Figure 3.1-30), developed to exclude influences of roughness and porosities of the surface. The sensors usually provide a measuring range from a few micrometers up to 1000 µm. They are used for differential measurements, in a similar way to comparator dials. Typical applications consist in the measurement of distances, diameters, linearity deviations, conical gradients, and distances between holes.

The advantages of contactless pneumatic sensors are that the workpiece surface is not damaged during measurement and that the compressed air used cleans greased, oiled, or otherwise contaminated surfaces, at least at the measuring points. The automatic alignment of drill-hole diameters during measurement can also reduce measuring times.

The disadvantages are the strict requirements for constancy and cleanness of the compressed air used, further processing of the measured values is possible only after conversion to an electrical quantity, and for rough surfaces (peak-to-valley heights >3–5 µm) contactless measurement no longer provides reliable measured values.

3.1.7
Further Reading

1 ADAM, W., BUSCH, M., NICKOLAY, B., *Sensoren für die Produktionstechnik;* Berlin: Springer, 1997.
2 DEUTSCHE GESELLSCHAFT FÜR ZERSTÖRUNGSFREIE PRÜFUNG, *Handbuch OF 1: Verfahren für die Optische Formerfassung;* Eigenverlag, 1995.
3 DUTSCHKE, W., *Fertigungsmeßtechnik;* Stuttgart: Teubner, 1993.
4 ERNST, A., *Digitale Längen- und Winkelmesstechnik;* Landsberg/Lech: Verlag Moderne Industrie, 1989.
5 GASVIK, K.J., *Optical Metrology;* Chichester: J. Wiley, 1995.
6 GEVATTER, H.-J., *Handbuch der Mess- und Automatisierungstechnik;* Berlin: Springer, 1999.
7 LEMKE, E.; *Fertigungsmeßtechnik;* Braunschweig: Vieweg, 1992.
8 PFEIFER, T., *Fertigungsmeßtechnik;* Munich: Oldenbourg, 1998.
9 SCHLEMMER, H., *Grundlagen der Sensorik;* Heidelberg: Wichmann, 1996.

3.2
Micro-geometric Features
A. WECKENMANN, *Universität Erlangen-Nürnberg, Erlangen, Germany*

Precision measurement of structures in the micrometer and sub-micrometer ranges is becoming more and more important. Because of the never-ending miniaturization it is central to the precision of production and metrology of microelectronics and micromechanics, but also to the measurement of the size distribution of microparticles, for example, in environmental protection. A number of measuring methods are available to perform these tasks. They range from conventional optical microscopy and its extension into the ultraviolet range, through electron microscopy, to the high-resolution near-field microscopy methods such as atomic force microscopy.

Optical microscopy includes conventional bright- and dark-field microscopy, confocal scanning microscopy, in the visible and ultraviolet spectral ranges, and interference microscopy. As a non-microscopic additional feature, far-field diffraction images of the objects are evaluated. Fundamental research into the interaction of the radiation used with the objects and theoretical modeling are important, additional aids in using these methods. Non-optical, high-resolution microscopy methods (scanning electron microscopy (SEM), atomic force microscopy (AFM), etc.) are currently used to examine and assess the microgeometry of the structures to be measured which cannot be resolved by light-optical methods, as a supplement to optical measuring methods. After further extensive research into the interaction of the scanning probes with the object structures and specific extension of microscope systems, eg, adding precision length measurement systems, high-resolution microscopy methods can also be used for calibration. So far this has not been possible because the principle of optical methods places a limit on the resolution that can be achieved.

3.2.1
Tactile Measuring Method

Tactile measuring methods for surface measurement are still the most important methods, especially in the area of metal-cutting and non-cutting machining operations in industry and research. It is the only operation that is anchored in national and international standards. Particularly the parameters and measurement conditions are fixed, so that the comparability of the measurement results can be secured. The surface roughness and topography greatly affect the mechanical and physical properties of parts. Properties such as fit, seal, friction, wear, fatigue, adhesion of coatings, electrical and thermal contact, and even optical properties such as gloss, transparency, etc., can be adjusted by manufacturing design. The surface laboratory is concerned with the assessment of roughness, waviness, texture, groove depth, and other special surface shapes. The contact stylus method is generally set-up off-line in the measuring room or in the workshop. Only in special cases are oil-proof calipers integrated into the processing equipment. The profile method is based on the linear sampling of the workpiece surface with a diamond needle whose tip has the shape of a cone or a pyramid (Figure 3.2-1). The radius of the tip is 2 and 10 µm and its angle usually 90°.

The static measuring force applied is less than 1 mN. Thereby, equidistant profile supporting points are measured directly to calculate various roughness and waviness characteristics. The commencement of this method dates back to about 1930. Nowadays, measurement systems with digital signal processing and profile evaluation are available. The instruments can be adjusted to fit the workpiece flexibly by modularly compiling the stylus instrument, feed mechanism, and evaluation system. Contact stylus instruments generally register a two-dimensional vertical profile cut in the workpiece surface. Latterly, its application has expanded by

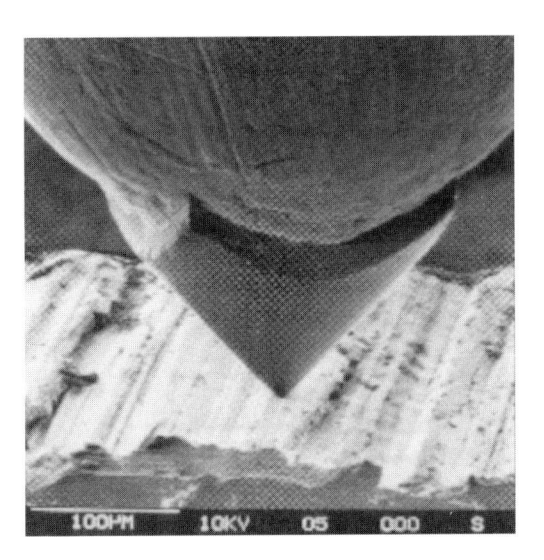

Fig. 3.2-1 Probe tip (courtesy: PTB)

introducing a successive cross traverse for the three-dimensional measurement of surface topography.

The amplitude resolution can be as good as 10 nm at any measurement point, and the best possible local resolution in the horizontal axis is 0.25 µm. The measuring range for contour measurements extends to 120 mm along the plane of the face and 6 mm in amplitude. The contact stylus instrument is traceable to the unit meter through reference standards.

Alignment of the cantilever is problematic. Additionally, the measuring instrument is sensitive to vibrations and oscillations. A further problem in some cases is a curved form of the surface of the workpiece.

For the adaptation of different workpiece geometries, a variety of different tactile profile meters exist, whose properties clearly determine the quality of the surface measurement. They can generally be traced back to the basic reference surface, skidded and double skidded system.

3.2.1.1 Reference Surface Tactile Probing System

In the skidless system (Figure 3.2-2), the stylus is located at the end of a probing arm that is guided over the surface of the object to be measured held in a linear guide in the vertical direction. The styli are rigidly connected with a reference plane that is usually located in the feed mechanism. The excursions of the stylus caused by the surface roughness are transmitted to a measuring transducer and converted to measuring signals, depending on the type of transducer, in analog or digital format. The measuring pick-up and the object to be measured are mechanically decoupled, and only the stylus itself slides over the surface of the object being measured. For that reason, skidless systems are extremely sensitive to vibrations.

3.2.1.2 Skidded System

The skidded system (Figure 3.2-3) uses the surface to be measured as a guide and has much smaller dimensions than the skidless system. The stylus contacts the surface to be measured with a skid and acquires the surface profile relative to the path of the skid with the probe tip. Depending on the measurement task, the

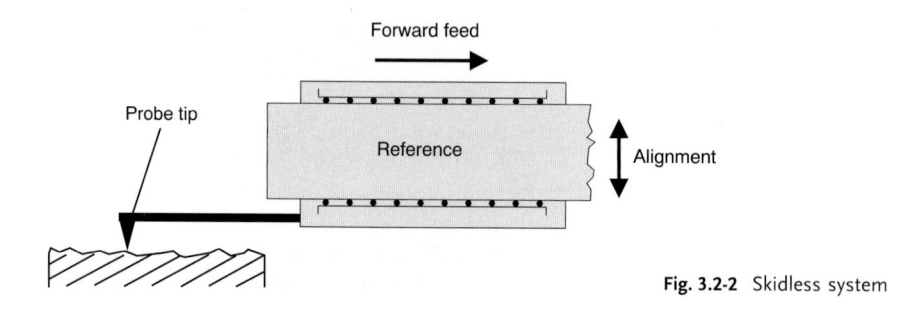

Fig. 3.2-2 Skidless system

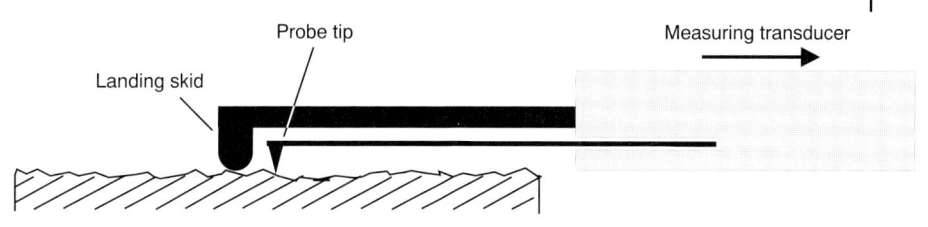

Fig. 3.2-3 Skidded system

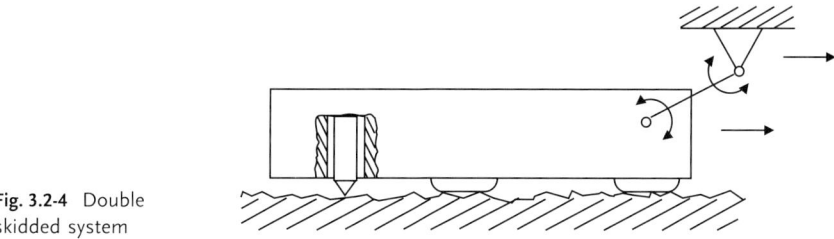

Fig. 3.2-4 Double skidded system

landing skid is mounted before, behind, or lateral to the probe tip. The unavoidable distance between the landing skid and the probe tip can lead to falsifications during the transfer of the profile, depending on the surface attributes of the workpiece to be measured. However, it is less precise because of the mechanical filtering that occurs while sliding over the surface. The skids act as an amplitude-independent, non-linear, high-pass filter and eliminates, depending on the probe and workpiece geometry, the macro-geometric form and waviness of the workpiece profile. This system is used for fast measurements in production.

3.2.1.3 **Double Skidded System**
The double skidded system (Figure 3.2-4) uses the surface under test as a reference, it is self-aligning, insensitive to vibrations, and requires large measuring surfaces because of its size.

The double skidded system can lead to considerable profile falsification owing to its landing skid, especially with profile tips that jut out.

3.2.2
Optical Measuring Methods

Optical 3D measuring methods permit fast, wide-area sampling point acquisition. In several measurements from different views, it is possible to measure all wearing zones and zones of the workpiece relevant to determining the form and surface characteristics of the workpiece with the required resolution. After transformation of the measured data into a common coordinate system, the sample is represented by a 3D set of sampling points. From the measured data it is possible to determine the form, surface, or wear characteristics. The advantages of this method are that the measuring process can be automated to a great extent and is therefore independent of the influences of the operator, it has a high measuring rate, and the surface of the measured object is acquired as a whole. Especially suitable for measured value acquisition for microgeometry are devices that operate on the principle of white-light interferometry or scattered light methods.

3.2.2.1 White Light Interferometry

Special white light interferometers permit wide-area form acquisition on optically rough surfaces. A measuring system called coherence radar is based on the principle of the Michelson interferometer, where the mirror in the measuring beam is replaced by the object to be measured. The light-emitting diode (LED) to be used as the light source causes white light interference that displays a typical modulation as a function of the phase shift between the measuring and reference beam, which is at a maximum when no phase shift exists, ie, the object being measured is in the reference plane (Figure 3.2-5).

Using a linear table, the object to be measured is pushed through the reference plane and the position of the linear table is stored as a vertical coordinate for each sampling point as soon as the maximum modulation of the interference signal acquired with a charge-coupled device (CCD) camera is detected. One advantage of this measuring method is that, unlike, for example, the triangulation method, illumination and observation are in the same direction, which makes measurements possible on structures with a large aspect ratio that are often encountered in microsystem technology. Both the topography of the measured object and, derived from it, the roughness of its surface can be measured.

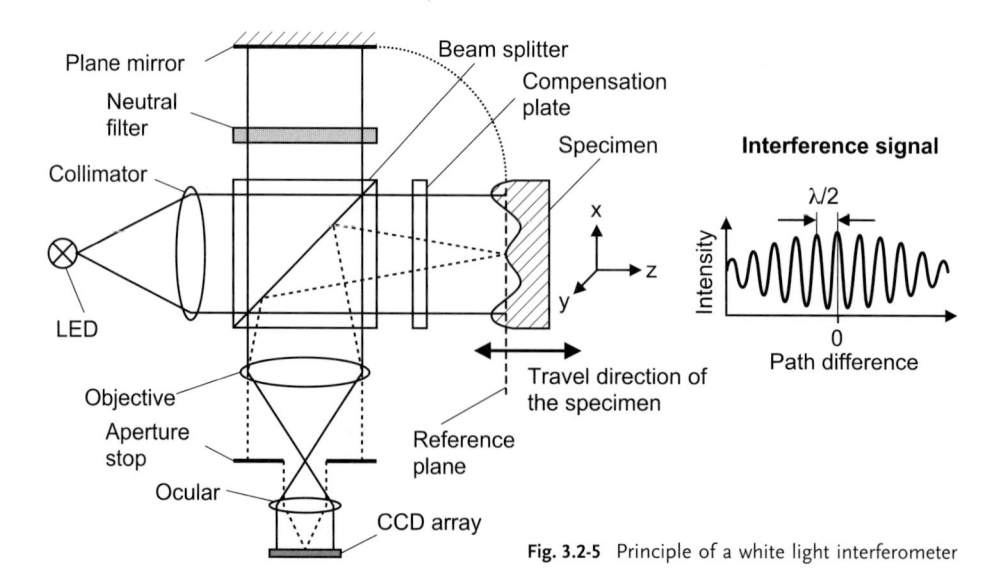

Fig. 3.2-5 Principle of a white light interferometer

Depending on the optical arrangement, measuring fields from about 50×50 mm to about 200×200 µm can be implemented. The lateral resolution depends on the measuring field and the number of columns or rows of the CCD camera (typically 512×512 pixels) and the pixel geometry; the resolution in the longitudinal direction is limited by the roughness of the workpiece surface and the traversing speed of the linear table (typically 1–2 µm at 4 µm/s). The measuring time is in the minutes range, depending on the maximum structure depth to be measured.

3.2.2.2 Scattered Light Method

The scattered light method is used to measure the roughness of workpiece surfaces. Light reflected from the workpiece has a spatial distribution that depends on the surface roughness. Smooth surfaces reflect incident light fully according to the law of reflection of geometric optics (angle of reflection with respect to the surface normal equal to the angle of incidence). On rough surfaces, portions of the scattered light are also reflected in other directions. Figure 3.2-6 shows the arrangement principle of a scattered light sensor. The collimated light of an LED is deflected on to the workpiece surface via a beam divider. The diameter of the measuring spot is about 1 mm. The scattered light is mapped with a lens on to a linear image sensor (photodiode or CCD line) so that the intensity of the light scattered in different directions can be measured at different locations on the detector. To assess the surface, the scatter value S_N is usually used. This is proportional to the second statistical moment of the measured intensity distribution and therefore describes its width. Larger S_N values indicate a greater proportion of scattered light, usually describing a rougher surface. One problem with the acceptance of this measuring method is that the measured S_N value does not correlate

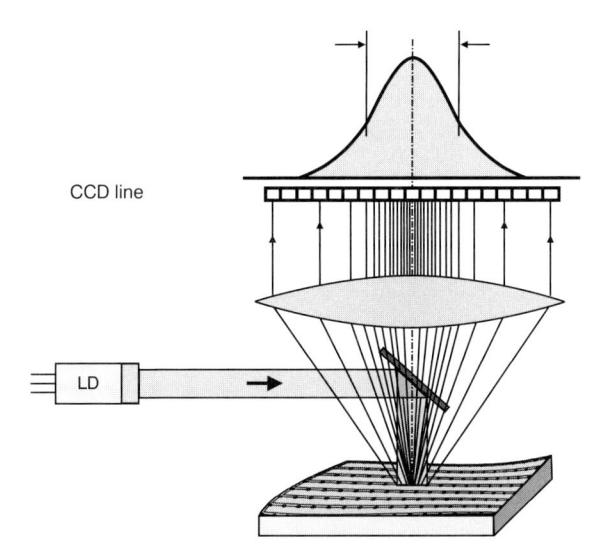

Fig. 3.2-6 Block diagram of a scattered light sensor

with the roughness quantities R_a and R_z which have been introduced into tactile roughness metrology and which are standardized.

3.2.2.3 Speckle Correlation

Speckle correlation differs between two methods: angular speckle correlation (ASK) and spectral speckle correlation (SSK). Both methods use the correlation coefficient of two takes of the surface as a measure for the roughness value R_q. This roughness value is equal to the standard deviation of the altitude values, so that a mathematical connection between this and the correlation of different speckle pictures exists as shown in statistical calculations.

Angular speckle correlation (ASK). Angular speckle correlation offers two advantages. On the one hand, the requirements for the laser system are small, since only one wavelength is necessary for the implementation of the measurement. On the other hand, owing to the difference in the angles of illumination, the measurement area can continuously be re-adjusted according to the measurement task. The disadvantages of an adjustable difference angle result in high requirements for the mechanical precision. Figure 3.2-7 displays a typical experimental set-up for ASK measurements with an adjustable difference angle. One of the illumination beams is faded out when taking the first picture, and for the second picture the other is faded out. It is necessary to move one of the pictures respective to the deviating illumination angle of the applied ASK, so that the offset opposite the second picture can be counter-balanced. The distance moved can roughly be

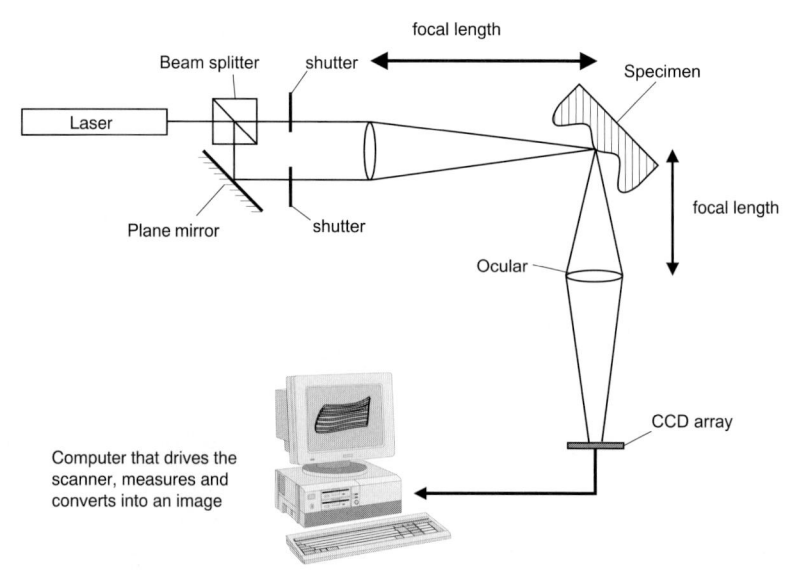

Fig. 3.2-7 Setup of an angular speckle correlation

calculated from the geometry of the setup and is always the same for a fixed setup. The exact value can be calculated in an evaluation program.

Spectral speckle correlation (SSK). The setup of SSK is simplified to the extent that no second illumination beam path is necessary. The adjustment of the measurement system during practical application is far easier and one can achieve an increase in stability. With the possibility of taking two pictures of the surface at the same time, the measurement time can be reduced. The disadvantage of this measurement system is the higher requirements for the laser system. At least two different wavelengths must be generated, so that an adaptation of the different roughness areas becomes possible. A larger number of wavelengths is, however, more advantageous.

The evaluation of the two pictures taken takes place via a two-dimensional cross-correlation coefficient. Experimental prerequisites for the correct evaluation consist in the observance of Shannon's theorem. This means that the spatial sampling frequency, in this case the reciprocal pixel size of the CCD camera, has to be at least twice as large as the spatial signal frequency. In other words,

$$d_{speckle} > 2d_{pixel} \tag{3.2-1}$$

$$d_{speckle} = \frac{4}{\pi} \cdot \frac{\lambda f}{2\omega_0} \tag{3.2-2}$$

where λ is the wavelength of the light used, f the focal length of the lens and ω_0 the diameter of the illuminated area on the surface.

3.2.2.4 Grazing Incidence X-Ray Reflectometry

The total reflection of X-rays from solid samples with flat and smooth surfaces was first reported by Compton in 1923, which can be assumed to mark the birth of the experimental technique of X-ray specular reflectivity. Since the angle of incidence is very shallow and almost parallel to the surface, measurement using X-ray total reflection is also called the grazing incidence experiment. If the surface is not ideally smooth but somewhat rough, the X-rays can be diffusely scattered in any direction. The experimental technique is known as X-ray diffuse scattering (X-ray non-specular reflection). Its development began immediately after the pioneering work in 1963 of Yoneda, who reported intensity modulation in X-ray diffuse scattering, known as Yoneda wings or anomalous reflection.

Nowadays, X-ray reflectometry based on total reflection has become a powerful tool for the analysis of surfaces and thin-film interfaces, and will continue with further progress. This is mainly due to the significant development of experimental techniques and instrumentation, especially the advent of synchrotron radiation and the progress achieved in detector technology. The advances in theoretical modeling and techniques for analyzing experimental data are also important.

Total reflection and the penetration capability of energy-rich X-rays are used for coating thickness measurement. The refractive index for X-rays is always <1. If the angle of incidence is made smaller, the X-radiation penetrates only up to a very small angle, the critical angle. If the angle of incidence is reduced still further, external total reflection on the interface occurs. The beam is reflected as by a mirror. In coating-substrate systems, part of the radiation is reflected and part of it penetrates the film. There are now two angles of total reflection at the air-coating and coating-substrate interfaces. The two partial beams interfere and form interference. Surface roughness and the optical densities of coating and substrate material affect the acuity of the resulting interference image. The most intense and sharpest interference images are obtained if the refractive index of the substrate material is less than the refractive index of the coating material.

The main limitations of the X-ray reflectivity technique are the limited range of the wave-vector transfer and the loss of the phase of the reflected amplitude. Nevertheless, an accuracy of approximately 0.2 nm has been reported in determining the thickness and roughness of a double-layer sample.

3.2.3
Probe Measuring Methods

Over the last decade, fundamental research into surface physics has given rise to a new class of analyzer, the scanning probe microscope. These devices allow the mapping of a surface in a lateral range of 150×150 µm down to atomic resolution according to similar measuring principles with slight technical variations. Figure 3.2-8 shows the principle of the structure of a scanning probe microscope.

Other members of this class are the magnetic force microscope, the optical near-field microscope and microscopes that work by a thermal or capacitive interface or with ion flows.

However, scanning probe microscopes are not only useful for characterizing surfaces with high spatial resolution. The sharp tips of the scanning tunneling,

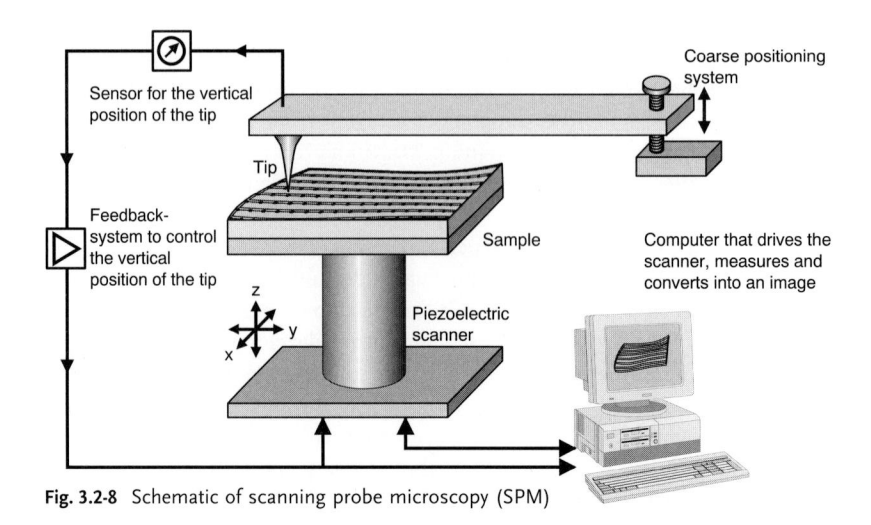

Fig. 3.2-8 Schematic of scanning probe microscopy (SPM)

scanning force, and lateral force microscopes can also be used as local sensors and as nano-tools for carrying out experiments or for making surface modifications on the atomic scale. In this way, time-stable atomic-scale structures can be generated, modified, and removed under environmental conditions. Chemical reactions can be induced locally with the AFM tip and crystal growth can be monitored in situ and in real time. Forces and interactions can be investigated on the (sub)atomic scale and the phenomenon of energy dissipation due to friction can be studied quantitatively on a microscopic scale.

3.2.3.1 Scanning Electron Microscopy (SEM)

In many areas of research it is important to obtain chemical, morphological information in the sub-micrometer range. Because of the limited resolution of optical microscopes (theoretically 0.15 µm), bundled electrons accelerated by electrical high voltage (up to 3 MV) in a high vacuum are used instead of light because they are strongly deflected by scattering at atmospheric pressure. Rotationally symmetric electrical and magnetic fields perform the same functions as lenses in an optical microscope, concentrating the electron beam coming from the hot cathode on to the object. The object to be measured is penetrated by the electrons to different degrees in the transmission electron microscope depending on the thickness and density of the electrons in such a way that the corresponding intensity distribution in the electron image represents the structure. The electron image is acquired on a photographic plate or fluorescent screen, yielding an approximately 200 000-fold magnification. In SEM (Figure 3.2-9), an electron beam (diameter about 10 nm) is moved over the object in a scanning pattern, ie, row by row. The electrons, both those scattered back and the secondary electrons that escape from the surface of the sample, are amplified by the scintillator and photomultiplier and provide the signal for brightness control of a synchronously controlled cathode-ray tube (large depth of field).

The resolution limit is determined by the diffraction phenomena at the aperture of the imaging system and the wavelength of the particles. With a 100 kV electron microscope, a resolution of 0.2 nm ($\lambda = 3.7$ pm, $A = 0.4$–0.8, error of lens aperture $C_s = 0.3$–1 mm) is achieved according to the equation

$$d_{\text{theor.}} = A \cdot \sqrt[4]{\lambda C_s} \ . \tag{3.2-3}$$

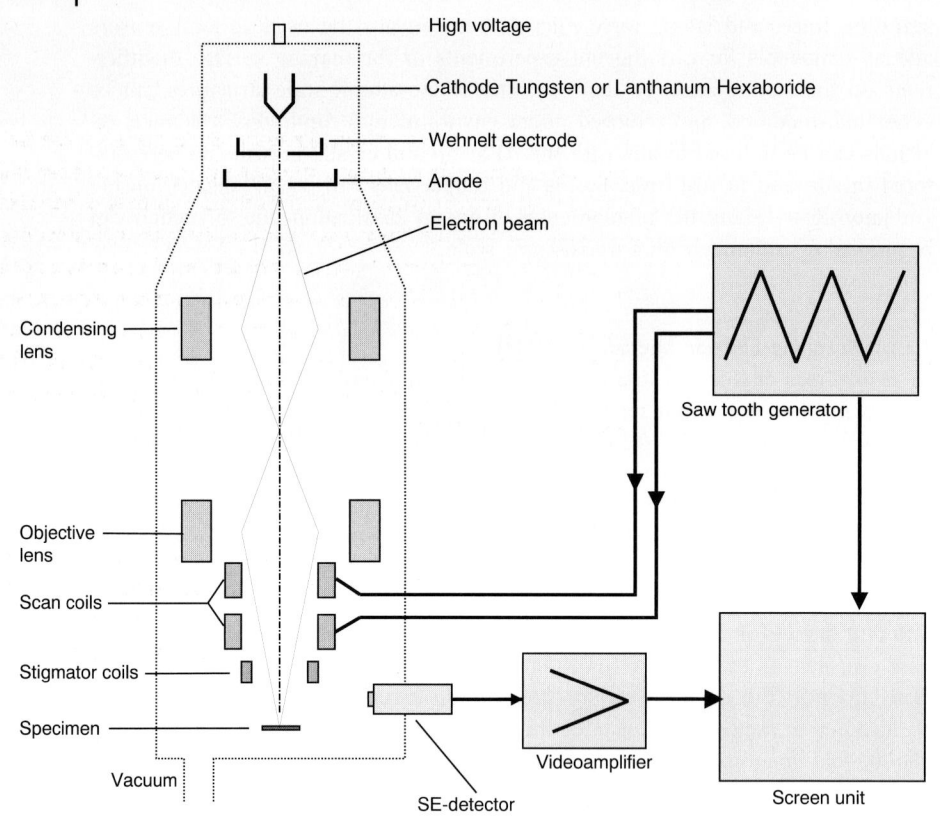

Fig. 3.2-9 Design of a scanning electron microscope

3.2.3.2 Scanning Tunneling Microscopy (STM)

In scanning tunneling microscopy, a conductive, atomically sharp needle is guided row by row over a conductive surface. If the probe is lowered near the surface to be measured, interaction due to the quantum physical tunneling effect occurs with the surface in the form of tunneling current (Figure 3.2-10). To obtain a measurable signal, the distance from the tip to the sample must only be about 10 Å. With highly sensitive amplifiers, it is possible to detect currents up to 2 fA (10^{-15} A). Influences from vibrations with amplitudes up to 1 µm (floor or sonic vibrations) and thermal drift of the components in the range of approximately 0.1 µm/cm are problems encountered in implementing this measuring method.

It is possible to choose between two different modes. If, during the measuring operation, the height of the tip is controlled in such a way that the tunneling current remains constant ($I_T(x,y)$ = constant), the probe displacement in the vertical direction provides a measure of the profile height of the surface at the measuring point (Figure 3.2-11).

In the second mode, the height of the needle tip is kept constant ($Z_T(x,y)$ = constant) and the variation in tunneling current is acquired during the

Fig. 3.2-10 Schematic representation of tip and sample interactions (tunneling effect)

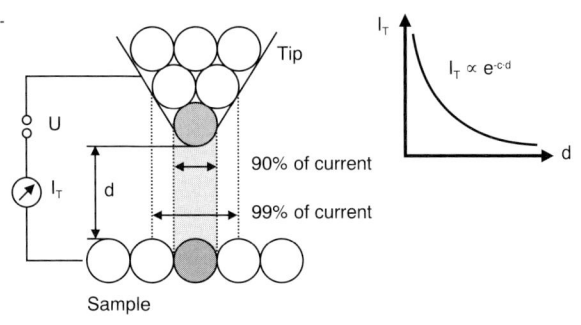

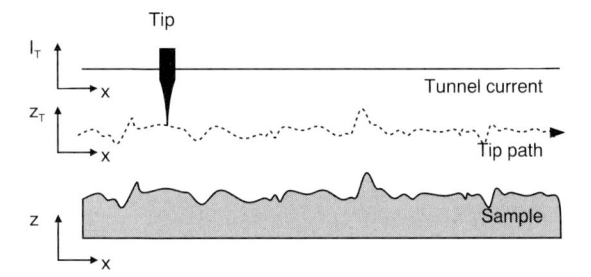

Fig. 3.2-11 Constant-current mode

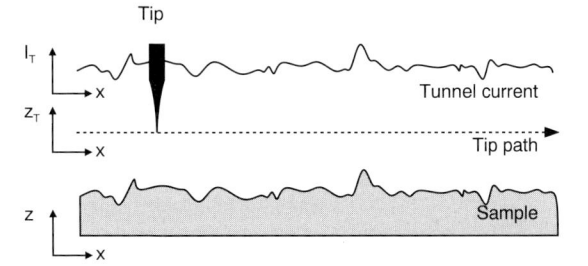

Fig. 3.2-12 Constant-height mode

measuring process (Figure 3.2-12). With this mode, however, there is a danger that the tip might come into contact with the surface because of irregularities of the sample and that incorrect measured current values could be obtained because of electrical contact.

The advantage of this measuring method is its very high resolution of about 0.01 nm. The disadvantage is its low measuring range (laterally maximum 100 µm, in the z-direction maximum 10 µm).

3.2.3.3 Scanning Near-field Optical Microscopy (SNOM)

SNOM is high-resolution optical microscopy implemented by scanning a small spot of 'light' over the specimen and detecting the reflected (or transmitted) light for image formation (Figure 3.2-13). This is the only similarity with confocal microscopy, where the focal point is scanned. Operation of conventional light microscopes suffers from the diffraction limit, which limits the optical resolution of the microscope to only approximately a half wavelength of the light being used. The resolution of the SNOM image is defined by the size and the properties of the aperture, not by the wavelength used. This means that SNOM provides an improvement in spatial resolution of at least one order of magnitude over conventional optical microscopes. However, the attainable resolution of approximately 50 nm is smaller than that for STM or AFM. SNOM utilizes tiny apertures of diameters in the range typically 50–100 nm, ie, smaller than half the wavelength of visible light. Typically such apertures are prepared in the metal coating at the apex of an optically transparent, sharp tip. Light cannot pass through such an aperture, but an evanescent field, the optical near-field, extends from it. The optical near-field decays exponentially with the distance, and is thus only detectable in the immediate vicinity of the tip.

The optical resolution limit for SNOM is governed by the light intensity passing through the aperture, usually by heating and pulling a fabricated fiber tip. A practical limit is usually encountered at aperture diameters between 80 and 200 nm but in ideal cases diameters down to <20 nm have been achieved.

If the aperture is brought close to the sample surface, the presence of the sample causes a disturbance of the optical near-field, which leads to the emission of light from the location opposite the aperture. Scanning the aperture at a distance of typically 10 nm from the sample with an accuracy of $\sim 5\,\text{Å}$, in order to prevent

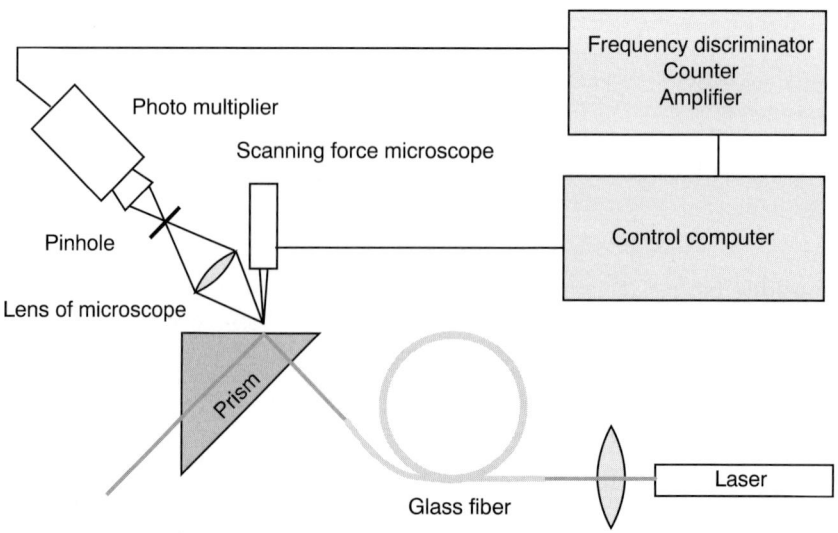

Fig. 3.2-13 Design of SNOM in combination with AFM

damage to the tip and/or sample, and simultaneously detecting emitted light in either the reflection or transmission mode produces a high-resolution optical image. By contrast, conventional optical microscopy relies on observation in the far-field where the achievable resolution is limited by diffraction.

SNOM instruments are technically closely related to scanning force and tunneling microscopes (SFM and STM) because probing involves scanning either the probe tip or the sample with tip-to-sample distance control.

3.2.3.4 Scanning Capacitance Microscopy (SCM)

Scanning capacitance microscopy (SCM) images spatial variations in capacitance (Figure 3.2-14). Like EFM (see Section 3.2.3.11), SCM induces a voltage between the tip and the sample. In the first mode, the cantilever operates in a non-contact, constant-height mode. A special circuit monitors the capacitance between the tip and the sample. Since the capacitance depends on the dielectric constant of the medium between the tip and sample, SCM studies can image variations in the thickness of a dielectric material on a semiconductor substrate. SCM can also be used to visualize sub-surface charge-carrier distributions, eg, to map dopant profiles in ion-implanted semiconductors.

In addition to measuring surface topography, the probe tip can also be used as a proximal probe. In the second mode, the probe tip is kept in contact with the surface to generate a topographic image. In addition, AC and DC voltages are applied between the tip and a semiconductor sample. Changes in the capacitance of the semiconductor beneath the probe tip are measured using a special sensor. Changes in the capacitance are mapped simultaneously with topography. These changes can be correlated with the dopant type and concentration of the semiconductor.

3.2.3.5 Scanning Thermal Microscopy (SThM)

In scanning thermal microscopy (SThM), a microscopically small wire loop as the detector is scanned over the sample surface, without coming into contact with the sample. Figure 3.2-15 shows the structure of a tip.

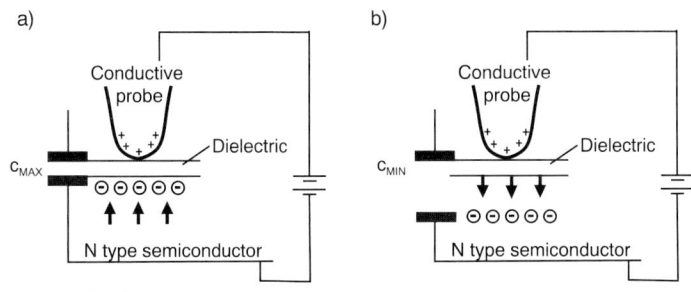

Fig. 3.2-14 Working principle of SCM

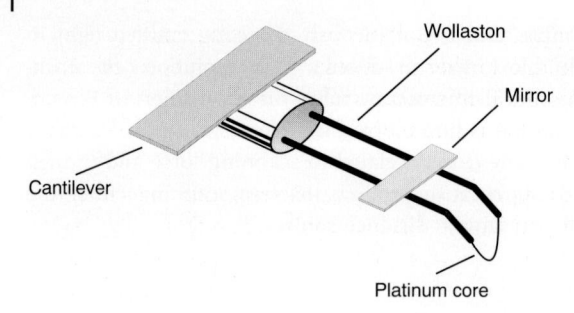

Fig. 3.2-15 Design of a tip for scanning thermal microscopy

(Figure labels: Wollaston, Mirror, Cantilever, Platinum core)

In addition to topography representation, which is analogous to that in AFM, thermal sample properties can also be acquired with a positional resolution in the nanometer range. If, for example, such a small current is conducted through the wire loop (see Figure 3.2-15) that no self-heating occurs, the local sample temperature can be derived from the resistance of the wire loop. If the current is increased so that a temperature difference between the tip and sample arises, it is possible to derive the local thermal conductivity directly from the measurable heat flow from the tip to the sample.

Thermal conductivity imaging is achieved using a special probe to measure both the topography and temperature. Using a Wollaston wire (which acts as a resistor in a Wheatstone bridge circuit), the surface is scanned with a second feedback signal that applies power to the probe keeping the temperature between the probe and sample constant. Variations in surface conductivity cause heat-flow changes in or out of the probe. These are recorded as a thermal conductivity map enabling individual phases of blends to be mapped.

The second possible design of a tip for SThM is a cantilever composed of two different metals, presented by Digital Instruments (a thermal element made up of two metal wires can also be used). The materials of the cantilever respond differently to changes in thermal conductivity, and cause the cantilever to deflect. The system generates an image, which is a map of the thermal conductivity, from the changes in the deflection of the cantilever. A topographical non-contact image can be generated from changes in the cantilever's amplitude of vibration. Thus, topographic information can be separated from local variations in the sample's thermal properties, and the two types of images can be collected simultaneously.

Temperature Resolution. Typical temperature resolution of 0.2 °C after temperature calibration of the probe is observed. Temperature calibration is probe dependent. The temperature resolution of the probe is currently limited by the electrical noise of the sensor.

Spatial Resolution. The measured spatial resolution of SThM is dependent on the characteristics of the sample, with the best observed resolution of 150–200 nm full width at half maximum (FWHM), as measured on electrically biased magnetoresistive stripes of magnetic data storage read elements. Note that three main factors affect the apparent spatial resolution of imaged temperature gradients: heat spreading in the sample due to thermal insulation properties of the background sample material versus the heat source; spatial extent of the heat source; and distance of the probe tip to the actual thermal source relative to the surface being imaged.

Samples which give the best apparent thermal spatial resolution (and therefore have the largest thermal gradients) have small heat sources in a thermally insulating substrate with the thermal source very close to the surface being imaged. Ideal samples also provide adequate electrical insulation between the circuit of interest and the probe sensor.

3.2.3.6 Atomic Force Microscopy (AFM)

Atomic force microscopy (AFM) probes the surface of a sample with a sharp tip, about 10 µm long and often less than 100 Å in diameter. The tip is located at the free end of a cantilever that is 100–200 µm long (Figure 3.2-16).

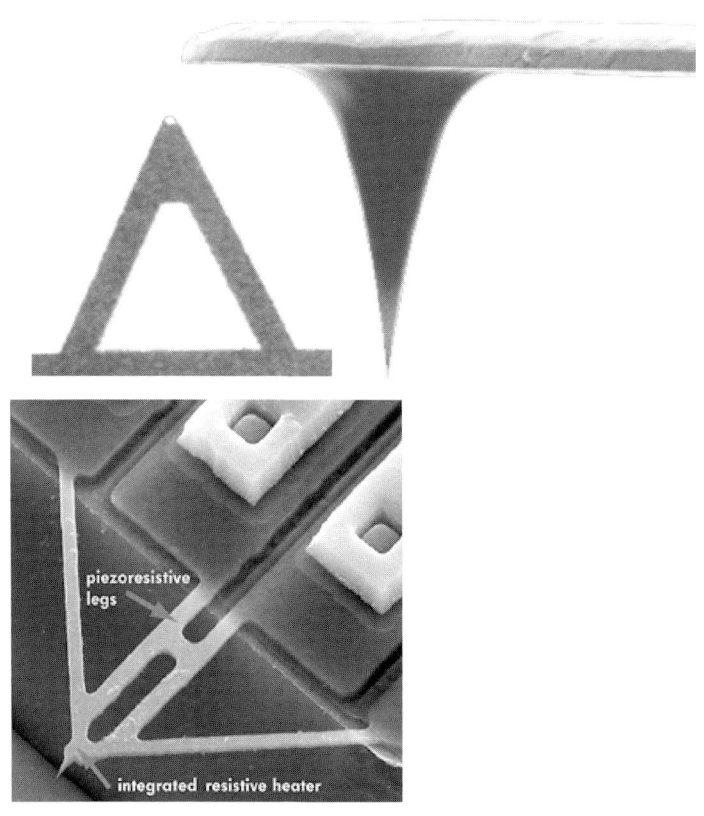

Fig. 3.2-16 Top and side view of a tip, tip height about 5–7 µm (courtesy: ThermoMicroscopes)

Forces between the tip and the sample surface cause the cantilever to bend or deflect. A detector measures the cantilever deflection as the tip is scanned over the sample, or the sample is scanned under the tip. The measured cantilever deflections allow a computer to generate a map of surface topography. In contrast to SEM, AFM can be used to study insulators and semiconductors as well as electrical conductors without special and expensive preparation of the specimen. The typical scan range of such an instrument is approximately 120 µm laterally and about 5 µm vertically.

Several forces typically contribute to the deflection of an AFM cantilever. The force most commonly associated with AFM is an interatomic force called the van der Waals force. The dependence of the van der Waals force upon the distance between the tip and the sample is shown in Figure 3.2-17.

Two distance regimes are labeled on Figure 3.2-17: (1) the contact regime and (2) the non-contact regime. In the contact regime, the cantilever is held less than a few ångströms from the sample surface, and the interatomic force between the cantilever and the sample is repulsive. In the non-contact regime, the cantilever is held tens to hundreds of Ångströms from the sample surface, and the interatomic force between the cantilever and sample is attractive (largely a result of the long-range van der Waals interactions). Both contact and non-contact imaging techniques are described in detail below.

Contact AFM. In the contact AFM mode (C-AFM), also known as the repulsive mode, an AFM tip makes soft 'physical contact' with the sample. The tip is attached to the end of a cantilever with a low spring constant (0.6–2.8 N/m), lower than the effective spring constant holding the atoms of the sample together. As the scanner traces the tip across the sample (or the sample under the tip), the contact force causes the cantilever to bend to accommodate changes in topography. Even if a very stiff cantilever is designed to exert large forces on the sample, the interatomic separation between the tip and sample atoms is unlikely to decrease much. Instead, the sample surface is likely to deform (nanolithography). In addition to the repulsive van der Waals force, two other forces are generally present during contact AFM operation: a capillary force exerted by the thin water layer often present in an ambient environment, and the force exerted by the cantilever itself. The capillary force arises when water wicks its way around the tip, applying a strong attractive force (about 10^{-8} N) that holds the tip in contact with the sur-

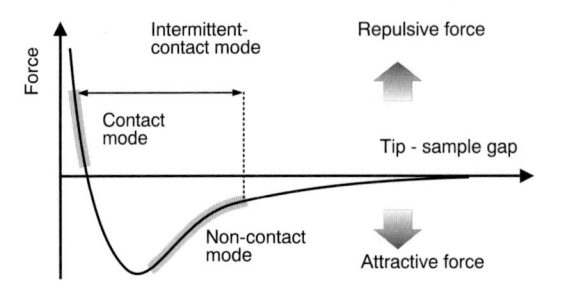

Fig. 3.2-17 Interatomic force versus distance curve

face. The magnitude of the capillary force depends on the tip-to-sample distance. The force exerted by the cantilever is like the force of a compressed spring. The magnitude and sign (repulsive or attractive) of the cantilever force depend on the deflection of the cantilever and on its spring constant. As long as the tip is in contact with the sample, the capillary force should be constant because the distance between the tip and the sample is virtually incompressible, assuming that the water layer is reasonably homogeneous. The variable force in C-AFM is the force exerted by the cantilever. The total force that the tip exerts on the sample is the sum of the capillary plus cantilever forces, and must be balanced by the repulsive van der Waals force for C-AFM. The magnitude of the total force exerted on the sample varies from 10^{-8} N (with the cantilever pulling away from the sample almost as hard as the water is pulling down the tip), to the more typical operating range of 10^{-7}–10^{-6} N. Usually AFMs detect the position of the cantilever with optical techniques. In the most common scheme, shown in Figure 3.2-18, a laser beam bounces off the back of the cantilever on to a position-sensitive photodetector (PSPD).

The PSPD itself can measure displacements of light as small as 10 Å. The ratio of the path length between the cantilever and the detector to the length of the cantilever itself produces a mechanical amplification. As a result, the system can detect sub-ångström vertical movement of the cantilever tip. Other methods of detecting cantilever deflection rely on optical interference or even a scanning tunneling microscope tip to read the cantilever deflection. Another technique is to fabricate the cantilever from a piezoresistive material so that its deflection can be detected electrically (strain from mechanical deformation causes a change in the material's resistivity). In that case, a laser beam and a PSPD are not necessary. Once the AFM has detected the cantilever deflection, it can generate the topographic data set by operating in one of two modes – constant-height or constant-force mode. In the constant-height mode, the spatial variation of the cantilever deflection can be used directly to generate the topographic data set because the height of the scanner is fixed as it scans. In the constant-force mode, the deflection of the cantilever can be used as input for a feedback circuit that moves the scanner up and down in z, responding to the topography by keeping the cantilever deflection constant. In that case, the image is generated from the scanner's motion. With the cantilever deflection held constant, the total force applied to the sample is constant. In the constant-force mode, the scanning speed is limited by the response time of the feedback circuit, but the total force

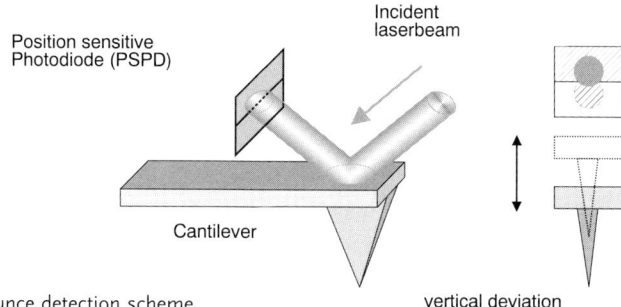

Fig. 3.2-18 The beam-bounce detection scheme

exerted on the sample by the tip is well controlled, generally preferred for most applications. The constant-height mode is often used for taking atomic-scale images of atomically flat surfaces, where the cantilever deflections and thus variations in applied force are small, also essential for recording real-time images of changing surfaces, where a high scan speed is essential.

Non-contact AFM. Non-contact AFM (NC-AFM) is a technique in which an AFM cantilever is vibrated near the surface of a sample. The spacing between the tip and the sample for NC-AFM is of the order of tens to hundreds of ångströms. This spacing is indicated by the van der Waals curve (Figure 3.2-17) as the non-contact regime. NC-AFM provides a means of measuring sample topography with little or no contact between the tip and the sample. The total force between the tip and the sample in the non-contact regime is very small, generally about 10^{-12} N. This small force is advantageous when studying soft or elastic samples. A further advantage is that samples such as silicon wafers are not contaminated with impurities through contact with the tip. Because the force between the tip and the sample in the non-contact regime is low, it is more difficult to measure than the force in the contact regime, which can be several orders of magnitude greater. In addition, cantilevers used for NC-AFM must be stiffer (typical 21–100 N/m) than those used for contact AFM because soft cantilevers can be pulled into contact with the sample surface. The small force values in the non-contact regime and the greater stiffness of the cantilevers used for NC-AFM are both factors that make the NC-AFM signal small, and therefore difficult to measure. Therefore, a sensitive AC detection scheme is used for NC-AFM operation. In the non-contact mode, the system vibrates a stiff cantilever near its resonant frequency (typically 100–400 kHz) with an amplitude of a few tens of Ångströms. Further, it detects changes in the resonant frequency or vibration amplitude as the tip comes near the sample surface. The sensitivity of this detection scheme provides sub-Ångström vertical resolution in the image, as with C-AFM. Changes in the resonant frequency of a cantilever can be used as a measure of changes in the force gradient, which reflect changes in the tip-to-sample spacing, or sample topography. In the NC-AFM mode, the system monitors the resonant frequency or vibrational amplitude of the cantilever and keeps it constant with the aid of a feedback system that moves the scanner up and down. By keeping the resonant frequency or amplitude constant, the system also keeps the average tip-to-sample distance constant. As with C-AFM (in the constant-force mode), the motion of the scanner is used to generate the data set. NC-AFM does not suffer from the tip or sample degradation effects that are sometimes observed after taking numerous scans with contact AFM. In the case of rigid samples, contact and non-contact images may look the same. However, if a few monolayers of condensed water are lying on the surface of a rigid sample, for instance, the images may look completely different. An AFM operating in the contact mode will penetrate the liquid layer to image the underlying surface, whereas in the non-contact mode an AFM will image the surface of the liquid layer. For cases where a sample of low moduli may be damaged by the dragging of an AFM tip across its surface, another mode of AFM operation is available; the intermittent-contact mode.

Intermittent-contact AFM. This technique (IC-AFM) was developed as a method for achieving high resolution without inducing destructive frictional forces. The cantilever is oscillated near its resonant frequency as it is scanned over the sample surface. The probe is brought closer to the sample surface until it begins to touch intermittently ('tap') on the surface. This contact with the sample causes the oscillation amplitude to be reduced. Once the tip is tapping on the surface, the oscillation amplitude scales in direct proportion to the average distance of the probe to the sample; for example, if the average separation between the tip and sample is 10 nm, then the oscillation amplitude will be roughly 20 nm peak-to-peak. The oscillation level is set below the free air amplitude and a feedback system adjusts the cantilever-sample separation to keep this amplitude constant as the tip is scanned in a raster pattern across the surface. Because the contact with the sample is only intermittent, the probe exerts negligible frictional forces on the sample and damage from these lateral forces is eliminated. Also, the oscillation amplitude is set sufficiently high (10–100 nm) that when the probe taps on the surface, the cantilever has sufficient restoring force (owing to the bending of the cantilever) to prevent the probe from becoming trapped in the contaminant layer by fluid meniscus forces or electrostatic forces.

3.2.3.7 Magnetic Force Microscopy (MFM)

Magnetic force microscopy (MFM) is the third generation of scanning probe techniques after scanning tunneling microscopy and scanning force microscopy. MFM has been designed to study the fringing field above magnetic materials. The principal idea of that method relies on the magnetostatic interaction between the magnetic sample and the magnetic sensor attached to the flexible cantilever, scanning in non-contact mode over the sample surface typically in the range of tens to hundreds of nanometers. A magnetic sample with a domain structure produces a complicated stray field over the surface. The aim of MFM is to map the stray field as close to the surface as possible. An interaction which appears when a sample is scanned by an MFM sensor is monitored via a deflection of a cantilever (Figure 3.2-19) or by vibrating the lever and measuring its resonance frequency.

A form of MFM operation affording line-by-line simultaneous acquisition of the topography and magnetic force gives the user a way of studying correlations be-

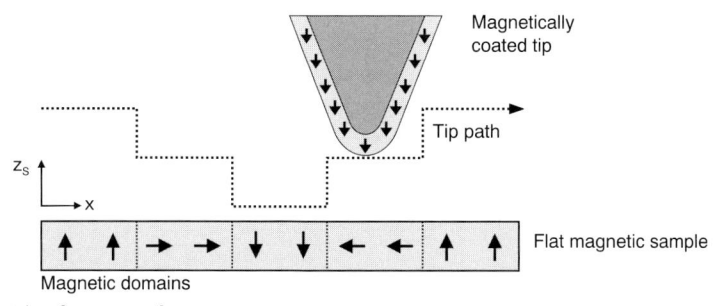

Fig. 3.2-19 Principle of magnetic force microscope

tween the morphology of the surface and the magnetic domain structure. Which effect dominates depends on the distance of the tip from the surface, because the interatomic magnetic force persists for greater tip-to-sample separations than the van der Waals force. If the tip is close to the surface, in the region where standard non-contact AFM is operated, the image will be predominantly topographic. As the separation between the tip and the sample is increased, magnetic effects become apparent. Collecting a series of images at different tip heights is one way of separating magnetic from topographic effects. MFM provides high sensitivity and a very high lateral resolution of 50 nm or better. This has mainly been achieved by using ferromagnetic thin-film sensors. The preparation of these sensors not only determines the resolution and sensitivity of MFM but also provides a way of studying magnetic features on a nanometer scale.

3.2.3.8 Lateral Force Microscopy (LFM)

Lateral force microscopy (LFM) measures lateral deflections (twisting) of the cantilever that arise from forces on the cantilever parallel to the plane of the sample surface. LFM studies are useful for imaging variations in surface friction that can arise from non-homogeneities in surface materials, and also for obtaining edge-enhanced images of any surface.

As depicted in Figure 3.2-20, lateral deflections of the cantilever usually arise from two sources, changes in surface friction (top) and changes in slope (bottom).

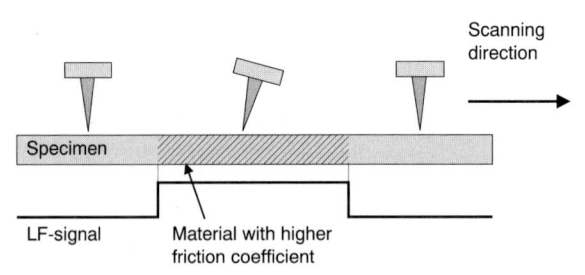

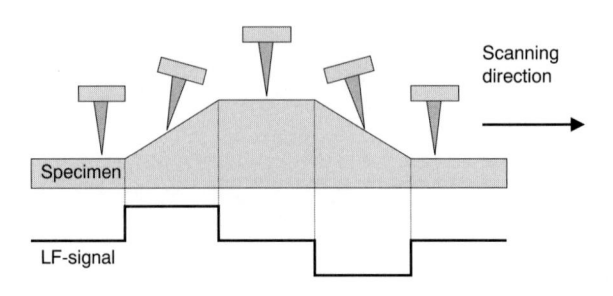

Fig. 3.2-20 Lateral deflection of the cantilever

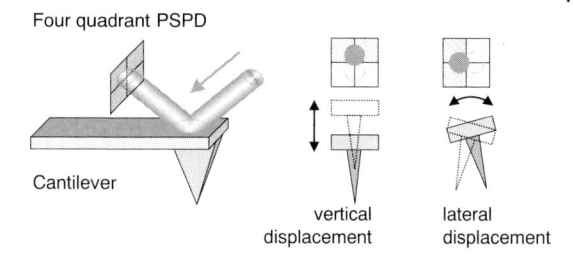

Fig. 3.2-21 Enhanced detection of tip position

In the first case, the tip may experience greater friction as it traverses some areas, causing the cantilever to twist more to a greater degree. In the second case, the cantilever may twist when it encounters a steep slope. To separate one effect from the other, LFM and AFM images should be collected simultaneously. LFM uses a position-sensitive photodetector to detect the deflection of the cantilever, just as for AFM. The difference is that for LFM, the PSPD also senses the cantilever's twist, or lateral deflection.

A comparison between Figures 3.2-18 and 3.2-21 illustrates the difference between an AFM measurement of the vertical deflection of the cantilever, and an LFM measurement of lateral deflection. AFM uses a 'bi-cell' PSPD, divided into two halves, A and B. LFM requires a 'quad-cell' PSPD, divided into four quadrants, A–D. By adding the signals from quadrants A and C, and comparing the result with the sum from quadrants B and D, the quad-cell can also sense the lateral component of the cantilever's deflection. A properly engineered system can generate both AFM and LFM data simultaneously.

3.2.3.9 Phase Detection Microscopy (PDM)

Phase detection microscopy (PDM), also referred to as phase imaging, is another technique that can be used to map variations in the mechanical and chemical surface properties such as elasticity, adhesion, friction, or viscoelasticity. Phase detection images can be produced while the instrument operates in any vibrating cantilever mode, such as NC-AFM, IC-AFM, or MFM mode. Phase detection information can also be collected while a force modulation microscopic (FMM) image (Section 3.2.3.10) is being taken. Phase detection refers to the monitoring of the phase lag between the signal that drives the cantilever to oscillate and the cantilever oscillation output signal (Figure 3.2-22). Changes in the phase lag reflect changes in the mechanical properties of the sample surface.

The system's feedback loop operates in the usual manner, using changes in the cantilever's deflection or vibration amplitude to measure sample topography. The phase lag is monitored while the topographic image is being taken so that images of topography and material properties can be collected simultaneously. One application of phase detection is to obtain material-properties information for samples whose topography is best measured using IC-AFM rather than C-AFM (Section 3.2.3.6). For these samples, phase detection is useful as an alternative to FMM, which uses C-AFM to measure topography.

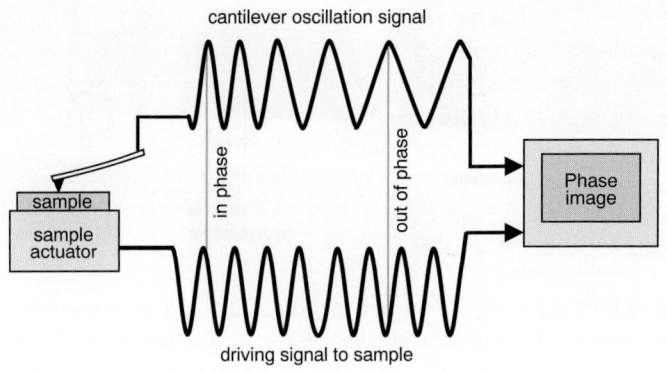

Fig. 3.2-22 Principle of a phase detection microscope

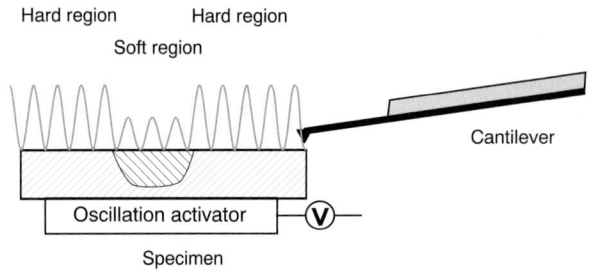

Fig. 3.2-23 The amplitude of cantilever oscillation varies according to the mechanical properties of the sample surface

The PDM image provides complementary information to the topography image, in the form of very dramatic high-contrast surface images, revealing the variations in the surface properties of an adhesive label. This technique offers a resolution in phase detection of 0.1°.

3.2.3.10 Force Modulation Microscopy (FMM)

This extension of AFM imaging includes the force modulation microscopy (FMM) characterization of a sample's mechanical properties. Like LFM and MFM, FMM allows the simultaneous acquisition of both topographic and material-properties data.

In the FMM mode, the AFM tip is scanned in contact with the sample, and the z feedback loop maintains a constant cantilever deflection (as for constant-force mode AFM). In addition, a periodic signal is applied to either the tip or the sample. The amplitude of cantilever modulation that results from this applied signal varies according to the elastic properties of the sample, as shown in Figure 3.2-23.

When the probe is brought into contact with a sample, the surface resists oscillation and the cantilever bends. Under the same applied force, a stiff area on the sample will deform less than a soft area, ie, stiffer surfaces cause greater resistance to the vertical oscillation and, consequently, greater bending of the cantilever. The system generates a force modulation image, which is a map of the sample's elastic properties, from the changes in the amplitude of cantilever modulation. The frequency of the applied signal is of the order of hundreds of kilohertz, which is faster than the z feedback loop which is set up to track. Thus, topographic information can be separated from local variations in the sample's elastic properties, and the two types of images can be collected simultaneously.

3.2.3.11 Electric Force Microscopy (EFM)

Electric force microscopy (EFM) applies a voltage between the tip and the sample while the cantilever hovers above the surface without touching it. The cantilever deflects when it scans over static charges, as depicted in Figure 3.2-24. The AC electrostatic force exerted on the probe tip is measured using a lock-in amplifier and recorded simultaneously with sample topography.

EFM maps locally charged domains on the sample surface, similarly to the way MFM (Section 3.2.3.7) plots the magnetic domains of the sample surface. The magnitude of the deflection, proportional to the charge density, can be measured with the standard beam-bounce system. EFM is used to study the spatial variation of surface charge carrier density. For instance, EFM can map the electrostatic fields of an electronic circuit as the device is turned on and off. This technique is known as 'voltage probing' and is a valuable tool for testing live microprocessor chips at the sub-micrometer scale.

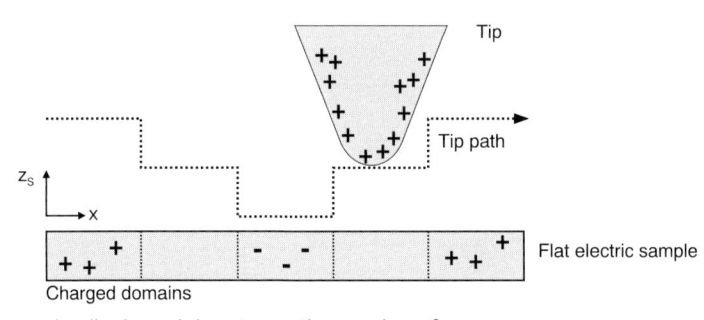

Fig. 3.2-24 EFM maps locally charged domains on the sample surface

3.2.3.12 Scanning Near-field Acoustic Microscopy (SNAM)

The scanning near-field acoustic microscope (Figure 3.2-25) is a further measuring instrument for determining mechanical sample properties. In the vertical direction, the resolution of the height differences by SNAM is limited to about 10 nm, and the lateral resolution is of the same order of magnitude as the curvature radius of the tip (about 100 nm) and is based on the free path length of the molecules in air.

A quartz resonator with a tip at the end is operated in a feedback loop at its resonant frequency. If the tip comes close to the surface, the attenuation of the oscillations of the resonator increases over a distance of a few micrometers because of hydrodynamic friction forces. The reduced vibration amplitude is detected with sensitive electronics and provides an image of the mechanical properties of the sample. The typical scanning speed of such a SNAM (up to 300 µm/s) is considerably higher than the scanning speed of a standard AFM (0.1–1 µm/s).

SNAM is therefore a non-contact method for exploring sample surfaces in air at scales down to a few millimeters and closes the gap between SFM and conventional tactile profile meter.

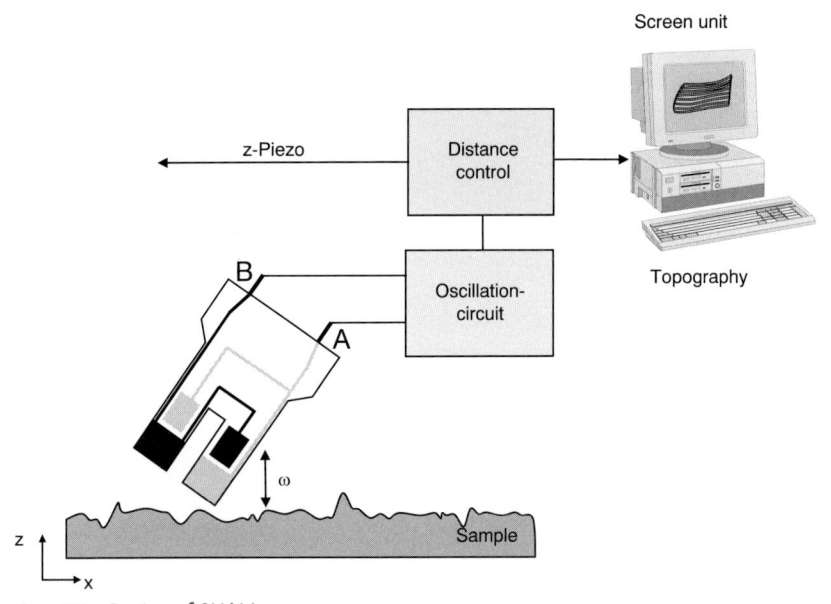

Fig. 3.2-25 Design of SNAM

3.2.4
Further Reading

1 Dagnall, H., *Exploring Surface Texture;* Leicester: Rank Taylor Hobson, 1986.
2 Hommelwerke GmbH, *Rauheitsmessung Theorie und Praxis;* Schwenningen: Schnurr Druck, 1993.
3 Sander, M., *Oberflächenmesstechnik für den Praktiker;* Göttingen: Feinprüf Perthen, 1993.
4 Thomas, T.R., *Rough Surfaces;* London: Imperial College Press, 1999.
5 Bodschwinna, H., Hillmann, W., *Oberflächenmesstechnik mit Tastschnittgeräten in der Industriellen Praxis;* Cologne: Beuth, 1992.
6 Dresel, T., Häusler, G., Venzke, H., *Appl. Opt.* **31** (1992) 919–925.
7 Koch, A.W., Ruprecht, M.W., Toedter, O., Häusler, G., *Optische Messtechnik an Technischen Oberflächen;* Renningen-Malmsheim: Expert-Verlag, 1998.
8 Pfeifer, T., *Optoelektronische Verfahren zur Messung Geometrischer Grössen in der Fertigung;* Ehningen bei Böblingen: Expert-Verlag, 1993.
9 Wiesendanger, R., Güntherodt, H.J., *Scanning Tunneling Microscope;* Vols. I–III. Heidelberg: Springer, 1992.
10 Fries, Th., *Rastersondenmikroskopie: Nobelpreistechnologie für die Anwendung;* 10 Feinwerktechnik – Mikrotechnik – Messtechnik (F+M) 101, 10, 1993.
11 *A Practical Guide to Scanning Probe Microscopy;* Sunnyvale: ThermoMicroscopes, 1997.

3.3
Sensors for Physical Properties
B. Karpuschewski, *Keio University, Yokohama, Japan*

3.3.1
Introduction

In this section, the possibilities of monitoring the physical properties of machined parts are discussed. Cutting processes with geometrically defined cutting edges such as hard turning have to be distinguished from abrasive processes such as grinding. In both cases workpiece material is removed in the form of chips due to the mechanical effect of the tool on the workpiece. Not only the geometry but also the number of cutting edges and their position relative to the workpiece are well known for cutting operations, whereas the situation for grinding processes is more complex. Here cutting edges are generated by single abrasives with irregular shape and size variation, which are connected by sufficient bond material. Owing to the large number of individual grains with changing micro-geometry, the conditions in the zone of contact can only be described by means of statistics. Although these major differences in the cutting conditions between grinding and cutting occur, the elementary process of material removal is still identical.

In Figure 3.3-1 the chip formation for cutting and grinding is shown schematically. The chip formation occurs owing to the formation of a pressure zone in front of the cutting edge rounding in the primary shear zone [1, 2]. This pressure zone effects the separation of the workpiece material into one part flowing as a chip over the rake face and another part, which is plastically and elastically deformed by the cutting edge rounding, the flank face, and the minor cutting edge

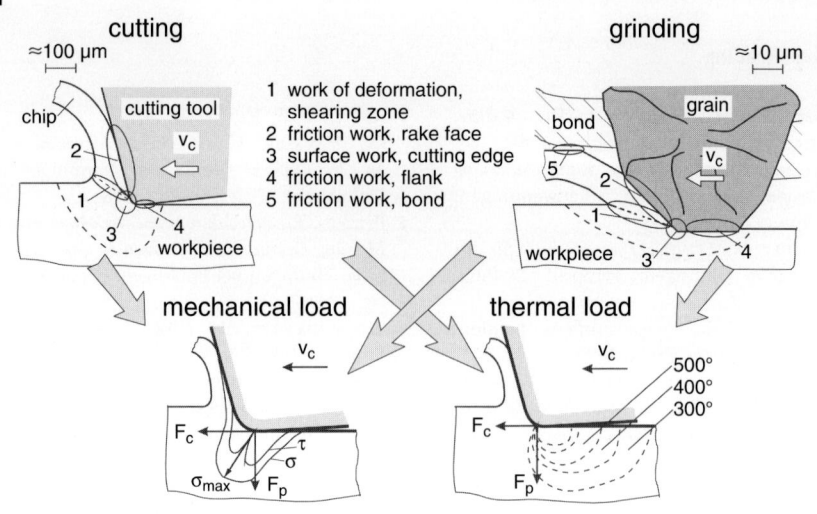

Fig. 3.3-1 Cutting and grinding chip formation and mechanical and thermal impact. Source: König [3], Wobker [4]

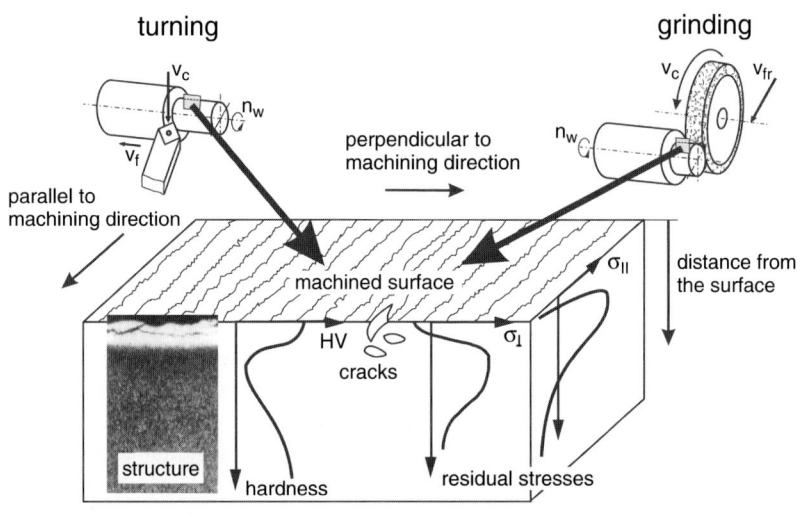

Fig. 3.3-2 Physical workpiece properties after turning or grinding

and pressed in the remaining workpiece material. Both processes generate a thermal and mechanical impact on the workpiece surface. These effects are the dominating influences for the physical properties. The mechanical impact is characterized by the generation of the contact area between the workpiece and tool, and abrasive grain and the resulting forces and stresses. The thermal load is determined by the heat distribution in the zone of contact, the temperatures that arise, and their temporal course [3–5]. Of course, the initial properties of the workpiece material also play an important role for the physical properties. These physical properties are shown in Figure 3.3-2 for a hardened steel material.

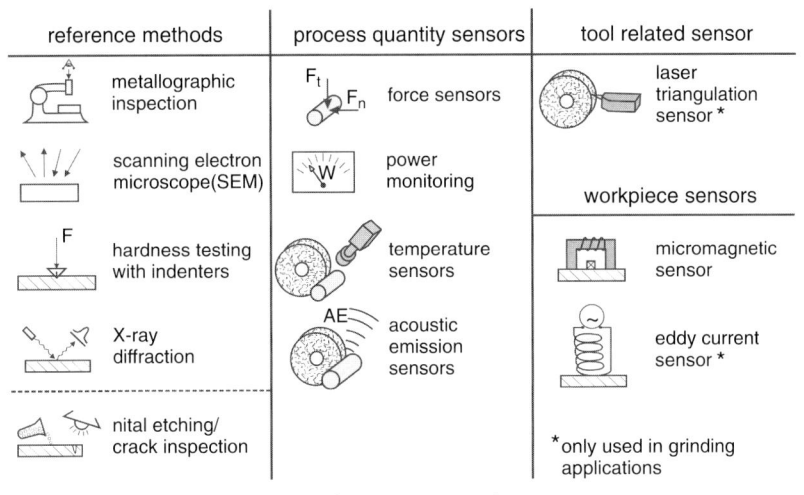

reference methods	process quantity sensors	tool related sensor
metallographic inspection	F_t, F_n force sensors	laser triangulation sensor*
scanning electron microscope (SEM)	power monitoring	workpiece sensors
hardness testing with indenters	temperature sensors	micromagnetic sensor
X-ray diffraction	AE acoustic emission sensors	eddy current sensor*
nital etching/ crack inspection		*only used in grinding applications

Fig. 3.3-3 Surface integrity characterization for cutting or grinding operations

All the physical properties of machined surfaces shown can be described as 'surface integrity'. This expression was introduced by Field and Kahles [6] more than three decades ago and is now a world-wide accepted technical term. Figure 3.3-3 shows an overview of systems for registering these physical properties.

3.3.2
Laboratory Reference Techniques

In the laboratory, high-resolution techniques such as hardness testing with indenters and metallographic inspection have proven their high standard. However, the main disadvantage of these methods is the measuring time, which limits their use to random sampling tests. In many cases the workpiece even has to be destroyed followed by extensive preparations to obtain information about the subsurface states or to investigate the cut-out segment of a larger part. In industrial applications, mainly methods are applied for detecting damage that has already occurred. Pure visual tests are very inaccurate and crack inspection is only suitable for crack-sensitive materials. Etching is the most widespread method of surface characterization, but still only a qualitative result can be obtained based on the experience of the inspection operator. X-ray diffraction using $\sin^2 \psi$ evaluation can be regarded as a standard technique to measure residual stresses [7].

3.3.3
Sensors for Process Quantities

During the interaction of tool and workpiece, material removal is initiated and a zone of contact is generated. The quantities which are measured during this interaction are called process quantities. Forces, power, temperature, and acoustic emission are the most common process quantities, which are discussed below.

3.3.3.1 Force Sensors

In this section only force sensors based on piezoelectric quartz force transducers are considered. Other possible sensor solutions for force measurements are discussed in Chapter 4. In turning of hardened steels or hard turning, the insert is usually fixed to a shank, which is mounted on a three-component piezoelectric force dynamometer. The hard turning operation is performed with low feed speeds and depths of cut and the use of tools with large cutting edge radii. In order to achieve low roughness values, only the cutting edge radius of the tool is used for machining, leading to a negative effective tool rake angle (see also Figure 3.3-1). The back force F_p, which is pushing the tool away from the workpiece, is the dominant force component with the steepest increase. The state of tool wear can thus be observed by measuring the cutting forces. With increasing cutting time, the cutting forces increase linearly. The development of surface integrity changes and back force increase due to hard turning performed with increasing tool wear is shown in Figure 3.3-4 [8]. Depending on increasing width of flank wear land VB_c, the back force and structural changes increase. These results can be regarded as typical and have been stated in many different investigations. Back force monitoring by piezoelectric dynamometers is thus a very efficient technique to monitor tool condition and to avoid any damage to the workpiece surface integrity. Wobker [4] and Schmidt [5] further improved this approach by calculating the friction power at the flank face, P_a, based on force measurements. If this friction power is related to the contact length between tool and workpiece, l_k, the specific friction power, P'_a, is calculated [5]. With this process quantity it is possible to predict the thermal load on the workpiece also taking the cutting edge micro-geometry into account.

In grinding there are different possibilities to integrate a piezoelectric dynamometer in the machine tool, which are described in Section 4.4.3. The tangential force is the most important component with regard to the surface integrity state,

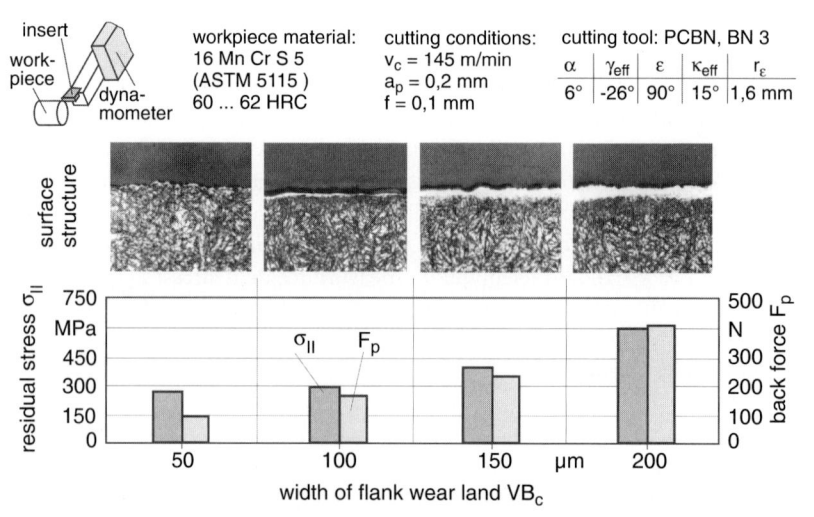

Fig. 3.3-4 Surface integrity state and back force as a function of tool wear. Source: Brandt [8]

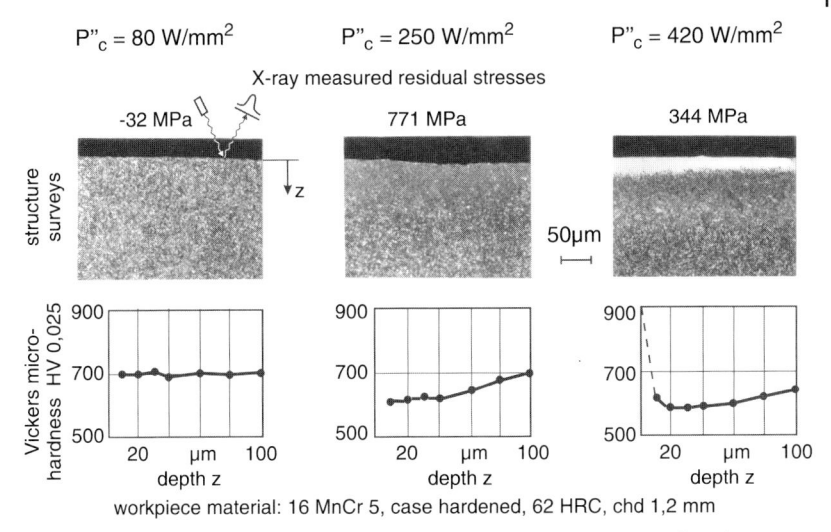

Fig. 3.3-5 Influence of the specific grinding power on the surface integrity state of steel

because the multiplication of tangential force and cutting speed results in the grinding power, P_c. If this grinding power is referred to the zone of contact, the specific grinding power, P_c'', can be calculated. This quantity is used to estimate the heat generation during grinding [eg, 1, 9]. Figure 3.3-5 shows representative structure surveys and Vickers micro-hardness depths of different plunge-cut ground workpieces made of case hardened steel. The specific grinding power as the main characteristic was varied by increasing the specific material removal rate, Q_w'. In addition to the graphical results, the X-ray measured residual stresses are also presented. The results reveal no thermal damage at the lowest related grinding power, followed by an increase in tensile residual stresses and an extended annealing zone for the second state.

The highest P_c'' causes structural deformations in the form of rehardening zones with sub-surface annealing and the previously explained reduction of tensile stresses. Brinksmeier has analyzed a wide variety of different grinding processes to establish an empirical model for the correlation between the specific grinding power based on force measurement and the X-ray calculated residual stress states [1]. The results show that it is not possible to predict the residual stress state only based on the specific grinding power without knowing the corresponding transfer function. The variations in the heat distribution due to different grinding wheel characteristics, process kinematics, and parameters are too widespread. Nevertheless, it can be clearly stated that a force measurement especially of the tangential force is a well suited method to control the surface integrity state of ground workpieces.

To summarize the examples presented, it can be said that a force measurement of only one decisive component is a very efficient method to avoid any kind of thermal damage on the machined surface either for turning or grinding. The only major disadvantage and the limiting factor for wide industrial use is the high investment required for this technique. Other solutions beside piezoelectric-based sensors will be introduced in Chapter 4.

3.3.3.2 Power Sensors

In cutting and especially in milling and drilling, power or torque sensors are often applied to the main spindle to monitor the process. It is the aim to avoid any overload of the spindle due to tool failure, eg, breakage of one cutting edge or of the whole tool (see Section 4.3.3). A direct correlation of the signals of these sensors with the surface integrity of machined surfaces is not the main purpose of their application. In turning of hardened steel, attempts were made to use the spindle power measured with a Hallsensor to find a correlation with the surface integrity state [5]. However, investigations revealed that the sensitivity of this sensor is not high enough to register changes of the workpiece physical properties.

In grinding, power monitoring is most often used. The main reason is the easy installation without influencing the working space of the machine tool and the relatively low costs. However, different investigations have clearly shown that the dynamic response of a power sensor at the main spindle is limited [eg, 10]. The power portion used for material removal is only a fraction of the total power consumption. However, still the mentioned advantages have promoted this sensor type for grinding applications. In [10] a result of power monitoring to detect grinding burn during internal grinding was published. Conventional abrasives were used to grind mild steel, and the detected high peak in the power signal over the grinding time must be attributed to a severe grinding burn. In most cases the signal increase is not spectacular, but is rather a steady increase over the grinding time due to continuing wear of the grinding wheel, especially when using superabrasives. A typical result is shown in Figure 3.3-6 for a grinding pro-

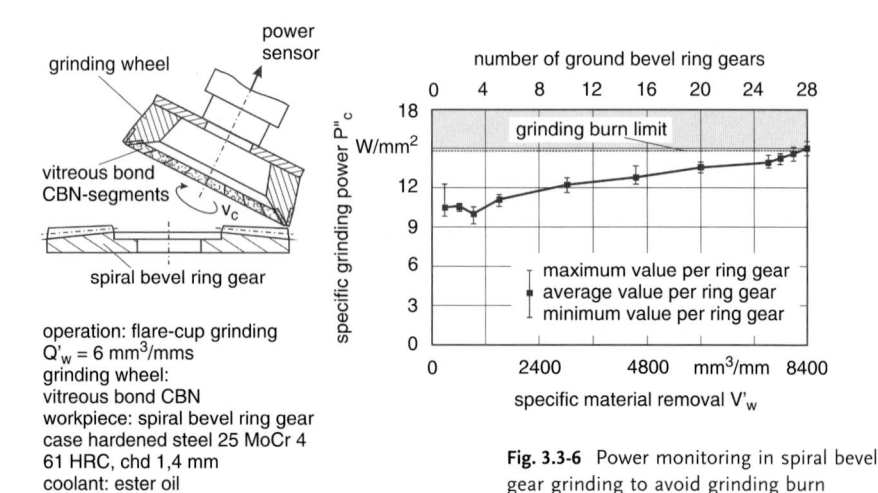

operation: flare-cup grinding
$Q'_w = 6$ mm³/mms
grinding wheel:
vitreous bond CBN
workpiece: spiral bevel ring gear
case hardened steel 25 MoCr 4
61 HRC, chd 1,4 mm
coolant: ester oil

Fig. 3.3-6 Power monitoring in spiral bevel gear grinding to avoid grinding burn

cess on spiral bevel ring gears, introducing a vitreous bond CBN grinding wheel [11]. Monitoring of the grinding power revealed a constant moderate increase in the material removal V'_w. At a specific material removal of 8100 mm^3/mm grinding burn was detected for the first time by nital etching. The macro- and microgeometry of the 28th workpiece was still within the tolerances, so the tool life criterion was the surface integrity state. For this type of medium- or even large-scale production in the automotive industry using grinding wheels with a long lifetime, power monitoring is a very effective way to avoid thermal damage of the workpiece and also to remove the environmentally harmful etching process. A similar system is monitoring the power consumption of the grinding spindle and also of the indexing head in a gear grinding machine. In addition, the rotation of the grinding spindle is also supervised by very sensitive inductive sensors to detect deviations especially in the entrance and exit of the grinding wheel in the tooth space, eg, due to distortions after heat treatment [12]. These results reveal that power monitoring can be a suitable sensor technique to avoid surface integrity changes during grinding. The most promising application is seen for superabrasives, because the slow increase in wear of the grinding wheel can be clearly determined with this dynamic limited method.

3.3.3.3 Temperature Sensors

In addition to forces and power, another important process quantity is the resulting temperature in the zone of contact. The mechanisms of chip formation as shown in Figure 3.3-1 lead to an almost total transformation of mechanical energy into heat. Thus all participating components in the zone of contact such as the workpiece, tool, chips, and, if used, coolant are thermally loaded. The resultant heat distribution is thus of major importance for the generated surface integrity state of the machined workpiece. The experimental effort to measure temperatures in the zone of contact is very high. Often the workpiece or the tool has to be destroyed to install the chosen sensor system in the zone of contact. All techniques for cutting and grinding can be distinguished in measurements based on heat conduction and heat radiation [2]. A detailed description of the most popular setups is given in Chapter 4. In the following, an example of heat radiation measurement in hard turning and heat conduction measurements in grinding is discussed.

Schmidt chose a heat radiation technique with an infrared camera, which measures the temperature on the workpiece directly underneath the insert (Figure 3.3-7, left) [5]. The measured temperatures on the workpiece are then used to calculate the maximum surface temperatures in the zone of contact through extrapolation by applying a differential approximation to the heat conductivity equation. In Figure 3.3-7 (right) the resulting maximum workpiece surface temperatures are shown for different cutting edge geometries and cutting parameters together with the contact length-related specific friction power at the flank face, P'_a, deduced from force measurements (see Section 3.3.3.1). High feeds and negative rake angles lead to an increase in the maximum workpiece surface temperature, whereas

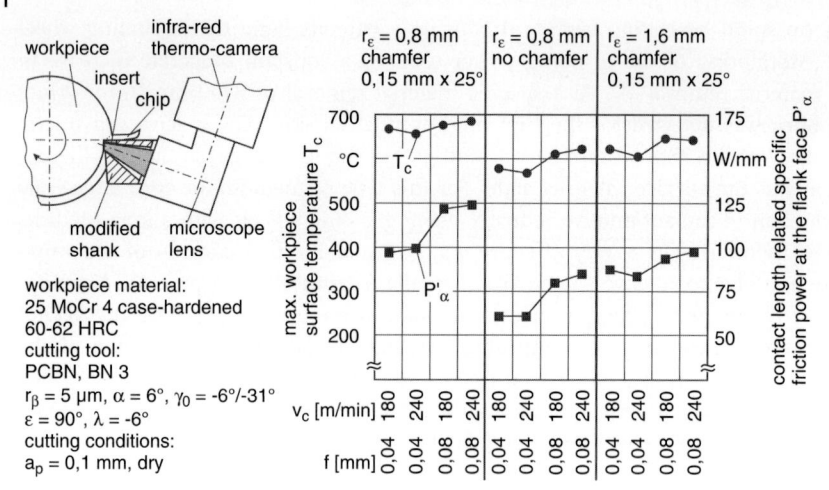

Fig. 3.3-7 Temperature measurement in hard turning based on heat radiation. Source: Schmidt [5]

the cutting speed has no significant influence. The contact length-related specific friction power at the flank face, P'_a, shows a good correlation with the measured and calculated temperatures, because the geometric conditions in the contact zone have been taken into account. The new defined quantity P'_a is in good agreement with the X-ray measured residual stress state on the workpiece surface after hard turning. The heat radiation-based temperature measurement was successfully used to verify and establish a new process quantity, P'_a, for surface integrity characterization based on the easier to apply force measurement.

In grinding, the same problems concerning temperature values and large gradients with respect to time and space are present, further intensified by the large number of geometrically undefined cutting edges and the almost general necessity for coolants. The possible sensor techniques based on heat conduction and heat radiation are explained in Section 4.4.3.5. An example of a successful temperature measurement with different types of thermocouples during surface grinding is shown in Figure 3.3-8 [13].

In these investigations a brazed thermal wire in a closed-circuit application and also thin-film thermocouples evaporated to the split workpiece were tested. The results reveal a systematic difference between the two sensor types. The temperatures determined with the thin-film thermocouples are 30% lower on average compared with the closed-circuit application, which was explained by imperfect insulation and a too large brazing point for the evaporated sensor [13]. Regardless of these differences, one major finding was the superior behavior of the vitreous bond CBN grinding wheel compared with a conventional corundum abrasive under the same grinding conditions.

The results have shown that temperature measurement in cutting or grinding is only possible with high technical effort. The modifications of workpiece or tool together with the financial and time investments restrict these measurements to fundamental research and industrial use is not possible.

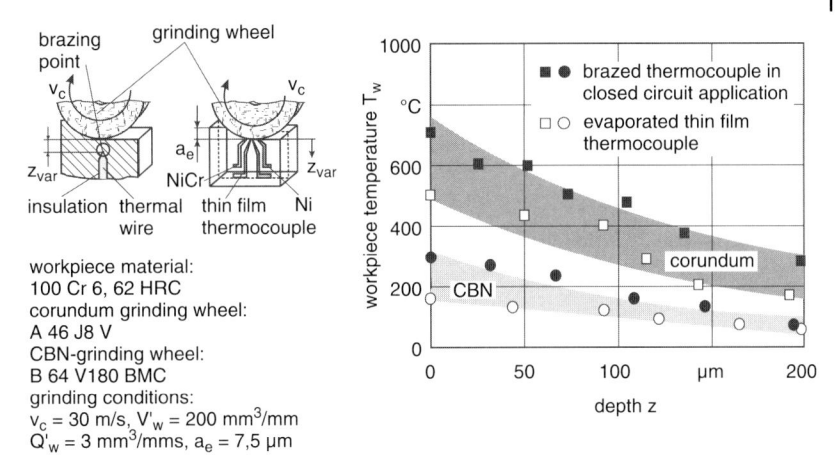

Fig. 3.3-8 Temperature profiles in surface grinding with different abrasives. Source: Choi [13]

3.3.3.4 Acoustic Emission Sensors

The application of acoustic emission (AE) sensors has become very popular in many kinds of machining processes over the last two decades. AE sensors combine some of the most important requirements for sensor systems such as relatively low costs, no negative influence on the stiffness of the machine tool, easy to mount and even capable of transmitting signals from rotating parts. First results on acoustic emissions were published in the 1950s for tensile tests. Since then, decades passed until this approach was first used to monitor cutting processes. The mechanisms leading to acoustic emission can be deformations through dislocations and distorted lattice planes, twin formation of polycrystalline structures, phase transitions, friction, crack formation, and propagation [eg, 14]. Owing to these different types, acoustic emission appears as a burst-type signal or as continuous emission. In cutting, the most important sources of acoustic emission are friction at the rake face, friction between workpiece and tool, plastic deformation in the shear zone, chip breakage, contact of the chip with either workpiece or cutting edge, and crack formation, as shown in Figure 3.3-9 [15]. The grinding process is characterized by the simultaneous contact of many different cutting edges, randomly shaped, with the workpiece surface. Every single contact of a grain is assumed to generate a stress pulse in the workpiece. During operation, the properties of the single grain and their overall distribution on the circumference of the grinding wheel will change owing to the occurrence of wear. Hence many different sources of acoustic emission have to be considered in the grinding process, as also shown in Figure 3.3-9. A change from austenite to martensite structures in ferrous materials also generates acoustic emission, although the energy content is significantly lower compared with the other sources. Hence every single effect has to be regarded as the origin of a wave front which is propagating through solid-state bodies.

Different types of signal evaluation can be applied to the AE sensor output. The most important quantities are root mean square value, raw acoustic emission sig-

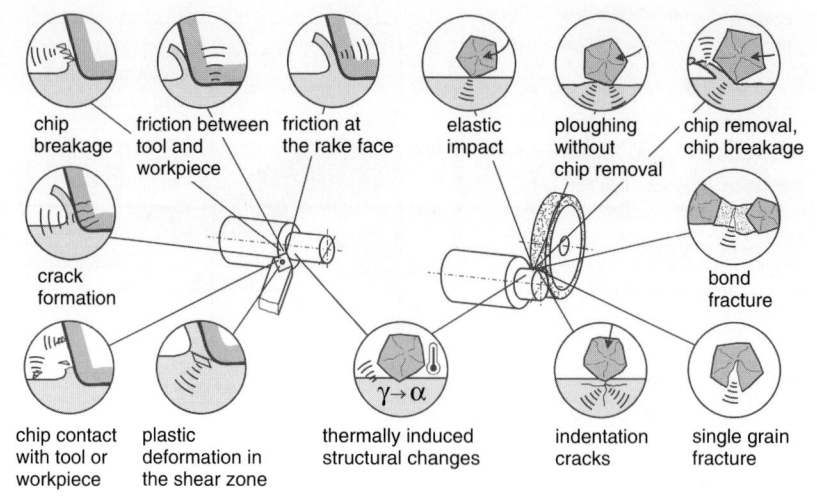

Fig. 3.3-9 Sources of acoustic emission in cutting and grinding

(chip breakage; friction between tool and workpiece; friction at the rake face; elastic impact; ploughing without chip removal; chip removal, chip breakage; crack formation; bond fracture; chip contact with tool or workpiece; plastic deformation in the shear zone; thermally induced structural changes $\gamma \rightarrow \alpha$; indentation cracks; single grain fracture)

nals and frequency analysis. AE sensors were often applied to cutting, especially turning operations [eg, 16]. In turning of hardened steel, one of the major concerns is to avoid any kind of surface integrity damage such as high residual stresses or white etching areas (see Figure 3.3-4). An AE sensor for monitoring purposes can be applied without any problems; usually it is mounted on the shank. Schmidt chose a position underneath the shank as close as possible to the zone of contact (Figure 3.3-10, left). The root mean square signal of the installed sensor was filtered in a very close frequency range between 150 and 250 kHz and transmitted to a digital oscilloscope for further analysis. The reason for this restriction is the aim to separate the effect of tool wear from the influence of other cutting parameters. By analyzing the frequency spectrum it was possible to identify the appropriate range [5]. Figure 3.3-10 (right) shows an example of the results achievable with this strategy.

The increase in tool wear leads to an almost linear decrease in the root mean square value, $U_{AE, RMS}$, in the chosen frequency range. The corresponding surface residual stress state, which was measured using the X-ray diffraction method, shows the opposite tendency. Hence it seems to be possible to monitor the surface integrity state of hard turned workpieces by analyzing the AE signal in a very close frequency range. Further investigations are needed to evaluate the influence of different system quantities such as machine tool, cutting insert, or workpiece on this frequency range.

In grinding, the application of an AE sensor is more complicated than in turning. Owing to the fast rotation of the grinding wheel and most often coolant supply, there are additional sources of noise, which have a significant influence on the measured signal. Possible positions of AE sensors in grinding are shown in Section 4.4.3. From the first beginnings of AE applications in grinding, attempts were made to correlate the signal with the occurrence of grinding burn. Whereas

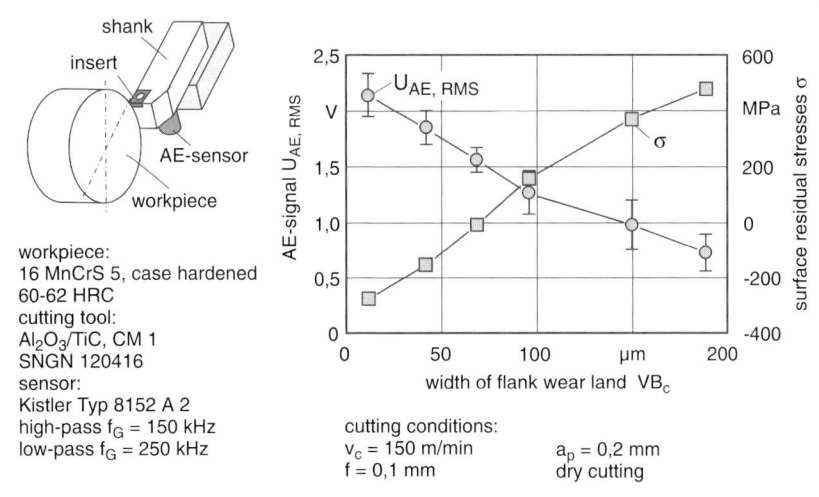

Fig. 3.3-10 AE analysis for surface integrity monitoring in hard turning. Source: Schmidt [5]

in early investigations only extreme damage was generated for detection [17], investigations on specially prepared workpieces [18] and real parts such as cams [19] have since been published. The studies of Klumpen [20] and Saxler [21] are directly related to the possibility of grinding burn detection with AE sensors. One fundamental result is that all process variations, which finally generate grinding burn, such as increasing material removal or infeed or reduced coolant supply, lead to an increase in the AE activity in grinding. Klumpen [20] could only identify grinding burn by applying a frequency analysis of the AE signal to determine the inclination of the integral differences. This must be regarded as a major disadvantage, because a frequency analysis is usually performed after grinding. The desired in-process solution would demand extremely high investment in analyzing hardware. For this reason, Saxler [21] concentrated on the AE signal in the time domain. In Figure 3.3-11 the major result of his work is shown.

Based on his investigations and theoretical considerations, he concluded that the AE sensor must be mounted to the workpiece to be most sensitive to the desired grinding burn detection. This is, of course, a major drawback for practical applications. An industrial test was conducted during gear grinding of planetary gears with an electroplated CBN-grinding wheel. With the aid of artificial neural networks he was able to achieve a dimensionless grinding burn characteristic value from the AE values of different frequency ranges in the time domain. The high efforts for training of the artificial neural network and the problems related to the sensor mounting on the workpiece side must be seen as limiting factors for wider industrial application. However, the results have clearly shown that AE systems can be regarded as suitable process quantity sensors in both cutting and grinding to monitor surface integrity changes.

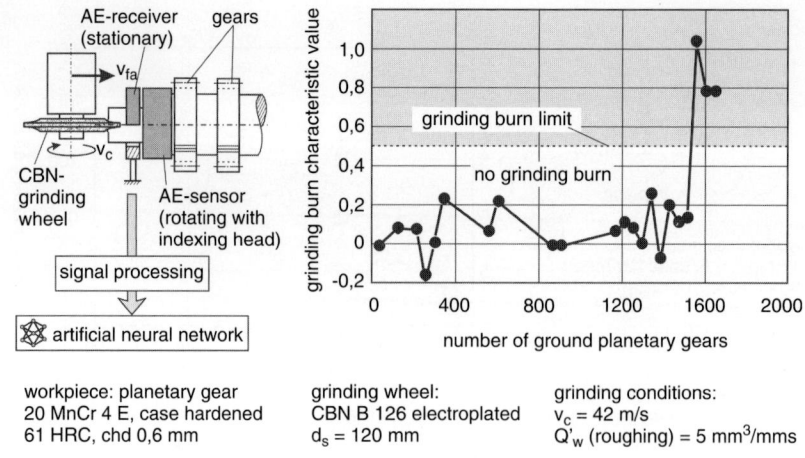

Fig. 3.3-11 Grinding burn detection with acoustic emission. Source: Saxler [21]

3.3.4
Sensors for Tools

The number of sensor systems which are available to be used on tools to acquire information about the surface integrity state is very limited. In cutting, no specific sensors are known which can be used on the tool to predict the achievable physical properties. In grinding, an appropriate sensor system must be able to measure the micro-geometry of the tool, which is of essential importance for the generation of surface integrity characteristics. The most promising system is based on laser triangulation.

Figure 3.3-12 shows the basic elements which are a laser diode with 40 mW continuous wave (c.w.) power and a position-sensitive detector (PSD) with amplifier and two lenses [22, 23]. The sensor is mounted on a two-axis stepper-drive unit to be moved in the direction normal to the grinding wheel surface and in an axial direction to make measurements on different traces on the grinding wheel circumference. For the determination of macro-geometric quantities such as radial runout, no practical limitations exist and the maximum surface speed may even exceed 300 m/s [22]. For micro-geometric measurements the investigations have revealed that the maximum speed of the grinding wheel should not exceed 20 m/s, based on hardware and software limitations. This means that for most applications the grinding wheel has to be decelerated for the measurement. This major drawback limits the practical field of application. The most interesting application for this sensor is seen in the supervision of superabrasives, especially CBN-grinding wheels. Their deterioration of the topography often leads to thermal damage of the workpiece. In [24] this sensor system was intensively tested during profile grinding of gears with an electroplated CBN-grinding wheel.

The measurement was done on the involute profile of the grinding wheel on 10 traces. A straight line approximated the measurement width, and additional tilting

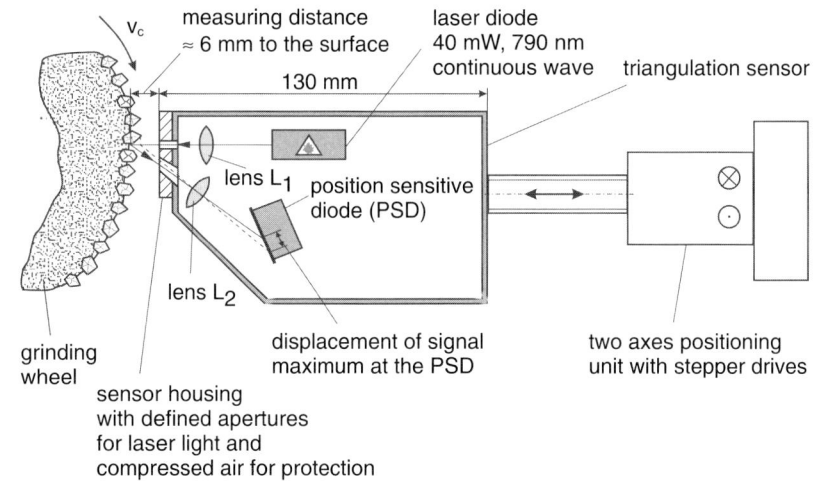

Fig. 3.3-12 Measurement principle of a laser triangulation system. Source: Werner [22]

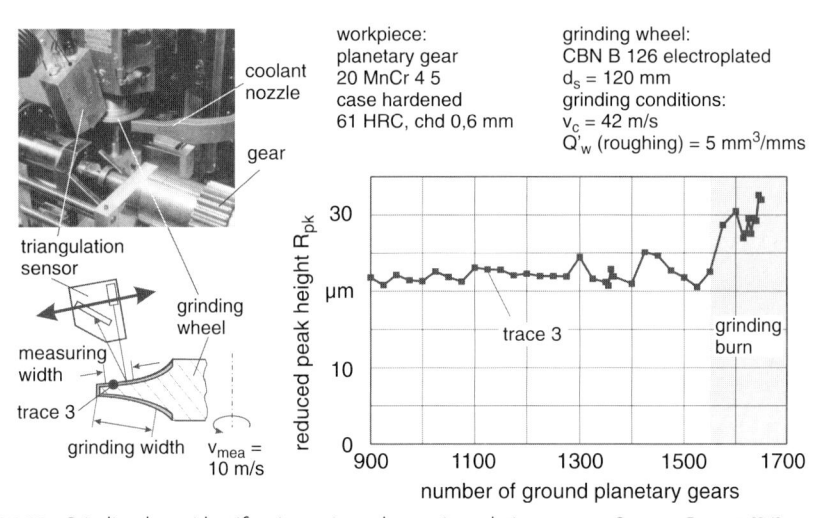

Fig. 3.3-13 Grinding burn identification using a laser triangulation sensor. Source: Regent [24]

of the sensor was not necessary. The measurement was done during the workpiece changing time. Figure 3.3-13 (left) shows the setup of the investigation. On the right a result of this sensor application is presented. The measured quantity is the reduced peak height, R_{pk}, deduced from the bearing ratio curve (see Section 4.4.4), which can be used to describe the change in the grinding wheel topography at the grains tips. As shown, the change in trace 3 can be clearly correlated with the occurrence of grinding burn, which was stated by nital etching and subsequent metallographic and X-ray inspection. Although this result is very promising, some problems have to be taken into consideration. Measurements and simulations have revealed that it is not possible to correlate the sensor roughness results definitely with a specific wear pattern. It is essential to identify the critical point where the unstable condition of the CBN-grinding wheel will start. Beyond this point, significant changes in the sensor quantities in different directions are found. Scanning of a larger area by shifting the sensor to different traces is always necessary to register this change definitely. This is possible with the installed positioning unit, but will increase the measurement time. The protection of the sensor against any kind of coolant or swarf is also essential. However, this sensor system is still the only serious method to correlate quantities measured in the working space of the machine tool on the grinding wheel with the occurrence of surface integrity changes. It is fast; even the repeated measurements of 10 traces at reduced speed will only take a maximum of 2 min, which is near the workpiece changing time of five gears in a special clamping unit. For expensive CBN-grinding wheels and high-value workpieces, this strategy offers high potential.

3.3.5
Sensors for Workpieces

The best way to investigate the influence of any cutting or grinding process on the physical properties of the machined workpiece would be to measure directly on the generated surface. However, so far only very few sensors are available to meet this demand. In the following, two techniques will be described which have the highest potential for this purpose, an eddy-current and a micro-magnetic method. Ultrasonic techniques, which are also used for crack detection, do not have a potential to monitor machined surfaces, because they are not able to resolve material changes in the first 100 µm from the surface owing to frequency limitations [1]. This is the most important area for surface integrity properties influenced by either cutting or grinding.

3.3.5.1 Eddy-current Sensors

The principle of eddy-current measurement for crack detection is based on the fact that cracks at the workpiece surface will disturb the eddy-current lines, which are in the measuring area of a coil with AC excitation [25]. All kinds of conductive materials can be tested. The penetration depth is determined by the excitation frequency. Both the conductivity and permeability of the workpiece can be investigated. The main application of this method is the monitoring of important material properties such as heat treatment to insure the proper hardness and hardness depth before succeeding operations [26]. In cutting, no application of this method for surface integrity characterization is known.

In grinding, an eddy-current sensor was introduced to monitor the occurrence of cracks. Figure 3.3-14 (left) shows the setup for this eddy-current-based measurement used for the determination of cracks generated during profile surface grinding of turbine blade roots.

Cracks generation is a special problem for the hard-to-machine materials of aircraft engine components such as nickel-, cobalt-, and titanium-based alloys. The

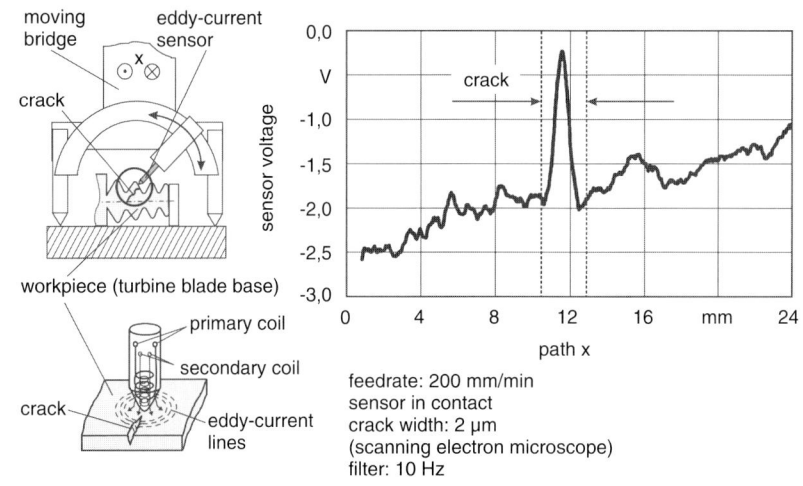

Fig. 3.3-14 Eddy-current crack detection after surface grinding of turbine blades. Source: Westkämper

purpose of the application was to determine cracks in the workspace of the machine tool immediately after grinding by moving a bridge with the sensor over the ground surface [27]. Figure 3.3-14 (right) shows the result of such a measurement. The crack was subsequently investigated with the aid of a scanning electron microscope and had a width of 2 μm. The eddy-current sensor could clearly determine this crack with a contact measurement. This size has to be regarded as the minimum resolution of the sensor. The sensor must be positioned in a perpendicular direction to the surface, because any tilting reduces the sensitivity. Thus an additional shift option was implemented in the moving bridge. Although the measurement speed was smaller than the grinding table speed, a check in the grinding machine may still be acceptable because of the high security demand on these workpieces. However, the limiting factor is the fact that only cracks in the surface of a ground workpiece can be identified, which is of interest only for a very limited range of materials.

3.3.5.2 Micro-magnetic Sensors

Changes in the physical properties on machined surfaces of ferrous materials can be determined with the aid of micro-magnetic techniques. The measuring principle is based on the fact that residual stresses, hardness values, and the structure in sub-surface layers influence the magnetic domains of ferromagnetic materials. Bloch walls separate adjacent ferromagnetic domains with different local magnetization directions. An exciting magnetic field causes Bloch-wall motions. As a result, the total magnetization of the workpiece changes. With a small coil of conductive wire at the surface of the workpiece, the change in the magnetization due to the Bloch-wall movements can be registered as an electrical pulse. The magne-

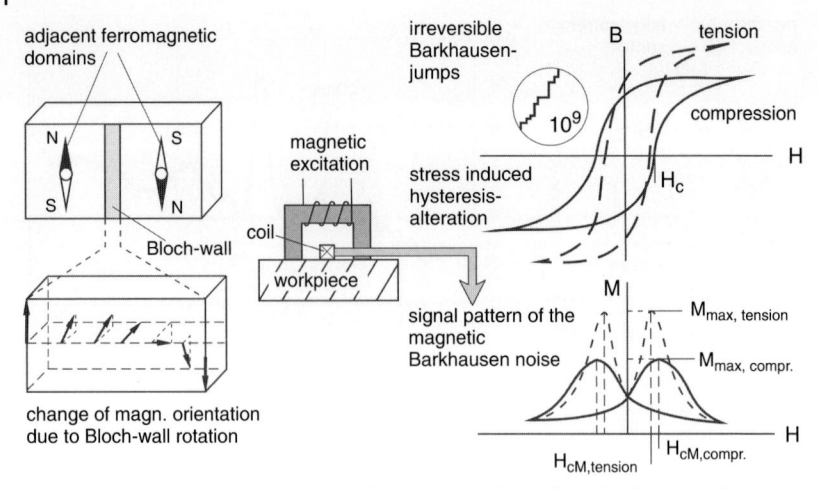

Fig. 3.3-15 Micro-magnetic structure and quantities deduced from the hysteresis loop

tization process is characterized by the well-known hysteresis loop (Figure 3.3-15). The presence and the distribution of elastic stresses in the material influence the Bloch walls to find the direction of easiest orientation to the lines of magnetic flux. Subsequently the existence of compressive stress in ferromagnetic materials reduces the intensity of the Barkhausen noise, whereas tensile stresses will increase the signal [9]. The most important quantities deduced from this signal are the maximum amplitude of the Barkhausen noise, M_{max}, and the coercivity, H_{cM}. The application of micro-magnetic technology to machined workpieces depends on the flexibility of the sensors.

In principle, a magnetization device and a signal receiver have to be adapted to the shape of the workpiece. The Barkhausen noise amplitude is detected with an air coil and the coercive strength with a Hall probe. The depth of penetration can be varied by different analyzing frequencies. Taking the magnetic properties of hardened steel into account, the minimum achievable penetration depth is approximately 15 µm. In any case the measurement time is very short and amounts to only a few seconds, which is one of the major advantages of this technique. This so-called two-parameter micro-magnetic setup was further improved by adding modules for the measurement of the incremental permeability, the harmonics of the exciting field, and eddy currents [24]. The major aim of this multi-parameter system was to separate further the influence of the initial material properties from the changes due to machining operations. This improved technique was tested on hard turned steel surfaces and on ground surfaces. In Figure 3.3-16, results from measurements on case hardened workpieces with varying tool wear of the PCBN insert are shown [5]. A total of 50 workpieces were tested. The micromagnetic investigation was done in a few minutes, including workpiece handling, whereas the X-ray investigation was performed with 20 h of pure measuring time. The results reveal a very good correlation of the two methods, especially taking into account that for the X-ray investigation only one measuring point at the sur-

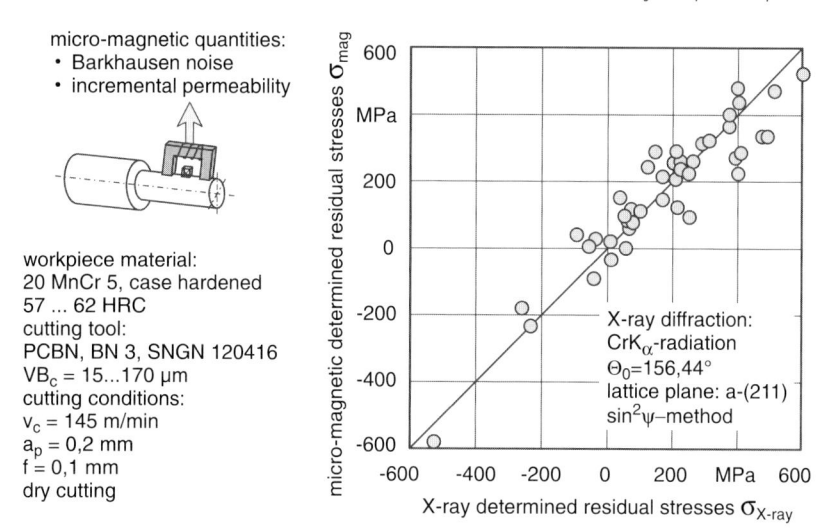

Fig. 3.3-16 X-ray and micro-magnetic determined residual stresses after hard turning. Source: Schmidt [5]

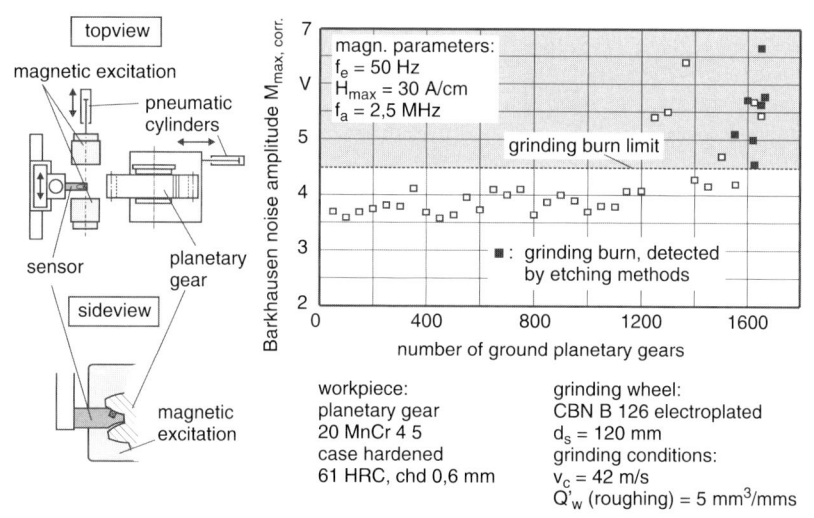

Fig. 3.3-17 Micro-magnetic surface integrity characterization of ground planetary gears. Source: Regent [24]

face could be measured, whereas the micro-magnetic value represents an average result on the circumference.

In grinding, the number of applications of this technique is higher than in hard turning [eg, 28, 29]. In [9], a detailed investigation of the potential of the two parameter micro-magnetic approach to characterize surface integrity states of workpieces also with different heat treatment was given. Subsequently this technique was transferred to practical application. Figure 3.3-17 shows the results of a large industrial test on planetary gears ground with electroplated CBN-grinding wheels [23, 24]. The geometric fitting of the micro-magnetic sensor and excitation is shown at the left in top and side views. In this case the excitation is separated from the measuring coil and Hall element. Pneumatic cylinders are used for gear approach and excitation clamping. On the right the results obtained over the lifetime of an electroplated wheel are shown. The Barkhausen noise amplitude is corrected to consider slight changes in the excitation field. It can be seen that all gears with identified grinding burn by nital etching are also recognized with the micro-magnetic setup. However, in addition, gears with high $M_{max, corr.}$ values appear, which do not show up as damage when nital etching is used. A possible explanation for this difference is the different penetration depths of the methods. Nital etching gives information only about the very top layer of the workpiece and sub-surface damage cannot be registered.

Micro-magnetic measurements can also reveal damage depending on the frequency range. It should be mentioned further that the measuring time required to scan all flanks of one gear is significantly longer than the grinding time. With intelligent strategies or an increased number of sensors in parallel use, this time can be shortened for suitable random testing. Totally automated measurement is possible. In addition to successful applications of the two-parameter method, tests with the improved multi-parameter system have also been performed for surface grinding [24]. Furthermore, in the laboratory the first tests on in-process measurements of surface integrity changes based on this micro-magnetic sensing have been conducted for outer diameter and surface grinding [24, 30]. In Figure 3.3-18 the first results of this approach during surface grinding of steel are presented. The sensor with integrated excitation moves on the surface behind the grinding wheel at the chosen table speed of 8 m/min. A conventional corundum grinding wheel was used on case hardened steel and cooling was provided by a 5% emulsion. Permanent contact is assured by spring loading. The X-ray measurement is

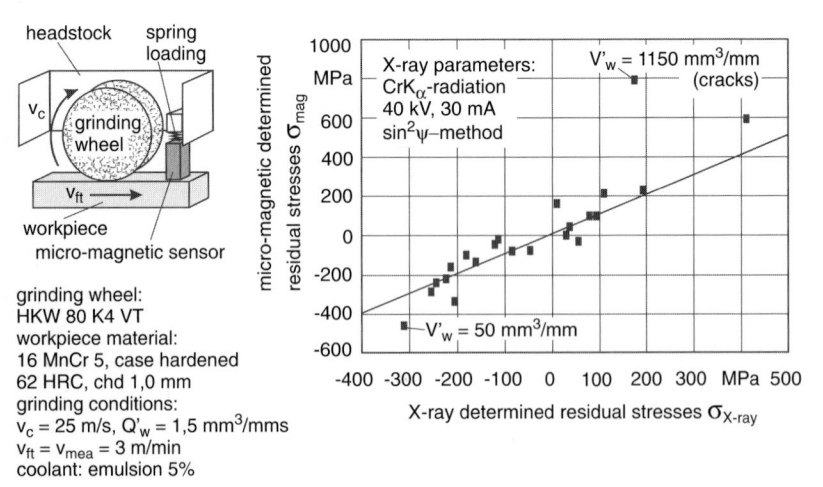

Fig. 3.3-18 Micro-magnetic in-process measurement of surface integrity during grinding. Source Regent [24]

made on one point of the ground surface and the micro-magnetic result represents the average over the whole workpiece length. The deviations in the area of compressive residual stresses and low tensile residual stresses are less than 100 MPa, and only in areas of significant damage with tensile stresses higher than 200 MPa do the deviations increase. This can be explained by the occurrence of cracks after grinding due to the high thermal load on the workpiece.

These cracks result from the high generated tensile stresses, which are measured inprocess, and lead to a significant release of the residual stress state for the later performed X-ray investigation. Although further investigations on the wear resistance of the sensor head, long-term influence of the coolant, maximum workpiece speed, geometric restrictions, and other parameters have to be conducted, this sensor offers the possibility of in-process workpiece surface integrity measurement for the first time.

3.3.6
Reference

1 BRINKSMEIER, E., *Habilitationsschrift;* Universität Hannover, 1991.
2 TÖNSHOFF, H.K., *Spanen – Grundlagen;* Berlin: Springer, 1995.
3 KÖNIG, W., BERKTOLD, A., KOCH, K.F., *Ann. CIRP* 42 (1993) 39–43.
4 WOBKER, H.-G., *Habilitationsschrift;* Universität Hannover, 1997.
5 SCHMIDT, J., *Dissertation;* Universität Hannover, 1999.
6 FIELD, M., KAHLES, J.F., *The Surface Integrity of Machined and Ground High Strength Steels,* Defense Metals Information Center Report 210; 1964, pp. 65–77.
7 MACHERAUCH, E., WOHLFAHRT, H., *HTM* 28 (1973) 201–211.
8 BRANDT, D., *Dissertation;* Universität Hannover, 1995.
9 KARPUSCHEWSKI, B., *Dissertation;* Universität Hannover, 1995.
10 INASAKI, I., *CIRP* 40 (1991) 359–362.
11 HAUPT, F., SEIDEL, T., KARPUSCHEWSKI, B., *Zahnflankenschleifen Bogenverzahnter Kegelradsätze mit CBN-Schleifscheiben,* VDI-Z 139, No. 9 (1997) 62–65.
12 ZF FRIEDRICHSHAFEN, REILHOFER K.G., *Patent-Offenlegungsschrift DE 4318102,* 1994.
13 CHOI, H.Z., *Dissertation;* Universität Hannover, 1986.
14 CREMER, L., HECKL, M., *Körperschall – Physikalische Grundlagen und Technische Anwendungen;* Berlin: Springer, 1967.
15 DORNFELD, D.A., KÖNIG, W., KETTELER, G., VDI-Ber. 988 (1993) 363–376.
16 BLUM, T., INASAKI, I., *J. Eng Ind.* 112 (1990) 203–211.
17 EDA, H., KISHI, K., et al., *Bull. Jpn. Soc. Prec. Eng.* 18 (1984) 299–304.
18 WEBSTER, J., MARINESCU, I., et al., *Ann. CIRP* 43 (1994) 299–304.
19 ROBERTS, D.A., LEETE, D.L., *Sens. Rev.* (1983) 72–75.
20 KLUMPEN, T., *Dissertation;* RWTH Aachen, 1994.
21 SAXLER, W., *Dissertation;* RWTH Aachen, 1997.
22 WERNER, F., *Dissertation;* Universität Hannover, 1994.
23 TÖNSHOFF, H.K., KARPUSCHEWSKI, B., REGENT, C., in: *30th International Symposium on Automotive Technology and Automation (ISATA), Florence, 16.–19. 6. 1997,* Vol. 30 Mec1, pp. 373–380.
24 REGENT, C., *Dissertation;* Universität Hannover, 1999.
25 CAHN, R.W., HAASEN, P., KRAMER, E.J. (eds.), *Materials Science and Technology, Vol. 3A and 3B: Electronic and Magnetic Properties of Metals and Ceramics;* Weinheim: VCH, 1994.
26 FÖRSTER, F., in: HÖLLER, P., et al. (eds.), *Nondestructive Characterization of Materials;* Berlin: Springer, 1989, pp. 505–515.

27 LANGE, D., in: *8. Internationales Braunschweiger Feinbearbeitungskolloquium*, 24.–26. 4. 1996, pp. 16/1–16/19.
28 FIX, R.M., TIITTO, K., TIITTO, S., *Mater. Eval.* **48** (1990) 904–908.
29 Shaw, B.A., Hyde, T.R., Evans, J.T., in: *1st International Conference on Barkhausen Noise and Micromagnetic Testing, Hannover*, 1–2. 9. 1998, pp. 14/1–10.
30 Tönshoff, H.K., Karpuschewski, B., et al., in: *Moderne Schleiftechnologie*, 14 May 1998, Furtwangen, pp. 7.1–7.13.

4
Sensors for Process Monitoring

4.1
Casting and Powder Metallurgy

4.1.1
Casting
H. D. HAFERKAMP, M. NIEMEYER and J. WEBER, *Universität Hannover, Hannover, Germany*

4.1.1.1 Introduction
The casting process represents the shortest route from the basic material, the alloyed melt, to the casting ready to be installed with optimized multiple functions. In contrast to this unique advantage exists the problem of the difficult control and diagnosis of the casting parameters which are responsible for the quality and the functionality of the casting. Only the melting parameters, chemical alloy composition and pouring temperature which can be set by inoculant or alloy wires and the heating capacity of the furnace before the casting are exceptions. The other parameters are subjected during the extremely short period of production in the casting process and solidification to dynamic and for the most part also reciprocal influences which are difficult to control. It is still said that for difficult and costly casting processes, eg, bell founding, you have to take off your hat before praying before the casting starts [1, 2].

With modern automated casting methods and the increasing use of computer-integrated manufacturing (CIM) systems in foundries, inaccuracy of the parameters must be avoided to guarantee a high quality of the casting products and to avoid a cost intensive interruption of production. The aims of perfect production and total quality management (TQM) require sensors which also control the mold filling and the solidification processes and thereby permit efficient process control and process control engineering [3, 4].

This demanding process control can only be realized with sensors which are adjusted to the severe conditions in a foundry such as high temperatures, difficult accessibility of the measuring point and the chemically aggressive effect of the melts. Because of the operating conditions, the sensors for casting process control, shown in Figure 4.1-1, can be divided into 'sensors without melt contact' and

4 Sensors for Process Monitoring

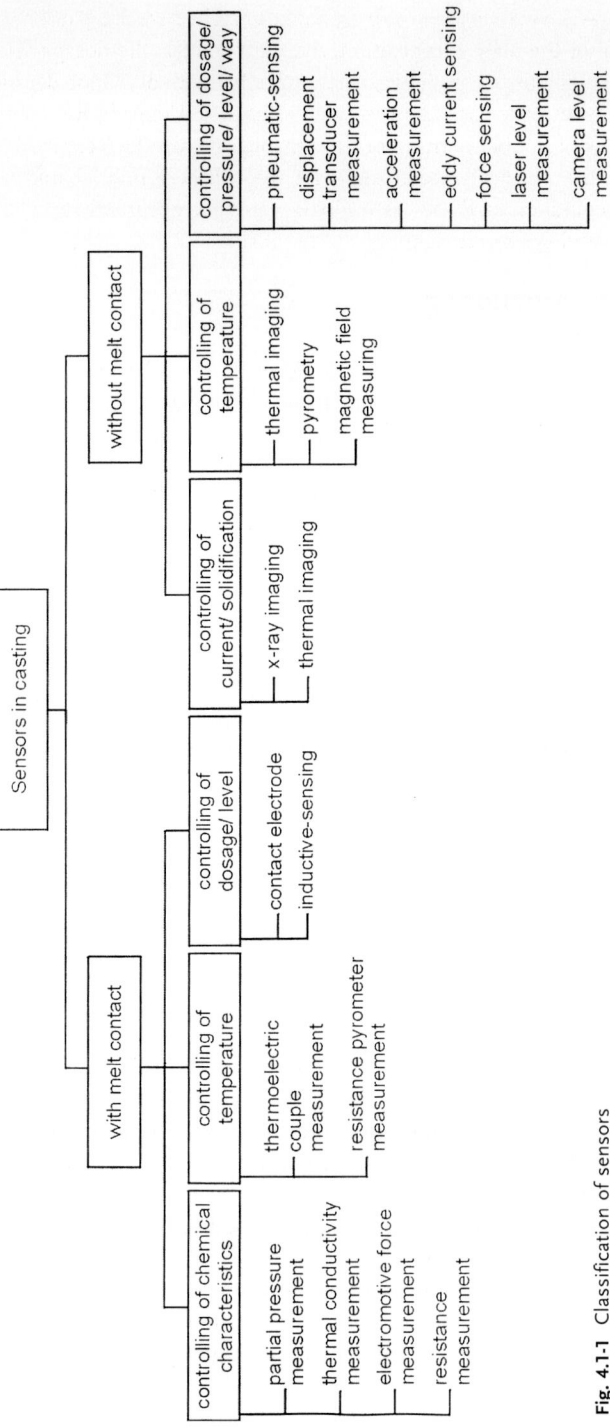

Sensors in casting

- **with melt contact**
  - controlling of chemical characteristics
    - partial pressure measurement
    - thermal conductivity measurement
    - electromotive force measurement
    - resistance measurement
  - controlling of temperature
    - thermoelectric couple measurement
    - resistance pyrometer measurement
  - controlling of dosage/level
    - contact electrode
    - inductive-sensing
- **without melt contact**
  - controlling of current/solidification
    - x-ray imaging
    - thermal imaging
  - controlling of temperature
    - thermal imaging
    - pyrometry
    - magnetic field measuring
  - controlling of dosage/pressure/level/way
    - pneumatic-sensing
    - displacement transducer measurement
    - acceleration measurement
    - eddy current sensing
    - force sensing
    - laser level measurement
    - camera level measurement

Fig. 4.1-1 Classification of sensors

'sensors with melt contact'. Further subgroups are distinguished by the particular control task, the control of the alloy composition, the temperature, the dosage and current of the melt and solidification. This division deliberately does not distinguish with regard to the separate casting processes as many of them do not allow a general summary without double naming of the sensors and also lack clarity.

In this classification, the physical measuring principle will be a final characteristic. The control and regulation of these casting parameters determine the quality of the casting products and the productivity of the foundries.

4.1.1.2 Sensors with Melt Contact

The functional groups of this type of sensor come directly into contact with the melt or the mold or are separated from the melt by protecting tubes. Normally the protecting tubes consist of thermodynamic permanent ceramics with high temperature stability as aluminium melts, for example, have a corrosive effect on the sensor material. Sensors with melt contact can be divided into types for the control of the chemical composition, types for the control of the temperature, and types for the control of the dosage or the level.

4.1.1.2.1 Sensors for Controlling Chemical Characteristics

The gas content, the chemical composition, and the purity of the melt are of decisive significance for the quality of the component. The chemical composition of the melt determines, in addition to the solidification characteristics of the casting which are influenced by the grain refining agent above all through the element content of the alloy, the mechanical properties of the component. The solvent power of metal melts for gases decreases with decrease in temperatures. Because of this, evolution of gaseous hydrogen and oxygen which are absorbed from the atmosphere and dissolved in the metal melts takes place and pores are formed in the casting. To guarantee a perfect component, the gas content must be controlled frequently before and during serial casting [5–7].

Partial Pressure Measurement

As hydrogen is the only gas which dissolves in aluminium melts, the hydrogen content can be simply controlled with the Chapel (continuous hydrogen analysis by pressure evaluation in liquids) and the Telegas or Alscan process. With the chapel process a porous graphite punch which is connected through a gas-tight ceramic tube to a pressure gage will be immersed in the melt and evacuated for a short time. The graphite punch reacts like a bubble into which the hydrogen diffuses out of the melt until the pressure in the probe and the hydrogen partial pressure in the melt are the same. If the state of equilibrium is reached the hydrogen content of the melt at a constant temperature can be calculated by using the Sievert laws [6–9]:

$$\log C_H = 0.5 \log p_{H_2} - A/T + B \tag{4.1-1}$$

where C_H = concentration of hydrogen dissolved in aluminium, p_{H_2} = partial pressure of segregated hydrogen, T = temperature, and A, B = Sievert constants, depending on the alloy composition.

The chapel process is easy to handle, reliable, fast and has been proved especially in Europe.

Thermal Conductivity Measurement

The Telegas or Alscan method has a ceramic probe below the melt level, out of which pure inert gas or nitrogen flows continuously into the melt and is then collected in a hood. While the blowholes are rising the dissolved hydrogen diffuses out of the melt until equilibrium of the gas circulation is reached. The hydrogen partial pressure is measured with a thermal conductivity-measuring cell [10–12]. The telegas or Al scan method is especially used in the USA. In contrast to the chapel process, the measurements must be carried out over a longer period, at least 15 min.

Electromotive Force Measurement

In the steel and copper industry, an electrochemical cell made of ceramic (Figure 4.1-2) is used for determining the oxygen content in the melt [13–16]. The gage heads contain a thermoelectric couple (see next section) and a voltaic cell which has a mixture of a metal and an oxide, eg, Cr/CrO, inside with a known oxygen partial pressure as a reference material.

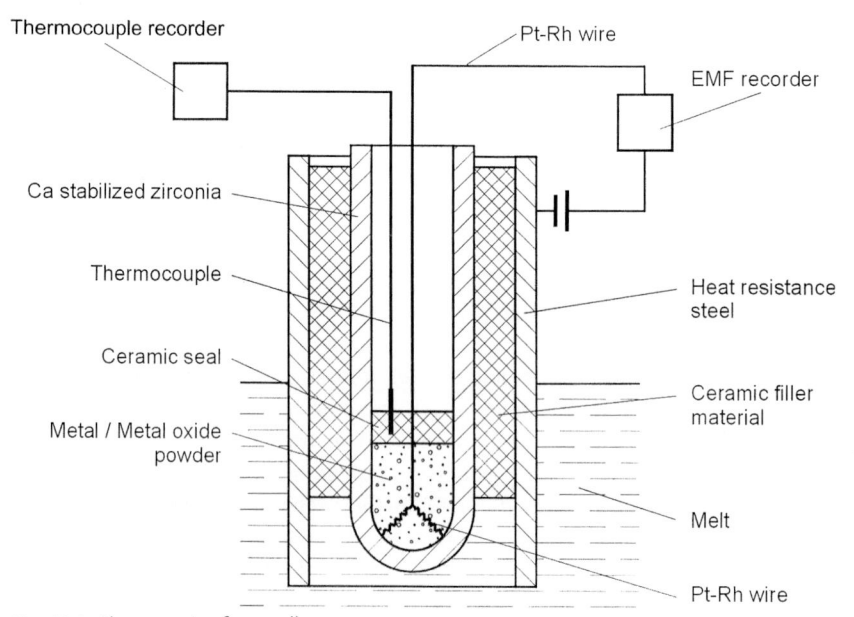

Fig. 4.1-2 Electromotive force cell

On immersing the gage head in the melt, an electromotive force between the reference material and the melt arises because of the oxygen ion conductivity of the partially stabilized ZrO_2. The relationship obeys the Nernst law:

$$E = -(RT/4F) \ln (p_{O_2}/p'_{O_2}) \qquad (4.1\text{-}2)$$

where E=energy, R=gas constant, T=temperature, F=Faraday constant, and p_{O_2}, p'_{O_2} =partial pressure of oxygen at the two electrodes.

The potential difference as a measure to calculate the oxygen activity of the melt can be used here. The temperature of the cell is an important factor in the measurement. A voltaic cell can be used at higher temperatures for the measurement of the oxygen content of solid or liquid metals, slag, and mattes. With this sensor the hydrogen, magnesium, and sodium contents can be determined when aluminium is melted [17, 18].

Resistance Measurement
The Liquid Metal Cleanliness Analyzer (LiMCA) is used to control the purity of the melt continuously. The measuring principle is mainly based on the registration of very small resistance modifications in the microohm range in liquid aluminium or magnesium caused by non-metallic inclusions. The robust and safe LiMCA sensor is used in light metal foundries and consists of a heat-resistant tube for sampling and two electrodes, one in a test-tube and the other in the surrounding melt [5, 19–21].

4.1.1.2.2 Sensors for Controlling Temperature

The temperature of the melt and the mold is of decisive significance for the correct mold filling and the cycle time of the serial casting, which implies the productivity of the company. Temperature sensors with melt contact are based on the principle of conduction, in contrast to the temperature sensors without melt contact. These sensors are also separated by protecting tubes or layers of aggressive melts. There is a division between thermoelectric couples and resistance pyrometers.

Thermoelectric Couple Measurement
Thermoelectric couples (Figure 4.1-3) are based on the thermoelectric effect (Seebeck effect). They consist of two wires of different metals with the ends soldered

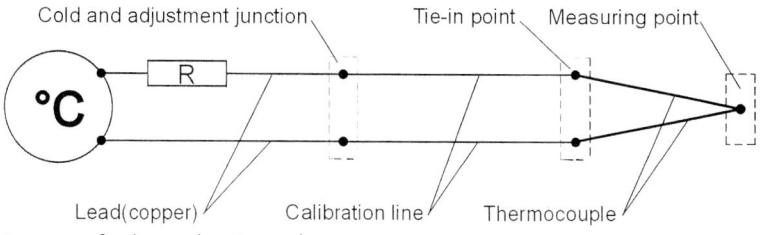

Fig. 4.1-3 Structure of a thermoelectric couple

or welded. A voltage arises when the two ends have different temperatures. This thermoelectric voltage depends on the metals used and on the temperature difference between the junction point and the connecting point (summing point) of the measuring instrument. The measurement of the thermoelectric voltage is carried out using high-resistance voltage measuring instruments. If necessary, possible disturbing secondary thermal effects at supplying parts must be eliminated through calibration lines. The measuring range is between −200 and 2500 °C depending on the metals. The following metal pairs are used: platinum/platinum rhodium, nickel/chrome nickel, iron/constantan, and copper/constantan [9, 22, 23].

Thermoelectric couples can also be produced without a protecting tube in very small sizes with a minimum diameter up to 0.5 mm and a free choice of the length. These so-called sheath thermoelectric couples are the most commonly used temperature sensors in light metal foundries because of their flexibility and reasonable price.

Resistance Pyrometer Measurement
The resistance pyrometer is based on the principle of a change in the electrical resistance with variation in the temperature of a conductor or semiconductor. Depending on the predominant electrical conducting mechanism, a difference is made between pyrometers with a positive (metals) and a negative (high-temperature conductors, negative temperature coefficient resistors, thermistors) resistance-temperature characteristic curve. Resistance pyrometers require analog or digital electrical connections for measurement and for higher demands measuring bridges and compensators are used. Similar to the thermoelectric couple, the advantages of these sensors are the reasonable price, the robustness, the flexibility, and the simple handling.

4.1.1.2.3 Sensors for Controlling the Dosage/Level
A correct dosage is decisive for quasi-stationary thermal economy of the mold and therefore significant for the quality of the casting. By reducing the cycle material the economy of the foundry is favored [24].

Contact Electrode Measurement
The easiest and most common way to control the dosage is realized with a contact electrode. When the melt touches the contact electrode a signal will be sent to the installation control which controls the dosage process [25, 26].

Inductive Sensing
In light metal furnaces, inductive level sensors which are protected from the melt by suitable austenitic or ceramic protecting tubes are used to control the level continuously. This principle is based on an induced voltage in a conducting loop in the sensor. This voltage causes an electric current which forms a magnetic field around the sensor. A signal is originated by the variation of the magnetic field by

the melt [27]. This type of level sensor is expensive, susceptible to wear and costly in maintenance.

4.1.1.3 Sensors without Melt Contact

The physical measuring methods and the technical realization of this kind of sensors are relatively complicated and complex, although necessary in order to guarantee continuous production and quality assurance. Since these sensors do not touch the melt, which is often chemically aggressive, and since they are not exposed to high thermal stresses, it is unlikely that they will fail. Sensors without melt contact can be divided into different types: sensors for controlling the current and solidification, for controlling the temperature, for controlling the dosage, pressure, level, and route.

4.1.1.3.1 Sensors for Controlling Current and Solidification

Precise knowledge of the melt current, the solidification and the thermal economy of the mold is an important factor in the design of casting dies. With this knowledge it is possible to attain perfect heating and cooling circuits, cycle times, and temperature distribution for directional solidification. For the continuous casting process the control of the position of the solidification contour is of great importance since the continuous cast velocity and the charging depend on this position. If the meniscus is not respected, liquid metal may flow over or run out [28–30].

X-ray Imaging

X-rays from a radioactive source, typically a rod-type emitter (eg, Co-60) in a lead protector, radiograph the mold. Since solid metals absorb X-rays better than melts owing to their higher density, the position of the solidification contour can be detected by a scintillation meter. The sprue, ie, the melt current during die casting, can be supervised and the position of the solidification contour can be directed in a continuous casting mold. Figure 4.1-4 shows a schematic diagram of this supervising method for continuous casting [30–35].

The complex protection of the workplace against radioactive radiation reduces the number of applications of this supervising method. X-ray processes and computed tomography (CT) are additionally used for nondestructive component testing and for the quality testing of safety components. Defects in casting, eg, inclusions, sink-holes, pores, cracks, etc., can be detected [37, 38].

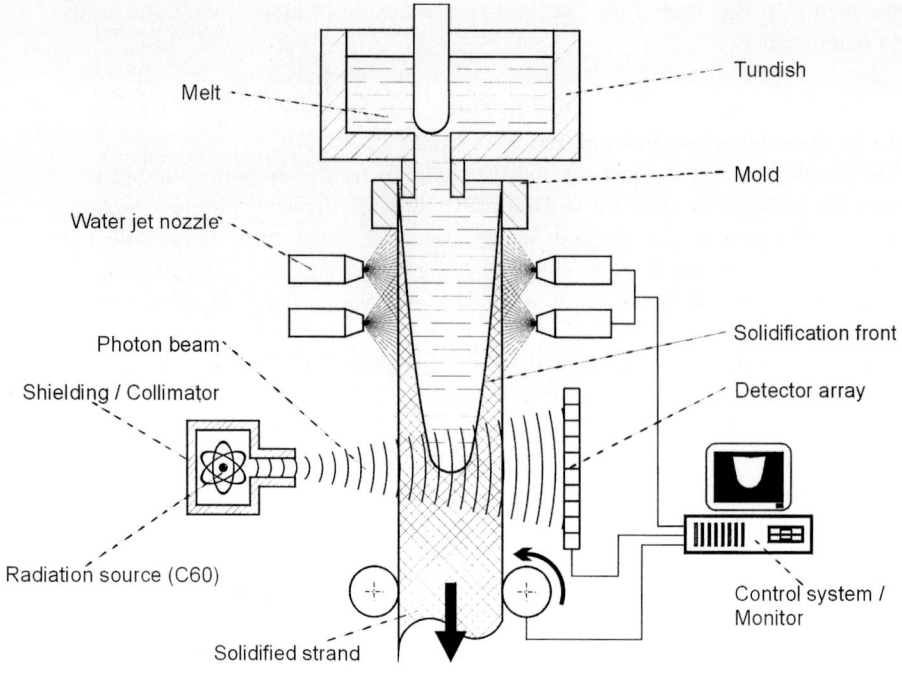

Fig. 4.1-4 Principle of X-ray imaging

Thermal Imaging
Cameras for thermal imaging visualize infrared radiation, ie, thermal radiation from object surfaces. Since the atmosphere is not transparent to thermal radiation over the whole radiation spectrum, these cameras are divided into near-, medium- and far-infrared cameras according to the sensitivity of their sensors [29, 30]. The flow of metal melts is examined for model molds consisting of a solid mold with die sinking and even face and which is closed by a moveable, transparent mold half of solid foam (aerogel) (Figure 4.1-5). Owing to its transparency to visible light and thermal radiation in the near-infrared range, the flow and solidification of steel, lead, aluminium, and magnesium melts, etc., can be observed [29, 39].

Since the assembly is complex and the use of the aerogel slab is difficult, thermal imaging for the examination of the flow of melts is only used in research or for the design of molds.

4.1.1.3.2 Sensors for Controlling Temperature
If the metallurgical melt flow is correct, up to 100% of rejects in die casting can occur owing to the wrong temperature of the mold. Non-contact temperature sensors permit a correct mold design and effective continuous control of the melt temperature at positions difficult to access or at temperatures that destroy contact sensors [40].

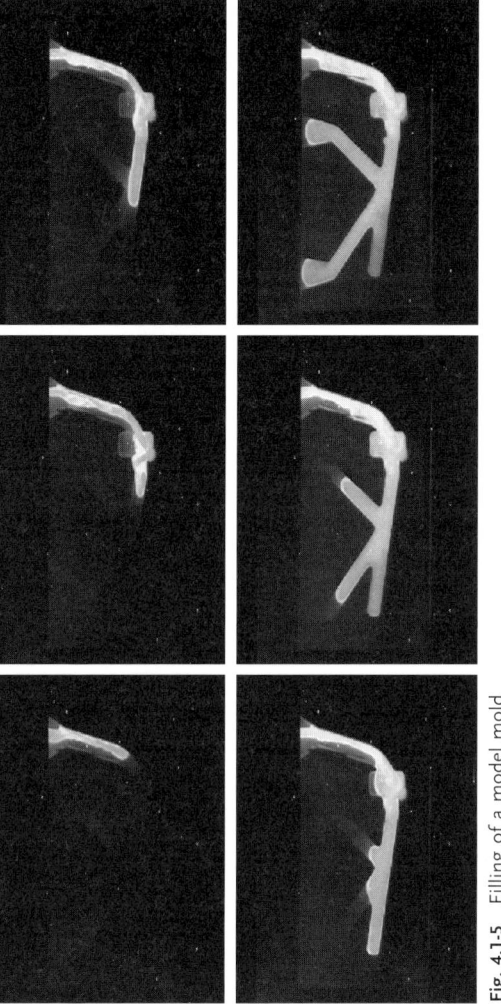

Fig. 4.1-5 Filling of a model mold

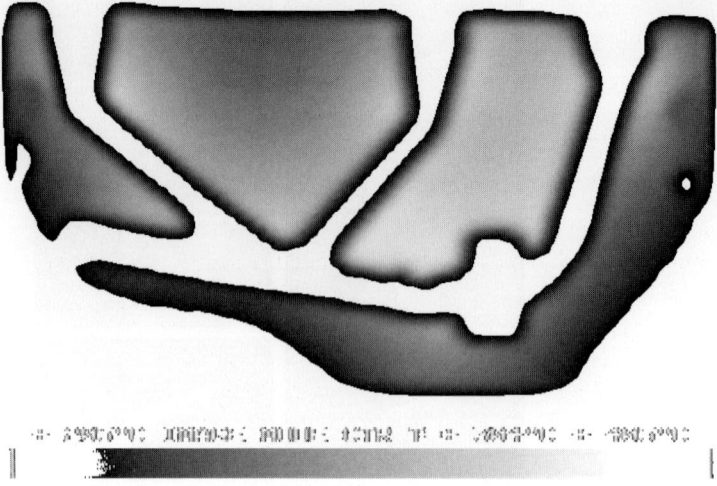

Fig. 4.1-6 Thermogram of a model mold

Thermal Imaging
For thermal imaging of the mold temperature, as shown in Figure 4.1-6, mainly far-infrared cameras are used due to the emission spectrum [29, 41]. With these examinations a relationship between the die casting temperature, the flow temperature of the cooling system, and the cast cavity could be found [40]. Further, thermal imaging is used for the verification of simulation results and mold designs [41, 42].

Another application of this type of camera is the supervision of the cast temperature for continuous casting. Additionally, conventional cameras are used for the observation of the billet surface, the billet orientation, etc. [43].

Pyrometry
Pyrometry is based on the same physical rules of thermal radiation and thermal imaging. In contrast to thermal imaging cameras, pyrometers detect the temperature only at intervals, but they are more economical, easier to use, and they have an excellent accuracy of up to ±1%. In general, foundries use total radiation pyrometers for low temperatures and ratio or two-color pyrometers for higher temperatures as emitted by iron and steel melts. Total radiation pyrometers can easy be tested as their signals are directly subject to the Stefan-Boltzmann law. For the two-color pyrometer two partial radiations in different wave ranges are considered. Ratio pyrometers measure the temperature of the object by the ratio of the radiation density of two different spectral regions. The advantage is that the transmission distance does not influence the measuring results [31, 44, 45].

In the steel industry, pyrometers have been used since the 1950s for the supervision of melt temperatures [31, 46]. Additionally, they are used for continuous casting for the control of the billet temperature, ie, for the control of the cooling system [47].

Magnetic Field Measurement

Magnetic field measurement is used for the high-precision heating of thixo billets. For this die casting process the cast material, the so-called thixo billet, is in a range between the solidus and liquidus temperatures, ie, the material is partly solid and partly liquid. This thixotropic state causes a change in the magnetic field which can be measured by a field-measuring sensor to an accuracy of down to 0.5% [48, 49]. This complex measuring process, which has to be calibrated for every alloy composition, is used over the whole cross-section of the thixo billet because of the highly required even distribution and measurement of the temperature.

4.1.1.3.3 Sensors for Controlling Dosage, Pressure, Level, and Route

The physical measuring method of this type of sensor is often the same so that it seems reasonable to combine these process parameters into one control group. Sensors in this group are mainly used for the supervision of die casting which, in spite of a more frequent use of sensors, is still called 'black box technology'. Further applications are continuous casting and break-mold casting [50, 51].

Pneumatic Sensing

Important machine parameters of die casting are the injection shot velocity and the pressure. The pressure is supervised by pneumatic sensors in the hydraulic system of the die casting machine. Pneumatic sensors are also used in the furnace gas chamber of dosage furnaces which have shown a high degree of reliability in the aluminium industry (Figure 4.1-7) [41, 52, 53].

Another application of pneumatic sensing is level measurement in dosage or blast furnaces. Figure 4.1-8 shows the functional principle of this sensor, which measures the pressure necessary for the exhaust of nitrogen bubbles from a ceramic tube on the bottom of the melting pot [54].

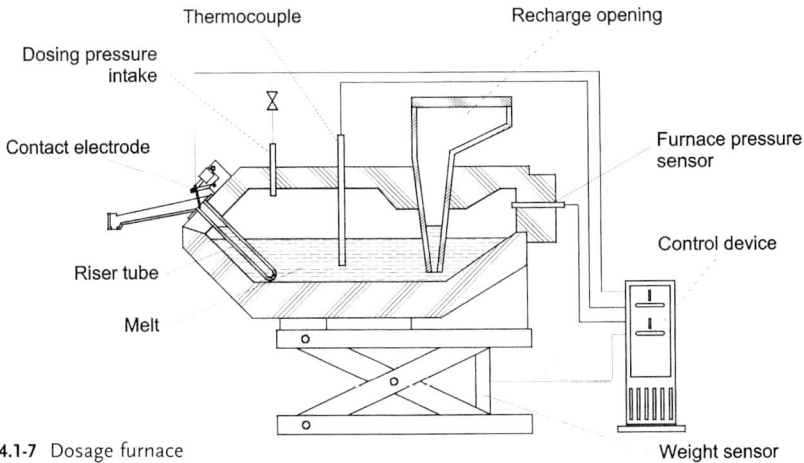

Fig. 4.1-7 Dosage furnace

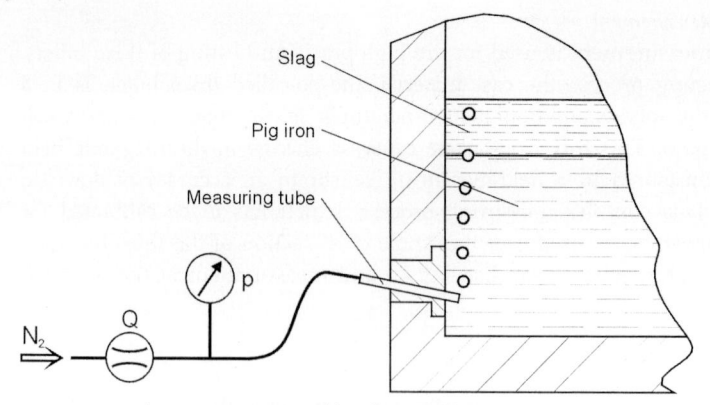

Fig. 4.1-8 Level measurement in a blast furnace [54]

For these control types, conventional pressure gages are used which are subject to the pneumatic or hydrostatic principle.

Displacement Transducer
The control of the injection shot velocity in die casting is the essential criterion for turbulence-free filling and therefore for components with only a few pores. The injection shot velocity is controlled in three phases depending on the piston displacement. Magnetic displacement transducers measure the piston position. The principle of this type of sensor is based on the influence of magnetic effects (eg, the Hall effect) which depend on the displacement [55]. The sensors are maintenance-free and extremely robust.

Acceleration Meter
In order to avoid the adhesion of the billet to the mold in continuous casting and to assure a clean billet surface, the continuous cast mold is set in an oscillating motion, vertical to the billet. This oscillation is supervised by seismic acceleration meters which represent a mass-spring damping system. The system consists of an inert seismic plate, a spring with a force proportional to the displacement and a damping component proportional to the velocity [22, 56, 57].

4.1.1.3.4 Eddy Current Sensing

Eddy current measurements represent another solution for the supervision of the level in a mold in the continuous casting process (Figure 4.1-9). According to Qui [58], the detection of sullage which must not enter the mold is another application when liquid steel is filled from the ladle into the tundish [59–61]. The changing level of the steel bath influences the number and course of the eddy current in liquid steel and the surrounding conductive objects. The resulting change in the electromagnetic field is measured [62].

Fig. 4.1-9 Principle of eddy current sensing

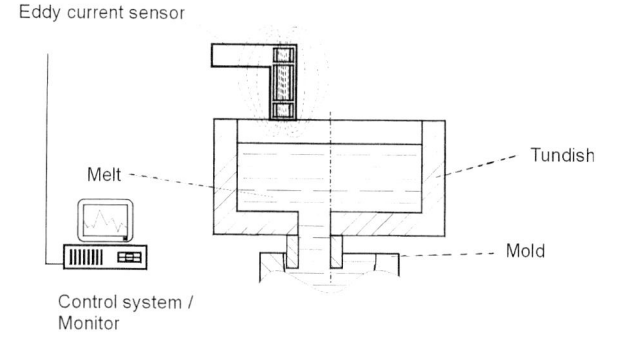

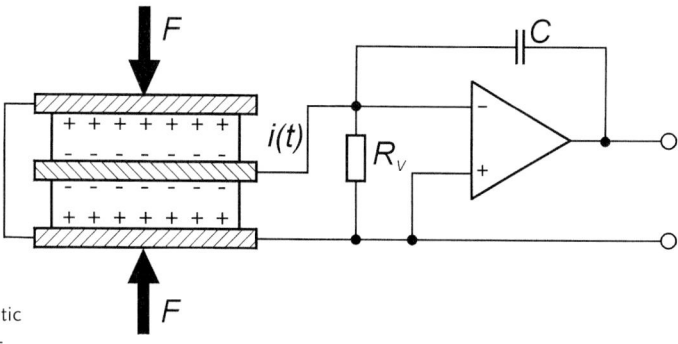

Fig. 4.1-10 Schematic diagram of a piezoelectric force gauge

Force Sensing

The most conventional way to measure the level continuously is furnace weighing with maintenance-free electronic load cells. In general, the level is indicated directly at the furnace by means of a signal lamp (see Figure 4.1-7) or it is indicated to the master computer. The load cell is cheap, maintenance-free and can be used for general purposes. Charging appliances are equipped with the same systems for balancing the material [63–66].

These load cells are based on the physical principle of piezoelectric force sensing technology (Figure 4.1.-10). When force is exerted on a piezoelectric crystal (eg, quartz, barium titanate ($BaTiO_3$)), negative crystal lattice points are offset against positive ones so that a difference in charge can be measured at the crystal surfaces as a function of force [22].

The function of sensors for the measurement of the internal pressure in casting chambers is subject to the same physical principle. With the measurement of the pressure development, important knowledge about the melt flow, mold filling and solidification during the filling process is achieved [67–69].

Laser Level Measurement

Laser sensors are used for the measurement of the meniscus in the continuous cast process and for level control of the launder and the sprue in automatic break-mold casting methods of aluminium and steel (Figure 4.1-11) [4, 70–73].

In laser level measurement, an emitter gives short light impulses at a high frequency (approximately 10 Hz) in the direction of the metal bath surface. From there a small proportion is reflected and sensed by a receiver. The transit time is a measure of the level [51].

Camera Level Measurement

Another system for level measurement in molding boxes works with a camera and secondary image processing so that the stopper control can keep the meniscus in the sprue at a constant level (Figure 4.1-12) [2, 74].

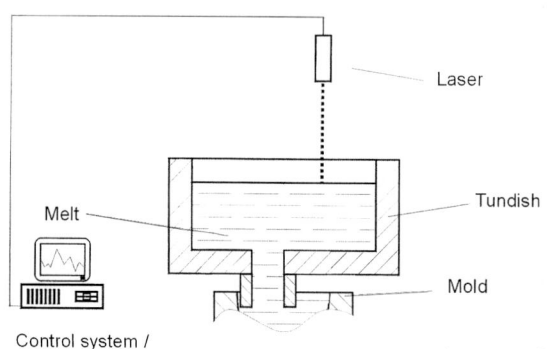

Fig. 4.1-11 Principle of laser level measurement

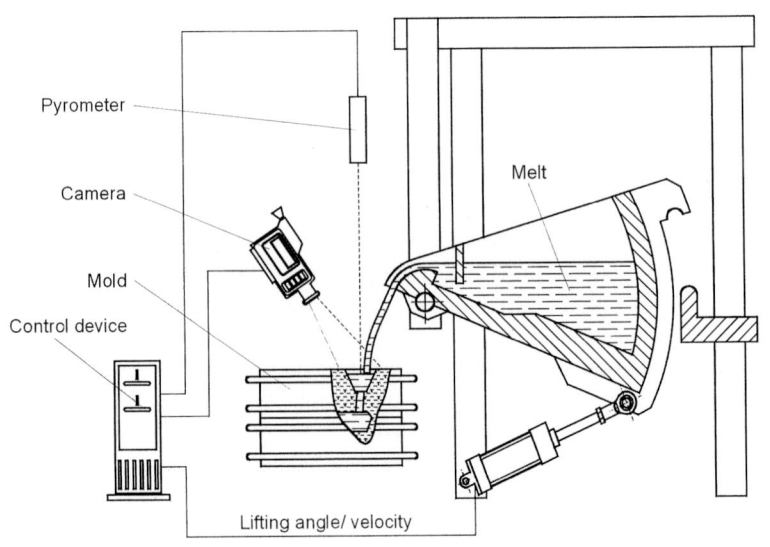

Fig. 4.1-12 Principle of camera level measurement

The cast behavior of types with many cores which in general differs widely depending on the mold can be limited by level control and the high requirements to achieve a constant hydrostatic pressure in the sprue can be fulfilled [74].

4.1.1.4 Summary

The quality of the casting and the productivity of a foundry depend on few but very important parameters which are difficult to control. This is mainly due to the fast dynamic processes during filling and solidification and to the sophisticated conditions in the foundries. The sensors specifically adapted to these requirements for the control of the chemical and physical properties of the melt and the perfect control of the machine and mold parameters such as cast velocity, pressure, and temperature allow optimum casting conditions. A sophisticated sensor technology creates the conditions for integral process control of automated casting processes, eg, die casting, which is still considered to be 'black box technology'. This sensor technology makes the integration of the casting process into production lines and CIM systems possible.

4.1.1.5
References

1 KAHN, F., *Giesserei* **80** (1993) 579–584.
2 NACKE, B., KESSLER, M., ANDREE, W., *Elektowärme Int.*, Ed. B (1995) 138–143.
3 STOLTENBERG, K., RÖHRIG, K., *Giesserei-Praxis* **80** (1993) 52–56.
4 MEZGER, F., *Giesserei* **82** (1995) 332–335.
5 LESSITER, M.J., RASMUSSEN, W.M., *Mod. Casting* **86** (2) (1996) 45–48.
6 EIGENFELD, K., WECHSELBERGER, O., KNOCHE, D., *Aluminium* **74** (1998) 244–247.
7 EIGENFELD, K., WECHSELBERGER, O., SCHAAN, KNOCHE, D., *Giesserei* **84** (1997) 45–47.
8 NOLTE, M., *Giesserei* **86** (1999) 72–75, 77–80, 83.
9 HASSE, S., *Giessereilexikon*; Berlin: Schiele & Schön, 1997.
10 CHEN, X.G., KLINKENBERG, F.J., ENGLER, S., HEUSLER, L., SCHNEIDER, W., *J. Miner. Met. Mater. Soc.* **46** (8) (1994) 34–38.
11 NEFF, D.V., in: *Proceedings of the 3rd International Conference on Molten Aluminium Processing*, Orlando, FL; 1992, pp. 387–405.
12 DASGUPTA, S., APELIAN, D., *Molten Aluminium Processing, 5th International AFS Conference*, Orlando, FL; 1998, pp. 233–258.
13 SEETHARAMAN, S., SICHEN, D., JAKOBSSON, A., *Sens. Model. Mater. Process.* (1997) 327–344.
14 *Sauerstoff-Messanlage Deltatherm III*; Künzer, 1993.
15 YAJIMA, T., KOIDE, K., TAKAI, H., FUKATSU, N., IWAHARA, H., in: *Proceedings of the 20th Commemorative Symposium on Solid State Ionics in Japan*; 1995, pp. 333–337.
16 DEKEYSER, J.C., SCHUTTER, F.D., POORTEN, C., VAN DER ZHANG, L., FRAY, D.J., in: *Proceedings of the 5th International Meeting on Chemical Sensors*; 1995, pp. 273–275.
17 FERGUS, J.W., *Giesserei-Praxis* **85** (1998) 443.
18 VANGRUNDERBEEK, J., LENS, P., CASTELIJNS, C., VERSTREKEN, P., *Light Met.* (1999) 1005–1009.
19 BUSSMANN, W., in: *Symposium der Deutschen Gesellschaft für Materialkunde*; 1995, pp. 189–197.
20 DUPUIS, C., DALLAIRE, F., MALTAIS, B., in: *128th TMS Annual Meeting, 'Light Metals 1999'*, Warrendale, PA; 1999, pp. 1069–1077.

21 Carozza, C., Lenard, P., Sankaranarayanan, R., Guthrie, R.I.L., in: *Proceedings of the International Symposium, 36th Annual Conference of Metallurgists of CIM*; 1997, pp. 185–196.
22 Beitz, W., Küttner, K.H., *Taschenbuch für den Maschinenbau*; Berlin; Springer, 1995.
23 Beckerath, A., *Sens. Rep.* **4** (2) (1989) 6–10.
24 Lindner, P., *VDI-Fortschr.-Ber.* 5, No. 582 (2000).
25 Malpohl, K., *CP+T* **15** (3) (1999) 8–12, 14.
26 Krause, H., Schiebold, K., Nielebock, E., Obieglo, B., *Giessereitechnik* **36** (1990) 19–20.
27 *Füllstandsmesseinrichtungen*; Wiehl-Bielstein: Carli Electro Automation, 1999.
28 Haferkamp, H., Bach, Fr.-W., Niemeyer, M., Viets, R., *Aluminium* **75** (1999) 945, 947–953.
29 Haferkamp, H., Bach, F.-W., Niemeyer, M., Viets, R., Weber, J., Breuer, M., Kruessel, T., in: *Proceedings of the IEEE International Symposium on Industrial Electronics*; 1999, pp. 1442–1447,
30 Regusewicz, F., *Arbeitskr. Aluminium-Automobil* (10) (1998) 151, 153–168.
31 Peacock, G.R., *Proc. SPIE* (1999) 171–189.
32 Fitting, D.W., Dubé, W.P., Siewert, T.A., Paran, J., *Review of Progress in Quantitative Nondestructive Evaluation*; New York: Plenum Press, 1995, pp. 2315–2321.
33 Chun, J.H., Hytros, M.M., Jureidini, I.M., Saka, N., Lanza, R.C., *Ann. CIRP* (1999) 147–150.
34 Hytros, M.M., Chun, J.H., Lanza, R.C., Saka, N., *J. Manuf. Sci. Eng., Trans. ASME* (1998) 515–522.
35 Dumitriu, B., Mitut, R., Bretthauer, G., Garbe, J., *Automatisierungstechnik* (839) (1997) 66–80.
36 Dumitriu, B., Mikut, R., Bretthauer, G., Werfel, G., Böttger, S., *Stahl Eisen* **119** (1999) 35–38.
37 Schmitz, V., Reiter, H., *Ing.-Werkst.* 4 (8) (1999) 43–45.
38 Kroth, E., *Giesserei* **85** (6) (1998) 35–38.
39 Haferkamp, H., Niemeyer, M., Pelz, C., Viets, R., Schaper, M., *Proc. SPIE* **3700** (1999) 164–170.
40 *Giesserei* **80** (1993) 451–456.
41 Muller, W., Feikus, F., *Trans. Am. Foundry men's Soc.* **104** (1996) 1111–1117.
42 Jaerke, P., Eigenfeld, K., *Giesserei* **74** (1987) 713–717.
43 Aigner, H., Angerer, R., Reisinger, J., *Berg- Hüttenmänn. Monatsh.* (140) (1995) 156–161.
44 Glueckert, U., *Erfassung und Messung von Wärmestrahlung*; Franzis-Verlag, 1992.
45 *Elektro* (22) (1991) 58–60.
46 Wanin, M., in: *Proceedings of the 3rd European Conference on Advanced Materials and Processes, Part 2*; 1993.
47 Lasday, S.B., *Ind. Heating* **60** (12) (1993) 36–38.
48 Juergens, R., Graeft, T., Gies, J., in: *Thermprocess Symposium*; 1999, pp. 45–61.
49 Graeft, T., Juergens, R., *Elektrowärme Int. Ed. B* **57** (B2) (1999) 69–73.
50 *Qual. Today* (1995) 50, 52.
51 Krueger, G., *Sens.-Technol. Anwendung* **509** (1984) 243–248.
52 Malpohl, K., *Casting Plant Technol.* **15** (3) (1999) p. 8–10, 14.
53 Croom, D.E., *Foundryman* **89** (1996) 440–445.
54 Danloy, G., Stolz, C., Crahay, J., Dubois, P., in: *58th Ironmaker Conference Proceedings*; 1999, pp. 89–98.
55 Sjögren, B., Fitze, R., *Giesserei* **83** (16) (1996) 92, 94, 96.
56 Chumakov, S.M., Sorokin, A.N., *Steel USSR* **28** (6) (1998) 27–29.
57 Miller, G., *Stahl Eisen* **111**, (12) (1991) 73–78.
58 Qui, D., *Scand. J. Metall.* (1997) 178–182.
59 *Ind. Heating* (1993) 36–38.
60 Klein, A., Wolf, M., *Comprehensive Machine and Process Condition Monitoring in Conventional Continuous Casting*; Iron and Steel Society, 1992, pp. 807–815.
61 Martin, J.F., Dusser, H., Nadif, M., *Steel Technol. Int.* **13** (1995) 183–185.
62 Rohac, J., Pisoft, V., *Stahl Eisen* **112** (3) (1992) 89–91.
63 Braunger, H.P., Hann, R., *Giesserei* **77** (3) (1990) 75–78.
64 Shingledecker, F., *Diecasting Eng.* **38** (4) (1994) 22, 24.
65 *Giesserei-Erfahrungsaustausch* **42** (1998) 546.
66 *Giesserei-Erfahrungsaustausch* **38** (1994) 321–322.
67 Kangas, S., Fritsche, S., *Diecasting Eng.* **43** (3), 38, 40, 42, 44–45.
68 *Giesserei* **83** (15) (1996) 37–38.

69 Niu, X.P., Tong, K.K., Hu, B.H., Pinwill, I., *Int. J. Cast Met. Res.* **11** (1998) 105–112.
70 *Aluminium* **76** (2000) 281–282.
71 Johansson, S., Lowback, G., *Sensors* **13** (8) (1996) 34–36.
72 Johansson, S., *Ind. Heating* **63** (9) (1996) 89–91.
73 Smith, J.R., *Foundry Trade J.* **166** (1992) 536, 538–539.
74 Nacke, B., *VDI-Ges. Energietech.* **1292** (1997) 299–313.

4.1.2
Powder Metallurgy
R. Wertheim, *ISCAR Ltd., Hardmetal Industrial Products, Tefen, Israel*

4.1.2.1 Introduction

Powder metallurgy (PM) is a metal processing technology in which metal powders are used to produce technological parts. It is an important commercial technology for the mass production of near-net shapes, eliminating or reducing the need for further machining processes. Certain metals or alloy combinations which cannot be produced by other methods can be formed by PM and sintering.

In the various PM manufacturing sequences, the powders are compressed into desired shapes, injected into molds or extruded through a nozzle to produce longer parts and profiles. After being shaped, the so-called 'green' product is heated to cause bonding of the particles into a hard, rigid mass.

Compression by pressing, injection molding, or extrusion is accomplished with suitable equipment using tools or molds designed specifically for the parts to be manufactured. The tooling, in pressing, consists of a die and at least one punch; in injection molding, of a die and a nozzle; and in extrusion, of die and injection equipment.

The green, very brittle product is transformed into a very hard part by sintering at a temperature below the melting point of the metal.

Figure 4.1-13 shows the four main conventional steps to produce metal parts after the metallic or ceramic powders have been produced: (A) the blending and mixing of the powder to the required particle size and various chemical compositions; (B) the compacting, in which the powder is pressed into the desired shape; (C) the sintering to the final or almost final size and shape; and (D) further possible steps: grinding, finishing, and coating.

In the following explanation, the use of monitoring, control, and sensors in the production of mainly hardmetal products made of carbides, nitrides, and oxides mixed with a suitable binder such as Co or Ni will be discussed.

4.1.2.2 Mixing and Blending of Metal Powders

The properties of the powder compound, the preparation and composition of the powder mix, and the shaping process are of significant importance in the production and performance of hardmetal products.

Figure 4.1-14, for example, shows a flow chart of the basic mixing procedure for the various shaping processes in the production of carbide cutting tools. In the

Particle Condition

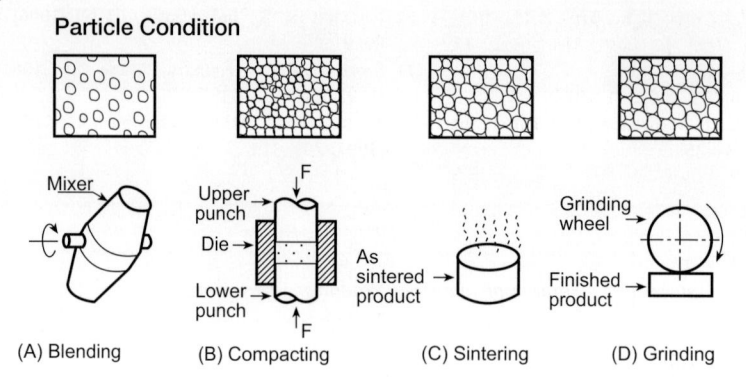

Fig. 4.1-13 The conventional powder metallurgy production sequence: (A) blending or mixing; (B) compacting; (C) sintering; (D) grinding [1]

simplest hardmetal composition, the basic mixture consists of tungsten carbide (WC) powder of a specified particle size and size distribution and cobalt (Co) powder; if necessary, addition of carbon black powder is used to correct the carbon content of the hardmetal. In order to determine the final hardmetal properties, cubic carbides of titanium (TiC), tantalum (TaC) and/or niobium (NbC) may be added to the mix or in the prealloyed form with the tungsten carbide. The name hardmetal is basically applied to all hard metallic materials, but in a narrower sense it is mainly associated with the above combinations of hard, distinctly brittle, metallic materials and a relatively soft ductile metal, predominantly from the iron group (Fe, Co, Ni), the so-called binder or binder metals. These binder metals (mainly Co) may be present in different amounts in a mixed crystal form in the binder phase.

For the subsequent wet milling, the required milling liquid, such as an alcohol, acetone, hexane, or other organic liquid, is added to the mixture. The purpose of the milling liquid is to protect the components of the mix from oxidation and also to insure optimum dispersion of the ingredients.

Powder milling is a crucial step, since adequate size reduction coupled with uniform distribution of all the ingredients can have a decisive effect on the sintering behavior. For wet milling, attritors or ball mills are used. In the stationary water-cooled container a stirrer rotates, giving a rotary motion to the milling medium, the charge, and the milling liquid. By means of a pumping system, the suspension being milled is circulated in order to insure uniform milling.

After milling, the suspension is sieved and dried for the next step. Depending on the subsequent forming process, a suitable procedure is selected as indicated in Figure 4.1-14.

The selected process or criteria depend on the specific requirements of the prepared powder. Therefore, for example, the powder mix for dry pressing or injection molding has to be brought into a granular form which has good flow properties, constant fill density, and a suitable granule size.

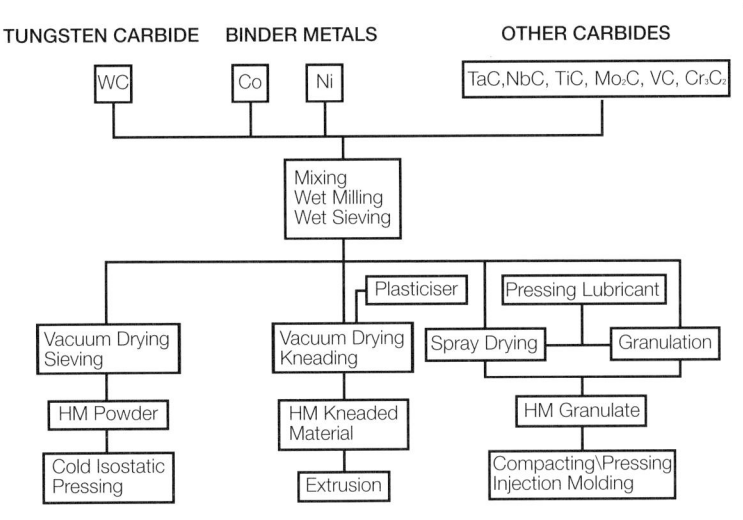

Fig. 4.1-14 Flow chart showing the preparation of carbide powder mix for various shaping processes [2]

The material from the wet milling process, consisting of powder mix, milling liquid, and dissolved or dispersed pressing lubricant, is processed by granulation or spray drying, into powder or powder mix. During spray drying, the suspension is forced from the pressurized container and atomized in a hot gas stream in the spray tower. This atomization results in the formation of spherical granules of variable diameter, eg, from a few micrometers to 200 μm.

4.1.2.2.1 Monitoring and Sensors in Powder Production

Monitoring the hardmetal powder includes the particle size, shape, distribution, and surface area. Features such as friction, flow or packing, composition, homogeneity, and contamination are essential for the subsequent compacting and sintering processes.

Determination of particle size by the evaluation of one of the geometric parameter depends on the shape, which can be spherical, flake, or irregular. The use of microscopy measurement techniques, such as optical, scanning electron, or transmission electron microscopy, are the most common sensors.

Screening is also used in obtaining sized powders. It provides a means for removing specific size fractions. The use of these methods is applicable for larger grain sizes and requires long screening durations.

Particle size analysis by sedimentation is mostly applicable to the smaller sizes. Particles settling in a liquid like water or air sensor device reach a terminal velocity dependent on both the particle size and the fluid velocity [3].

Size analysis by sedimentation uses a predetermined settling height and places a dispersed powder at the top of a tube. The amount of powder settling at the bottom (as a function of time) allows the calculation of particle size distribution. Obviously, the fastest settling particles are the largest whilst the smallest can take a considerable time to settle. Sensors and automatic instrumentation for performing sedimentation and separation can use gravity forces or centrifugal force devices, light blocking, or X-ray attenuation methods.

Air classification sensors achieve a separation of powders into selected size fractions using a cyclone or a spinning disk and cross-current airflow.

For automatic sensing of particle size, a low-angle Frauenhofer light scattering system using monochromatic (laser) light is used (Figure 4.1-15). Intensity and angular extent are affected by the particle sizes passing in front of the photodiode-array detector. A computer providing the particle-size distribution analyzes the data. An electrical conductivity-sensing device provides another option for measuring the number and size of particles suspended in the fluid. Conductivity measurement is achieved by making the fluid conductive and applying a small voltage across an opening.

A light-blocking sensor based on a light cell and a photocell is also used to determine particle size. The amount of light blockage due to the light beam interruption by moving particles is detected by the photocell, indicating particle-size distribution.

A large number of other sensors are used in the powder production steps, eg, mixing, blending, or spray drying. Most of these are not built into the production sequence itself to provide a direct feedback signal, but are mainly used as measurement sensors in open-looped systems.

4.1.2.3 Compacting of Metal Powders

Compacting of powders before sintering can be performed to give a low- or high-density component, or simultaneous pressing and sintering can be used to give the final product. Powders with good sintering densification can be shaped using low pressures as used in some compacting applications and in the injection molding process. During compacting, the powder is deformed into a high-density com-

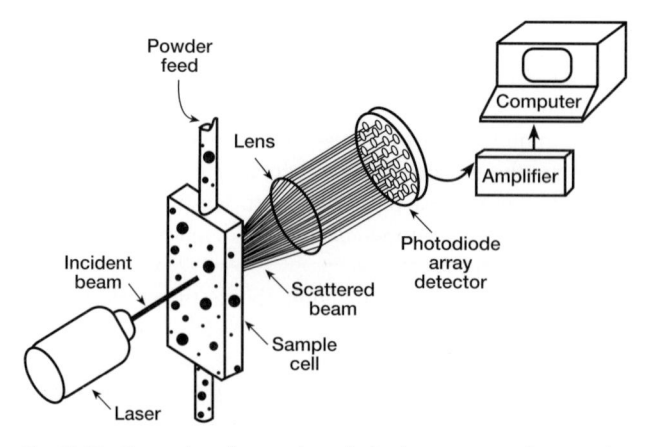

Fig. 4.1-15 Sensor based on a photodiode detector to analyze powder-particle size [3]

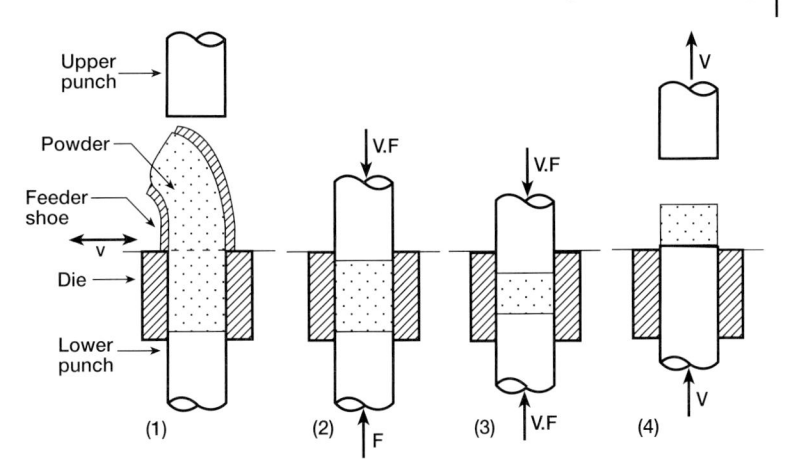

Fig. 4.1-16 Pressing powders: (1) filling the die cavity with powder; (2) initial and (3) final positions of upper and lower punches during compacting; (4) ejection part

ponent that approaches the final geometry. The means of delivering the high pressure to the powder, the mechanical constraints, the powder properties, and the rate of applying pressure are the main parameters determining the density which are analyzed during the process.

Conventional compacting of powder is normally performed with the pressure applied along one axis as shown in Figure 4.1-16. The steps during the pressing cycle start by filling the die with a very precise amount of powder which is controlled by the movement of the feeder shoe.

The lower punch position during filling determines the required volume. After filling the cavity, the lower punch drops to the pressing position and the upper punch is brought into the die. Both punches are moved and loaded to generate stress within the powder mass. At the end of the movement, the powder compound experiences the maximum stress. Finally, the upper punch is removed and the lower punch is moved upwardly to eject the compact.

In Figure 4.1-16, the pressure is transmitted from both the top and bottom punches. Alternatively, a single-action pressing can be performed when pressure is transmitted from only one punch. The applied pressure results from the punch movement forming a smaller volume of the powder particles and causing a decrease in pore space. During pressing, more particles are in contact with each other and the density is higher up to a very specific required value as shown in Figure 4.1-17. As shown, the initial rate of densification with the compacting pressure is high. With continued deformation the slope of the powder density versus pressure curve declines, reflecting work hardening. At the onset of the compacting cycle (1), voids exist between the particles. With increased pressure better packing and decreased porosity are achieved. The worked part following pressing (3) is defined as a 'green compact', which indicates that it has not been fully processed. The green density is much greater than the starting bulk density, which gives adequate strength for handling.

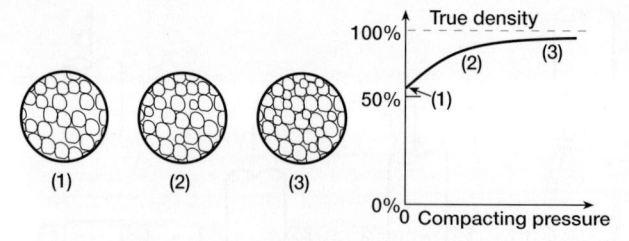

Fig. 4.1-17 Effect of applied pressure during compacting on the density: (1) initial powders after filling; (2) repacking; (3) compacting of particles [1]

Fig. 4.1-18 Pressing setup with a controlled process and an automatic handling system

4.1.2.3.1 Compacting Equipment

Presses used in conventional PM compacting are mechanical, hydraulic, or a combination of the two. Because of differences in part complexity, presses can be distinguished as pressing from one direction, referred to as single-action presses; or pressing from two directions, which can be either a double or multiple action. Current available press technology can provide up to 10 separate action controls to produce complex geometric parts.

Figure 4.1-18 shows a typical pressing setup with a controlled process. The system shown is used for double-action pressing of carbide inserts for cutting tools. The positioning of the upper and lower punches is adjusted according to the required powder volume and the compacting ratio. A computer that records and optimizes the process parameters is connected to the compacting system.

4.1.2.3.2 Sensors and Control

In compacting, the pressed height is determined by the powder fill and the compacting pressure. To maintain control of the final compact height and shape, both the apparent and final densities must be known. To achieve more uniform density of the pressed compact, the double-action pressing system is used.

Figure 4.1-19 shows a scheme of a controlled pressing system in which the lower punch can be adjusted according to the required filling height. The filling height or volume is reduced during pressing to the required size, which corresponds to the green density, and the final component size after sintering. A high green density normally results in improved final properties. However, as the compacting pressure increases, the mechanical locking of the component in the die cavity also increases. Thus, the ejection forces increase with increasing compacting pressures.

If mechanical motions taken off a cam system deliver the pressure, then the compact dimensions are the main controlled parameters. Any variations in the powder fill create small-density variations between parts. Generally, if the pressure is delivered from a hydraulic source, the pressing is usually slower than when using a mechanical pressing system. The most important controlled variable to maintain high accuracy is the force or pressure value. This can be implemented by using a force sensor in both the mechanical and hydraulic presses. The controller analyzes the maximum force developed in each compacting step and decides accordingly the quality of the part and adjustment of the filling up position.

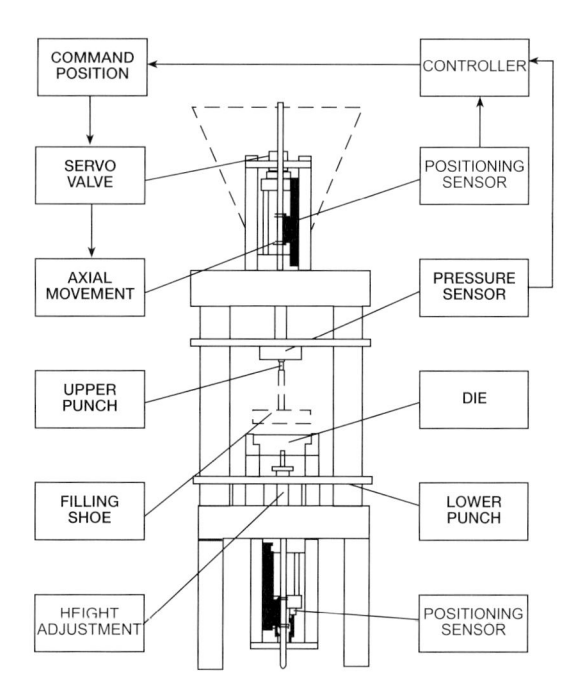

Fig. 4.1-19 Typical pressing system with sensors for punch positioning and pressing pressure

The use of strain-gage sensors, load cells, pressure transducers, piezoelectric cells or similar devices are used as sensors for the force-controlling system.

High-precision positioning of punches is done, for example, with an optical NC linear encoder with a resolution of up to 1 µm. The positioning sensor provides the signal to the controller which decides the command level transferred to the servo valve. New presses are equipped with proportionally controlled electric motors as a basis for high-precision compacting. A special capacity sensor is used to detect the absence of powder in the filling system to avoid damage due to lack of powder in the feeding system. All data related to pressure and punch positioning are processed in a computer in accordance with the known density and power properties.

4.1.2.4 The Sintering Process

In the sequence of the individual production steps shown in Figures 4.1-13 and 4.1-14, from raw material to the finished hardmetal product, the sintering operation is the process that confers the mechanical, physical, and chemical properties of the material. These properties are of essential importance for real applications. The sintering process therefore plays one of the most important parts in the manufacture of components, and determines the hardness, strength, and dimensional accuracy of the products.

Sintering is basically a heat treatment and a metallurgical process performed on the compact to bond the metallic particles, thereby increasing strength and hardness.

During this process, the loosely bound powder aggregates after pressing become denser through a change in position of the arrangement and structure of the atoms. From this, it is clear that the sintering process is the sum of the predominantly physical processes, which results in the complete or almost complete filling of the pores. Where the powder mixture is composed of different elements, the process leads to a material structure which is almost completely expressed by the appropriate phase diagram. Sintering of hardmetal is composed of multiple steps, which are dependent on temperature, time, and other influencing factors. There is no simple rule which can describe the complete process. However, the technical literature includes sintering equations, expressed in the form of relationships which describe by simple mathematical forms the dependence of the final material properties or the volume of pores on production parameters, such as density of the pressed body, the sintering temperature, or the time of sintering.

Material transport mechanisms can include solid-phase sintering of homogeneous or heterogeneous powder and liquid-phase sintering. The predominant process in sintering hardmetal is permanent liquid-phase sintering, which means that liquid is present during practically the whole process of isothermal sintering.

On the other hand, in most of the sintering processes of hardmetal, the final stage of the treatment is usually carried out at temperatures between 0.7 and 0.9 times the binder metal's melting point. In this case, the terms *solid-state sintering* or *solid-phase sintering* can be used because the binder metal remains unmelted at

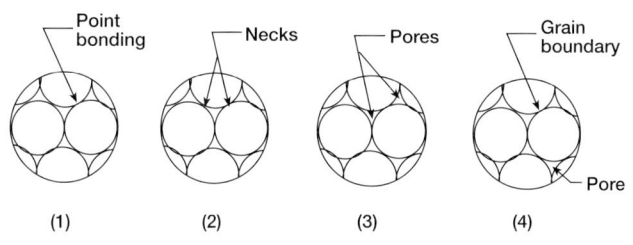

Fig. 4.1-20 Sintering on a microscopic scale: (1) bonding of particles at contact points; (2) contact points grow into 'necks'; (3) reduction of pores between sizes; (4) development of grain boundaries between particles [1–3]

these temperatures. The green compact consists of many distinct particles, each with its own individual surface, and so the total surface area is very high. Under the influence of heat, the surface area is reduced through the formation and growth of bonds between the particles, with associated reduction in surface energy. The finer the initial powder size, the higher is the total surface area and the greater the driving force behind the process to provide higher strength.

Figure 4.1-20 shows on a microscopic scale the changes that occur during sintering of metallic powders. Sintering involves mass transport to create the necks and transform them into grain boundaries. The principal mechanism by which this occurs is diffusion; other possible mechanisms include plastic flow. Shrinkage occurs during sintering as a result of pore-size reduction. This depends to a large extent on the density of the green compact, which is dependent on the pressure during compaction. Shrinkage is generally predictable when processing conditions are closely controlled.

4.1.2.4.1 **Sintering Furnaces**
Various types of sintering equipment are available. Continuous furnaces, vacuum batch-type furnaces, and hot isostatic pressing (HIP) equipment are used in the PM industry.

For PM applications with medium-to-high production rates, the sintering furnaces are designed with mechanized flow-through capability for the workparts shown schematically in Figure 4.1-21.

The heat treatment consists of three steps accomplished in three consecutive chambers. The three main steps in these continuous furnaces are (1) preheating, in which lubricants and binders are burned off, (2) sintering, and (3) cooling. The treatment is illustrated also in Figure 4.1-21 showing schematically also the sintering temperatures as a function of time or position in the furnace.

In modern sintering practice, various sensors control the atmosphere in the furnace. The purposes of a controlled atmosphere are to (1) protect from oxidation, (2) provide a reducing atmosphere to remove existing oxides, (3) provide a carbonizing atmosphere, and (4) assist in removing lubricants and binders used in pressing. Common sintering furnace atmospheres are inert gas, nitrogen based, disso-

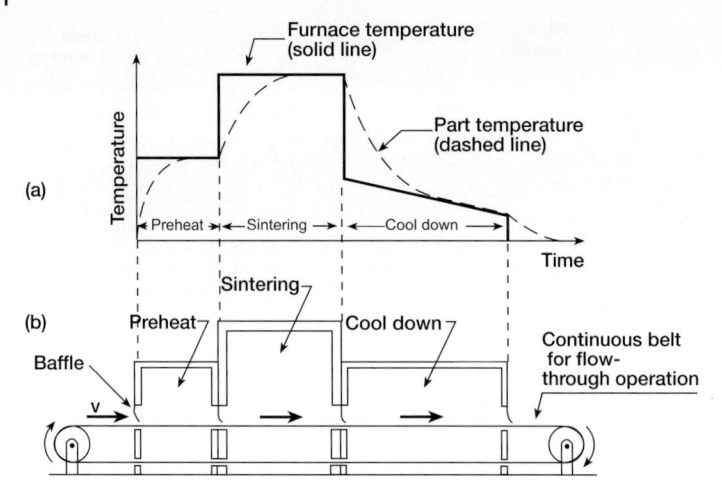

Fig. 4.1-21 (a) Typical heat treatment cycle in sintering and (b) schematic cross section of a continuous sintering furnace [1]

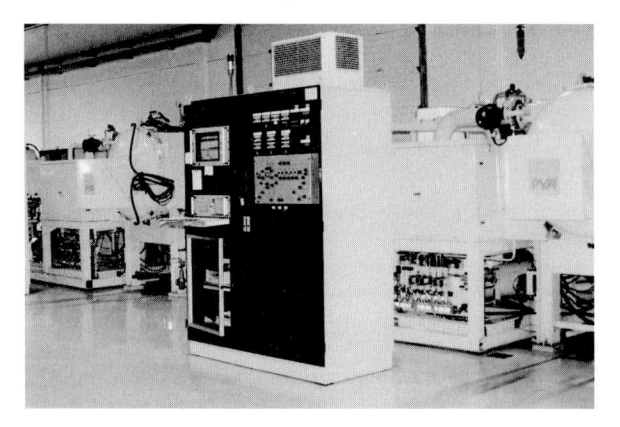

Fig. 4.1-22 Batch-batch type sintering furnace and control panel

ciated ammonia, hydrogen, and natural gas based. Vacuum atmospheres are used for certain metals, such as stainless steel and tungsten.

In order to ensure high product quality and repeatability, the furnaces are equipped with fully controlled automatic sintering programs. To arrive at a highly reproducible quality standard for series production, the required processing parameters for the furnace cycle, eg, the time-temperature cycle for heating, sintering, and cooling the pressure relationship or protective atmosphere, are recorded by printouts or stored in a computer. These new types of furnaces can be obtained in either a continuous or discontinuous (batch) format.

Figure 4.1-22 shows modern sintering furnaces with a fully controlled sintering process. The control panel provides information about the programmed temperature and pressure values together with the gas flow features at inlet and outlet positions. Flow rates, temperatures and pressure are controlled and recorded continuously. Sensors are used at various positions in order to follow and control the complete sintering process steps.

These modern furnaces permit direct sintering of pressed parts without a preliminary pre-sintering since an effective separation of the wax can be achieved by a suitable heating cycle and the evaporation of the organic pressing medium. Through the correct regulation of the temperature in the heating zones, marked temperature gradients in the reaction chamber (even during the heating phase) can be avoided. The temperature transfer from the heating unit to the parts occurs predominantly via radiation. Temperature measurement is carried out using thermocouples depending on temperature range and atmosphere, which are usually manufactured from chromel-alumel or platinum-rhodium alloys. The earlier method of measuring temperature with an optical pyrometer (radiation pyrometer) is now only used as an additional control technique. Owing to the porous nature of the structure of pressed parts in relation to other materials, there is a higher inclination to react with the surrounding atmosphere. Therefore, the sintering atmosphere plays a very important part in the process. Various gas combinations and pressure profiles can influence the properties of the sintered product. Of all the technically pure gases manufactured, hydrogen is the most widely available, since pure, dry hydrogen is a suitable protective gas for the sintering of WC-Co hardmetals and is frequently used as a sintering atmosphere. However, because of the possible build-up of oxy-hydrogen gas, and thus the danger of an explosion, special safety measures are necessary.

4.1.2.4.2 Monitoring the Sintering Process

Figure 4.1-23 shows the basic sintering furnace design for the production of WC+Co parts in batches. The sintering process steps and the main controlled variables can be described using the system elements shown.

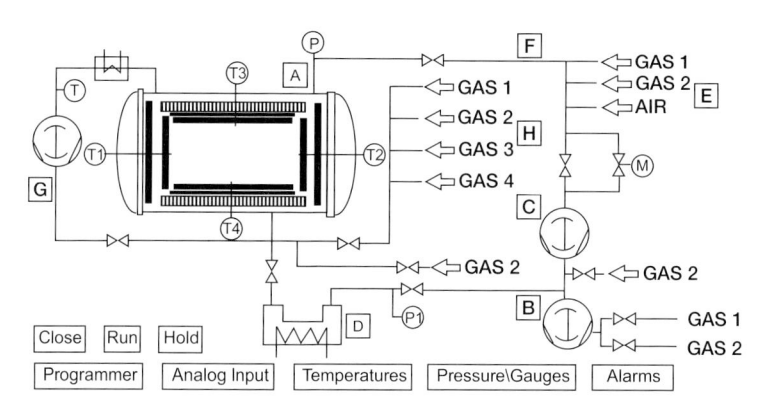

Fig. 4.1-23 Typical non-continuous sintering furnace for WC+Co production in batches

The 'green' compacts are loaded on graphite plates inside the furnace. In order to avoid adherence, an intermediate ceramic oxide is used. The quantity of products within the furnace depends on the total chamber volume.

After closure of the furnace door, the first step is to pump all gases down to a pressure of around 0.020 Torr (mmHg), which is essential to avoid any undesirable reactions. The gases are pumped through line A by pumps B and C to approximately 1 and 0.01 Torr, respectively. The pressure is measured and controlled continuously. The pumping time depends on the furnace volume and workpiece weight. At the same time, the condenser D is pumped to the same vacuum pressure. As the sintering temperature reaches a very high level, it is essential to avoid any air remaining in the condenser and in the furnace itself. In order to insure low pressure, a leak test is carried out. The criteria are, for example, a pressure change of 10 μmHg in 1 min.

The second step is dewaxing, which aims to remove the organic binder from the green workpieces. During this process, an inert gas such as argon or nitrogen is supplied into the furnace F. The heating then starts with a certain acceleration from room temperature to approximately 600 °C. During this process the gas pressure difference between the entrance to the furnace and the graphite case with the products is more than 2 Torr to insure the exit of the paraffin from the material. The pressure (P) is measured in entrance position A and at the exit point from the condenser P1. In order to insure constant and uniform temperature in all parts of the hot zones, the temperature is measured and controlled continually in several positions: frontal ($T1$) and rear ($T2$) doors and top ($T3$) and bottom ($T4$) of furnace. The normal temperature tolerance is ±3 °C. The normal duration of step 2 is 5 h depending on the binder composition and constant. Prior to step 3, the gas flow of argon or nitrogen is stopped. The third step is a continuous heating to approximately 1200 °C whilst the vacuum pump insures the constant pressure of about 0.02 mmHg. Pressure and temperatures in various points are controlled continuously.

During step 4, the liquid-phase reaction starts at a temperature of above 1300 °C while argon gas flows with a relatively high pressure of about 20 mmHg to insure the full reaction around the grain boundaries and minimizing the pore size.

After finalizing the sintering process, the next step (5) is the cooling process, which is done by using various gases to accelerate the process. Pump G is activated to supply the protective gases from H during the cooling procedure.

Temperature and pressure are controlled as shown in Figures 4.1-24 and 4.1-25. Figure 4.1-24 shows, for example, the programmed temperature in the furnace from room temperature up to more than 1350 °C and down to room temperature in five main steps. The average measured values of $T1$, $T2$, $T3$ and $T4$ are very close to those programmed. Figure 4.1-25 shows schematically the measured gas pressure in the sintering furnace. During Step 1 of pumping, the pressure starting from 1 bar is reduced to values below 0.04 Torr and the temperature remains at room values. During steps 2 and 3 the temperature is raised to 600 and 1200 °C while the pressure during dewaxing (step 2) is, for example, 20 Torr and during heating (step 3) again below 0.04 Torr. Pressure sensors for the higher

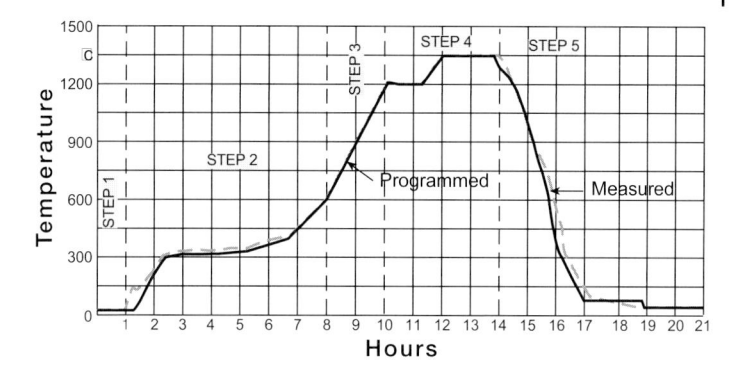

Fig. 4.1-24 The programmed and measured temperatures during controlled batch sintering

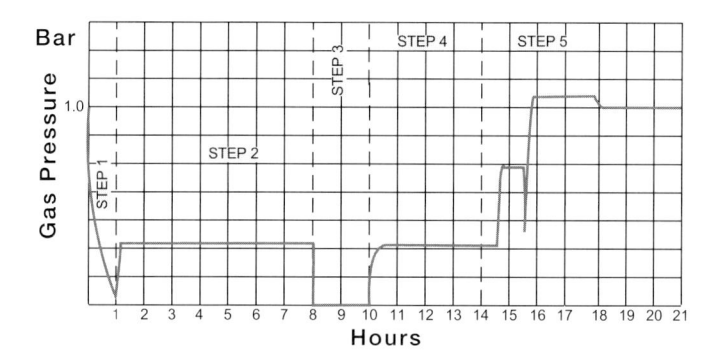

Fig. 4.1-25 General behavior of the gas pressure during batch sintering

range of vacuum technology are normally capacitance-, diaphragm-, or piezo-type devices, whereas, for the lower range, the sensors today are cold cathode gages or compaction gages.

During the sintering step (4), the temperature is kept almost constant in accordance with the material composition, and the gas pressure is adjusted to insure cobalt reaction along grain boundaries and to minimize the pore volume.

Sensors to control the gas dosing are used to regulate the pressure in the vacuum system. Gas-flow valves also close automatically in the event of power failure. During sintering of other materials such as iron powders or stainless steel, completely different temperatures and pressures are used.

4.1.2.5 References

1 GROOVER, M.P., *Fundamentals of Modern Manufacturing*; Prentice-Hall, Englewood Cliffs, NJ, 1996.
2 ANON., *Powder Metallurgy of Hardmetals*; Lecture Series, EPMA European Powder Metallurgy Association, Shrewsbury, 1995.
3 GERMAN, R.M., *Powder Metallurgy Science*; Metal Powder Industries Federation, NJ, 1994.

4.2
Metal Forming

E. Doege, F. Meiners, T. Mende, W. Strache, J. W. Yun, *Institute for Metal Forming and Metal Forming Machine Tools, University of Hannover, Germany*

The profitable use of advanced monitoring systems is more and more integrated in modern mass manufacturing processes since reliable equipment is available. The idea is to improve the metal forming process due to the high availability of tools and machines by decreasing machine setup and failure times. Therefore, it is important to employ new sensor technologies in metal forming systems for the observation of process signals.

4.2.1
Sensors for the Punching Process

In the last 30 years, enormous improvements have been achieved in the stamping process concerning economic production, accuracy and possible shape of the parts [1]. Today's tools are more sophisticated and more expensive. The costs of a modern multi-stage tool can be more than $ US 100000 and requires constant process monitoring to achieve high availability of the tool. This aspect is very important for the trend of just-in-time production. Also, customer requirements for 100% quality control can be fulfilled with indirect quality control by the process signals. Therefore, the demand for tool safety devices and process control units is increasing constantly [2]. Traditional limit switches [3] are not sufficient. The manufacturers' expectations for modern process control systems are as follows:

- complete quality assurance and documentation (100% indirect product quality control);
- protection of expensive and complex multi-stage tools against breakage and subsequent damage;
- machine overload protection;
- detection of feeding faults;
- extended production time with no supervision (ghost shifts);
- decrease of setup times and support with stored parameters;
- fewer production stoppages by premature recognition of process disturbances;
- permanent process monitoring to support the user with process information to permit optimal process setting;
- higher press speeds to increase productivity;
- control of existing tools;
- no sensor handling in the tooling room.

To fulfil all these requirements, the process control system should have sensors which are sensitive enough to recognize the disturbances and they must guarantee easy handling in daily production (no cables in the tool room). The signal processing must also be very sophisticated to detect breakage, wear, and process trends.

4.2.1.1 Sensors and Process Signals

The most common process signals for the monitoring of the punching process and the press load [5] are forces and acoustic emissions. Both signals include process information, which can be controlled or analyzed by process monitoring devices. In addition, these devices need one of the following signals as a reference basis for the monitoring or analysis:

- time;
- crankshaft angle;
- slide path.

Normally the time signal is used as the reference base for process monitoring/control. However, the time base depends on the press speed. If the press changes speed and if the signal is controlled by the window or the tolerance band technique [6, 7], the time-based signal will vary and could cause a press stop. In this case a crankshaft resolver with a high resolution at the lower dead center will be used. A linear distance sensor at the press slide can alternatively be used as a basis for the process signals. The disadvantage of the linear distance sensor is the low resolution at the lower dead center, because of the sine shape of the slide path. In Figure 4.2-1 the typical process signals of a punching process are shown.

Acoustic Emission Sensors
Short-term disturbances (tool breakage or cracks in the product material) can be easily detected by acoustic emission sensors [6–8]. This sudden change in the press load produces an acoustic emission signal up to 150 kHz, traveling through the tool and the machine. Most importantly, acoustic emission sensors should be placed as close as possible to the metal forming process to avoid disturbances (the

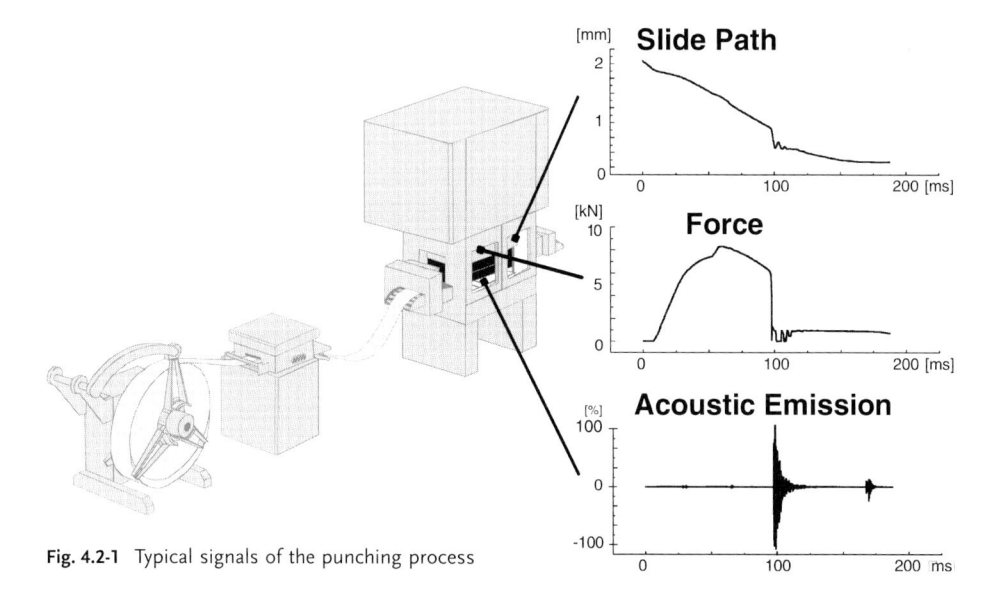

Fig. 4.2-1 Typical signals of the punching process

Acoustic Emission Sensor

Structure of the Sensor

A - Piezoelectric Element
B - Damping Mass
C - Diaphragm

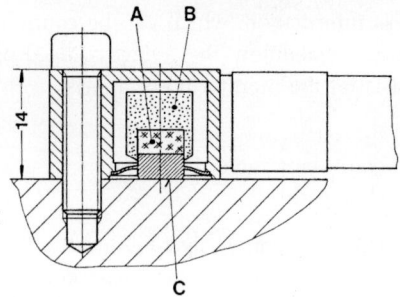

Fig. 4.2-2 Acoustic emission sensor (Kistler Instrumente AG)

machine's vibrations). Each mechanical contact (gap) between the forming process and the acoustic emission sensors filters the acoustic emission spectrum as a low pass. Therefore, the acoustic emission sensors should be placed into the tool [6] or next to the tool. In Figure 4.2-2 a piezoelectric sensor is shown with a very wide transmission band, which enables the sensor to measure acoustic emission signals in the 100 kHz range, because the piezoelectric element is mounted in a damping mass with no seismic mass (no resonance).

Force Sensors
The most important process signals are the signals of the force sensors (see Figure 4.2-3), which are placed in the structure or on the surface. Piezoelectric force sensors or piezoelectric transverse measuring pins are mostly used in the structure. On the surface the common devices are piezoelectric or resistive strain gages. A later calibration of all these sensors is necessary, because the strain and the sensitivity of the sensors depend on the surrounded structure of the machine or tool. Existing monitoring systems are mostly based on simple force monitoring. The force signal is mainly used for process monitoring. When the adjusted force limit is exceeded, the machine will automatically be stopped by the emergency stop.

4.2.1.2 Sensor Locations

In Figure 4.2-4 the most common sensor locations for the punching process control are shown. Acoustic emission sensors must be placed very close to the process. Typical locations for the acoustic emission sensors are the upper and the lower tool or the slide and the table. A greater distance to the process will increase the noise signal by the press.

The force signal is normally measured by sensors which are placed in the press frame, the connecting rods, the slide or directly in the tool [9]. Some presses are equipped by the press manufacturer with sensors to protect the press against

Sensors inside the Structure

Transverse Measuring Pin **Force Sensor**

Sensors on the Surface

Strain Sensor **Strain Gauges**

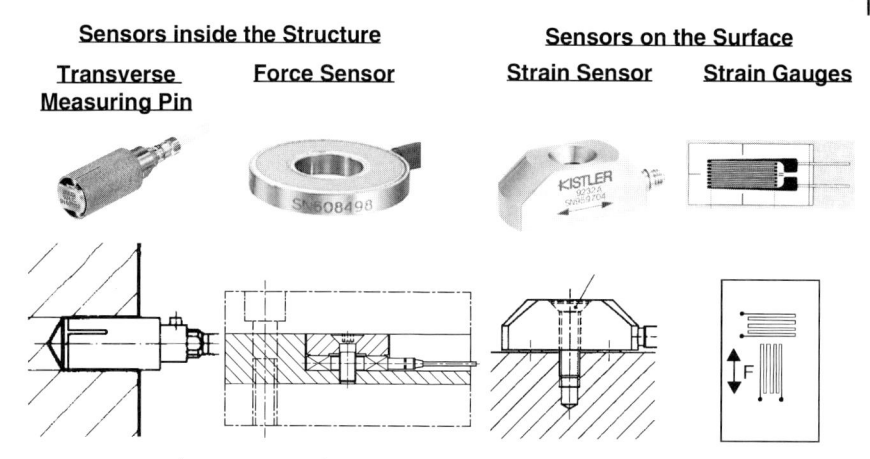

Fig. 4.2-3 Force and strain sensors for process control (Kistler Instrumente AG)

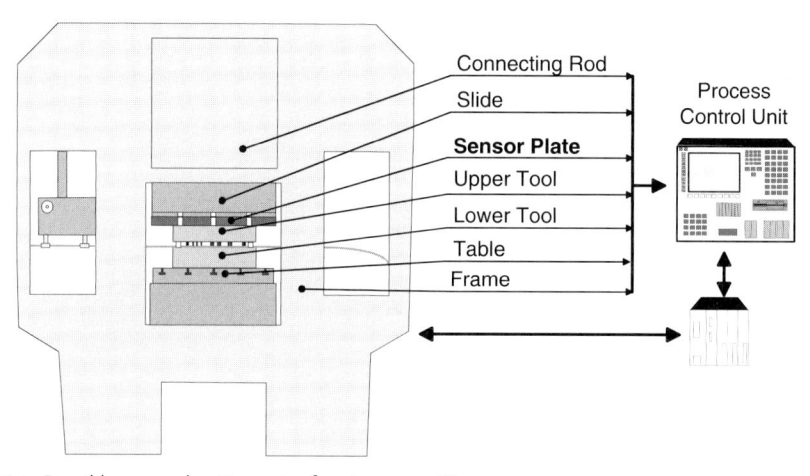

Fig. 4.2-4 Possible sensor locations at a forming press [9]

force overload. The distance between these sensors, which are placed in/on the press frame or the connecting rod, and the forming process is too large to detect more than the force overload. The signals of these force sensors and the acoustic emission sensor underlie many disturbances, eg, the press drive and vibrations. The best sensor signals can be obtained when the force sensors are directly placed in the tool (see Figure 4.2-5). The second best solution for the signal quality is to place the force sensors directly above or under the tool. See the sensor plate and table locations in Figure 4.2-4.

4.2.1.3 Sensor Applications

In this section, sensor applications, which are close to the forming process, will be described in detail. The integration into the top plate of the upper tool is shown in Figure 4.2-5. With this application a single forming operation can be perfectly monitored. The influence of a neighboring forming operation on the measured force signal is very low. Typical sensors for this application are piezoelectric transverse measuring pins or force rings, because the sensors are placed in the structure. The total or a part of the forming operation force is transmitted and measured by the sensors. A disadvantage is the large number of expensive sensors in a tool and the bad tool handling in daily production. The very rough environment in the tool shop also complicates the handling of the tools with sensitive sensor cables.

Better tool handling and lower sensor costs can be achieved when the sensors are integrated into machine parts or remain at the press structure. One solution, which was presented by Terzyk et al. [6], is the integration of force sensors into the slots of the press table. In Figure 4.2-6, two slot force sensors are shown,

Fig. 4.2-5 Force sensors integrated into the upper tool [6]

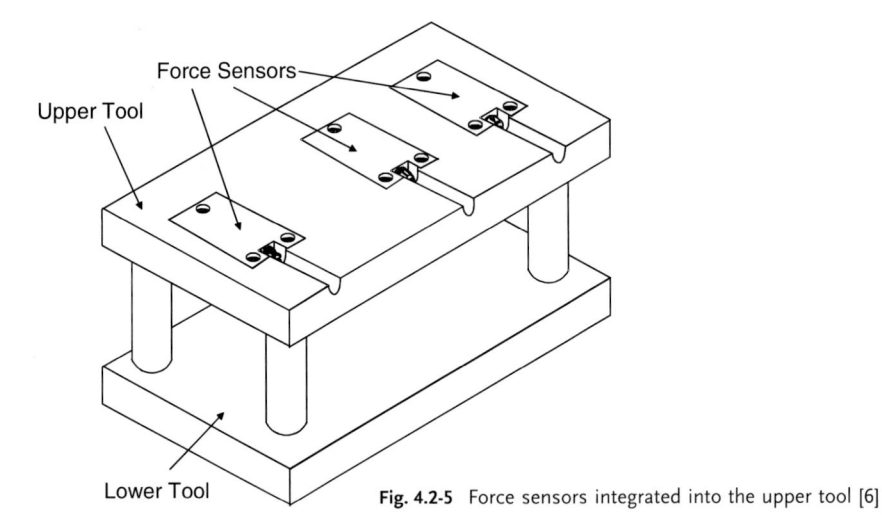

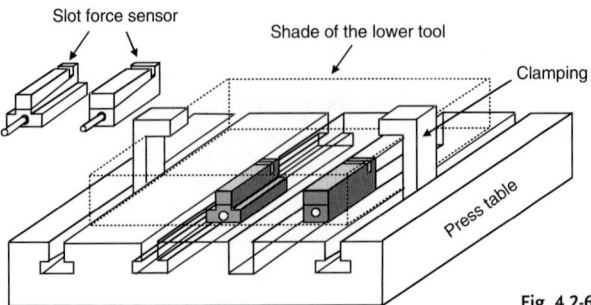

Fig. 4.2-6 Table slot force sensors [6]

which are placed under the lower tool. The advantage of this solution is the high flexibility and the integration into existing processes, because the shape of the table slots is standardized.

On the other hand, the slots must be cleaned and must have straight surfaces. These sensors cannot be placed in the center of the tool, because there are holes in the table in this area for scrap transportation.

A good combination of process-sensitive signals and good handling is achieved by a multi-sensor plate, which is placed between the slide and the upper tool. In Figure 4.2-7 the scheme of the multi-sensor plate is presented. The multi-sensor plate consists of a frame plate, which has the same shape as the slide, and several sensor cassettes, which contain force and acoustic emission sensors. The following requirements are the basis for the development of the system:

- easy handling in the production workshop;
- short distance to the process;
- integration of several force sensors for detailed process monitoring;
- connection devices for additional sensors;
- improved process control by a combined force/acoustic emission monitoring;
- modular design for high flexibility;
- integration into existing tool-press systems.

Easy handling is solved by using a modular cassette system, which is fixed by a frame plate and two guiding rails to the press slide (Figure 4.2-7). During a tool change the sensors will remain at the slide. All cables between the cassettes and the docking station are integrated in the frame plate. Because of the modular design, the multi-sensor plate can easily be adapted to the requirements of the user. The number and the locations of the standardized sensor cassettes can be changed. The docking station houses the charge amplifier and the connectors for additional sensors and is mounted on the frame plate. The frame plate has a height of 25 mm and the same shape as the slide, so that the tools can be fixed to the slide in the usual way.

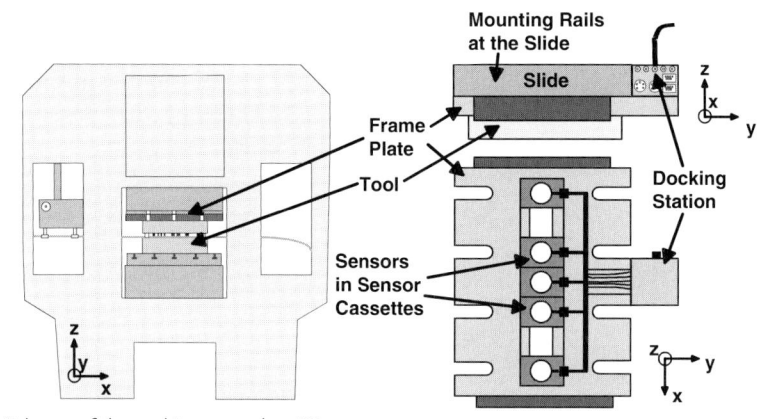

Fig. 4.2-7 Scheme of the multi-sensor plate [9]

4 Sensors for Process Monitoring

A multi-sensor plate with four cassettes and the docking station for a 500 kN press is shown in Figure 4.2-8.

Some typical process signals measured with the multi-sensor plate are presented below. A production tool with 11 forming operations (cutting, deep drawing, stamping) separated into four modules will be analyzed by means of the multi-sensor plate. The workpiece, the tool setup, the force and acoustic emission signals are shown in Figure 4.2-9. The two force cassettes of the multi-sensor plate are placed above the first and above the last (fourth) module to demonstrate the local resolution of the system. The acoustic emission cassette is placed in the middle of the tool.

The measured signals contain information on the cutting/forming process, on the blank holder and on the tool stop reaction. The contact of the blank holder occurs at point A in the force diagram and at point 1 in the acoustic emission diagram. Characteristic cutting operations can be identified at B/2 and C/3. The resulting cutting impact is very significant in the acoustic emission signal (peaks 2 and 3). Owing to an incorrect slide height (too tight), the upper tool is running on the stops of the lower tool (impact at D/4). The tight tool mounting causes an increase in the force signals up to point E. The force signal above the first module is higher than that above the last module, because the stops are in the first two modules of the tool (four modules). The lower dead center is reached at point E (highest force signal). The lift-off of the stops and of the blank holder occurs at the moments F/5 and G/6. The force curve is evidence for the incorrect adjusted slide height (too tight).

The correlation between the force signals and the acoustic emissions in the diagrams is significant and the combination of the two signals permits the identification of different cutting/forming operations.

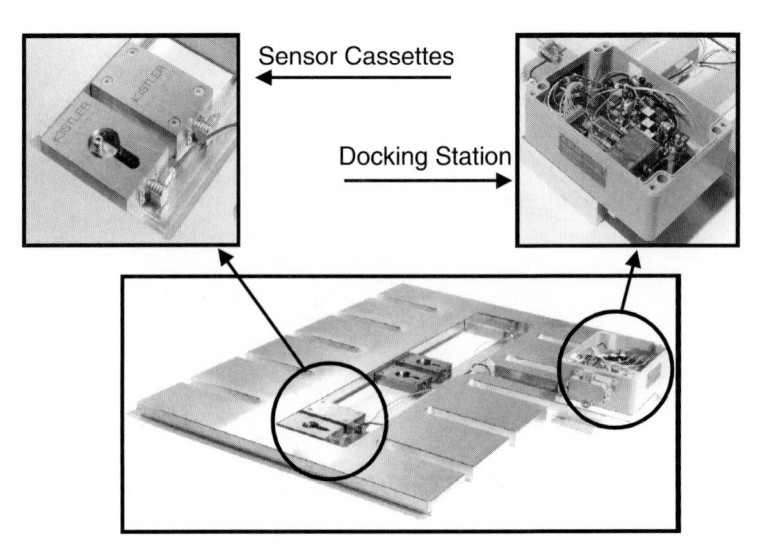

Fig. 4.2-8 Multi-sensor plate for a 500 kN press

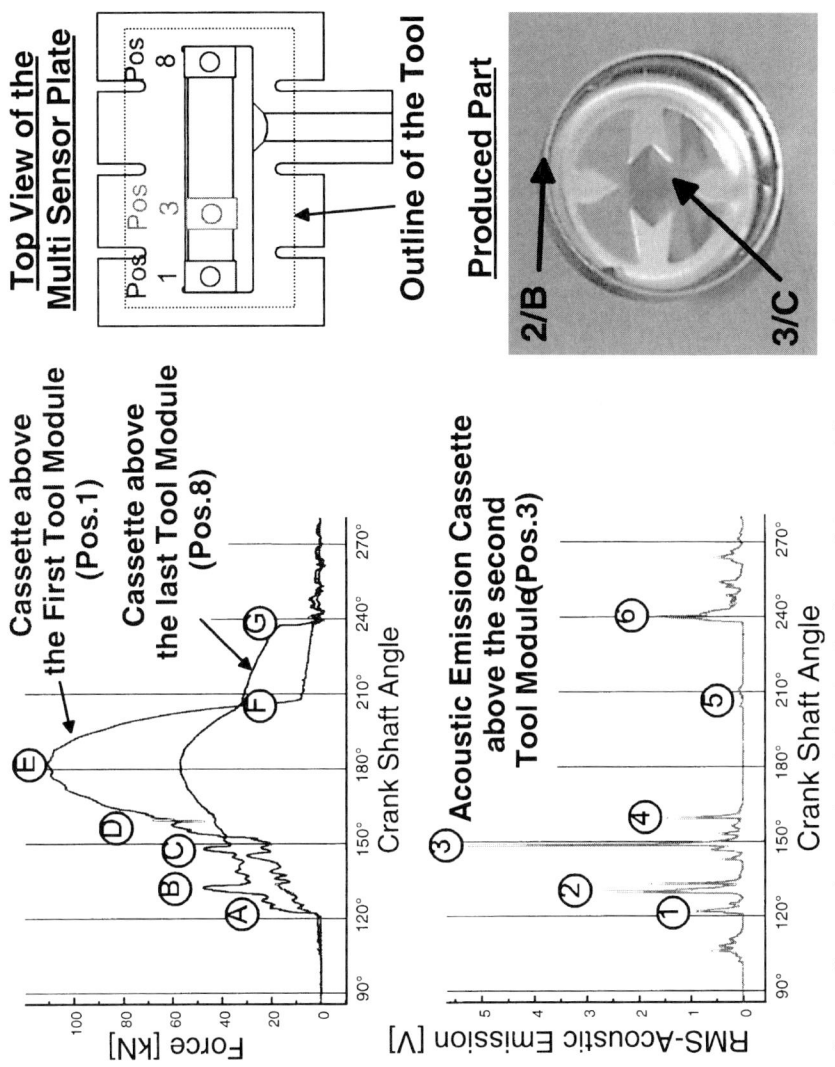

Fig. 4.2-9 Force and acoustic emission signals of a modular metal forming tool measured with the multi-sensor plate for a 500 kN press

The sensor signals should be significant so that the user can 'see and understand' the complex forming operation. Especially for the tool setup the stored signals of previous setups can be very helpful by using the same setup and therefore saving time and achieving the same product quality.

A tight slide position causes unnecessary high press forces in the lower dead center and product defects. This load decreases the tool and the machine lifetime and increases the energy consumption. The signals in Figure 4.2-10 were measured in a press shop with a production tool. At the normal slide height a force signal of a cutting operation is measured before the lower dead center. At the tight setup of the tool (0.6 mm lower) a significant second peak occurs at the lower dead center. The stored signals of the force cassettes enable the user to setup the tool properly with less load for the machine and the tool.

Another important aspect for the monitoring of the punching process is the detection of tool breakage. For this detection acoustic emission sensors should be used, because the reaction of the tool on overload and breakage is more significant in the acoustic emission signal than in the force signal. The force compresses the punch and energy will be stored in the punch. After the breakage (overload), the stored energy is released as acoustic emissions to the environment like a compressed spring. These acoustic emissions have significant amplitudes and can easily be detected in a 'silent moment' of the process.

In Figure 4.2-11 the signals of force and acoustic emissions of a normal punch and of a breaking punch are shown. There are only slight differences in the force signals. Especially the 'small valley' around 60 ms cannot be found in the signal

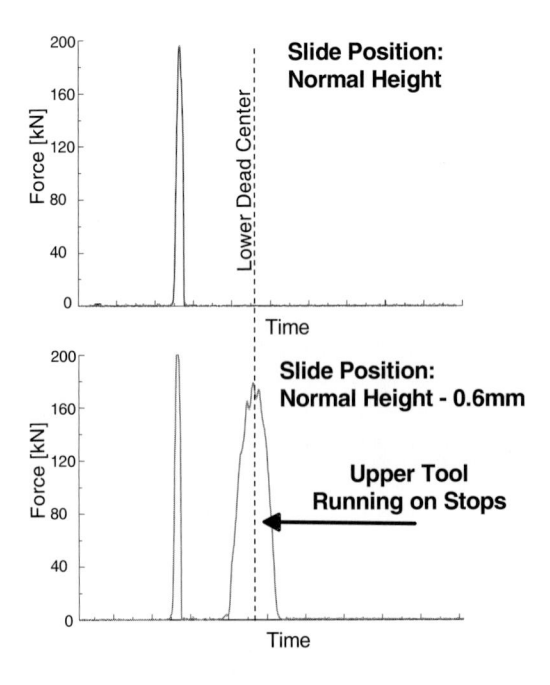

Fig. 4.2-10 Force signals of normal and incorrect tool setup

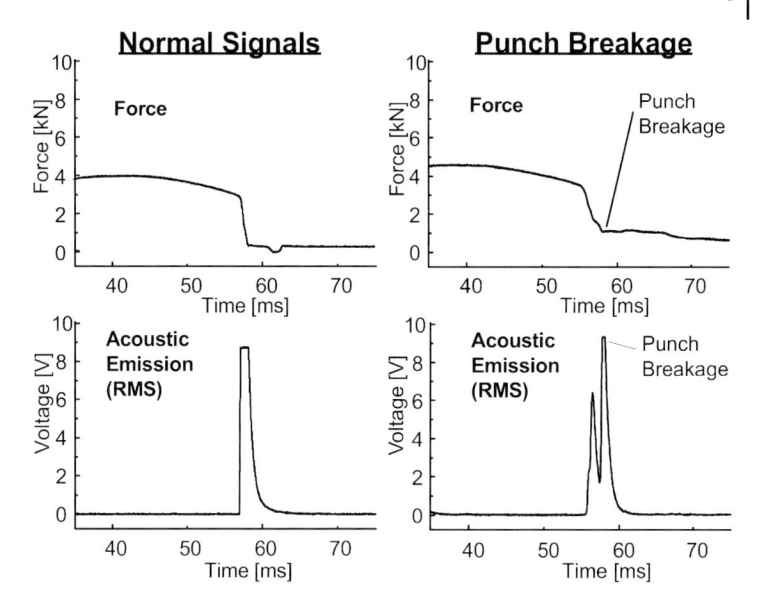

Fig. 4.2-11 Force and acoustic emission signals of a breaking punch

of the breaking punch. This is the moment when the punch moves upwards in the mold. The acoustic emission signal is more significant. The punch causes a second peak at the moment of breakage. This event can easily be detected with a narrow tolerance band [6] around the 'normal' curve.

4.2.2
Sensors for the Sheet Metal Forming Process

Sheet metal forming is a complex process which is affected by a manifold of influences. The high demands to quality and cost efficiency at the production of sheet metal components are increasing continuously. These high requirements can only be met with optimum designed and faultless manufacturing processes. Hence it is necessary to have fundamental knowledge about the behavior of the used materials and machines as well as the possibilities for the control of the actual process parameters. Furthermore it is of great importance to control the course of events during the forming operation because the process affecting parameters cannot be kept constant for any space of time. Material and tool properties as well as machine parameters are subjected to variations which are affecting the process stability adversely. Improvements can be achieved by the on-line measurement of indirect and direct process describing parameters and their transfer to a process monitoring system.

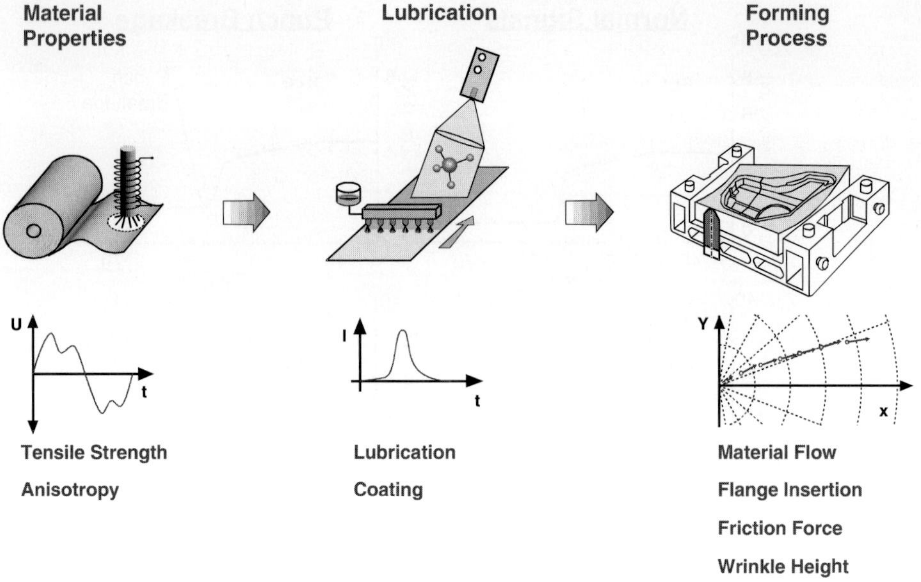

Material Properties	Lubrication	Forming Process
Tensile Strength, Anisotropy	Lubrication, Coating	Material Flow, Flange Insertion, Friction Force, Wrinkle Height

Fig. 4.2-12 Deep drawing process chain and monitorable signals

4.2.2.1 Deep Drawing Process and Signals

Sheet metal forming processes are affected by a manifold of influences. Figure 4.2-12 shows the different succeeding stages of the deep drawing process.

Examples of monitorable signals are material properties such as tensile strength, anisotropy, ductility, lubrication dose, wear of tools, and adjustment of forming machines. Changes in these parameters cause several failure modes such as cracks, wrinkles, etc., and also long process starting times, production insecurity and deviations from the required quality [11]. Differences in material charges often lead to a change of ductility, formability, and surface properties. The periphery affects the forming operation by variations within the straightening process, the accuracy of blanking, and the blank position in the drawing die. Also, changes in lubrication, tool wear, and different tool positions with respect to the press are unavoidable. The forming press affects the drawing result by changes in ram tilt, deflection of the press table, frame deformation, and deviations in the adjustment of punch speed and die cushion force.

4.2.2.2 Material Properties

In sheet metal forming, the working accuracy depends on mechanical properties such as tensile strength, normal anisotropy, and hardening exponent. These parameters fluctuate from coil to coil and charge in the ranges of the specified tolerances. The increasing standard quality requirements for the process and products demand a testing method which is capable of monitoring these material properties on-line and prior to the deep drawing process. Applying the magnetoinductive testing method, a sensor is inserted into the process chain after reeling off the blank from a coil. The sensor head is held at a defined distance above the sheet material and an exciting signal is brought to a magnetic coil to induce a magnetic field in the material (Figure 4.2-13).

The signals received mainly depend on the microstructural composition such as grain size, grain orientation, alloying elements, and dislocation density. Further, the resulting electromagnetic properties are correlated with mechanical parameters which were determined previously in tensile tests. The dependences on the properties are shown in Figure 4.2-14.

By using correlation statistics, a multiple regression equation allows the prediction of mechanical properties directly from the magnetoinductive measurements

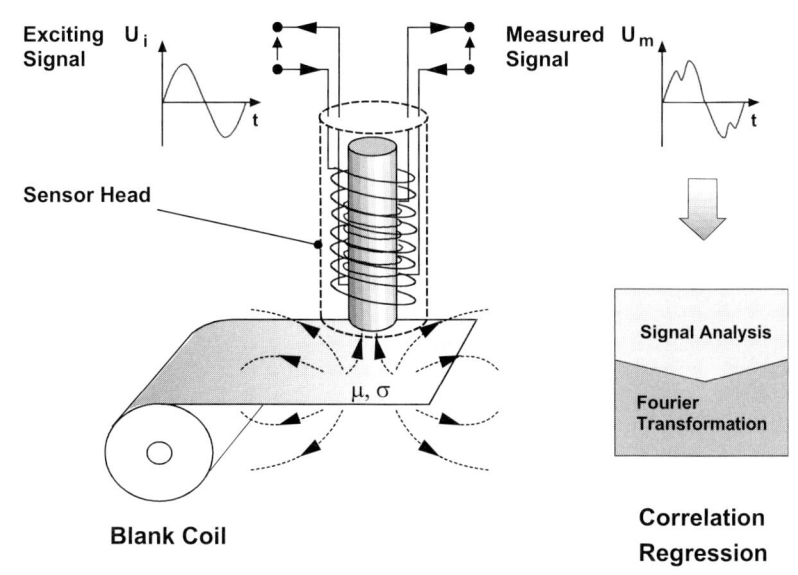

Fig. 4.2-13 Determination of the magneto-inductive signal parameters [12]

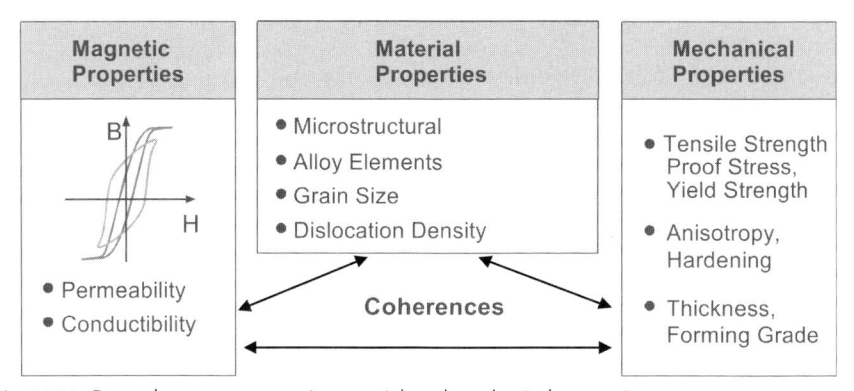

Fig. 4.2-14 Dependences on magnetic, material, and mechanical properties

and nondestructive testing method to avoid wrinkles and cracks caused by tolerance deviations in the sheet quality.

4.2.2.3 Lubrication

The lubrication properties affect the formability during the deep drawing process. The importance of control and analysis of the lubrication properties has significantly increased in pressing processes owing to the introduction of new generations of automatic transfer presses. The control of incoming material gives possibilities to reject the material before further processing them. The yield of the pressing process will increase, giving savings of material and production costs [13].

Pressforming processes require uniformity of oil films on the metal surface. During deep drawing the oil film separates the sheet metal from the die to allow the material to flow constantly between blankholder and die. The use of oil avoids cold welding of the steel on the active tool surfaces which can cause galling and passing of the friction force limit. As a result, the deep drawing process fails owing to cracks in the material. Galling means the formation of cold welds between blank sheet material, especially stainless steel and aluminium alloys with die material at high local pressures. During sliding these welds shear off and cause scratches in the material. Another important process parameter is the necessary blankholder force. This force is affected directly by the friction coefficient which depends on the quantity of the lubricant as shown in Figure 4.2-15.

In deep drawing processes, the blankholder force will be kept at a defined level to reach a defined surface pressing. Differences in the amount of lubrication cause deviations from the acceptable tolerance zone. For an increased amount of

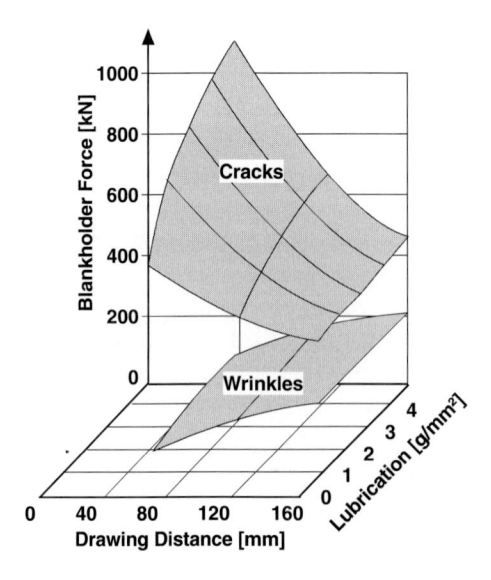

Fig. 4.2-15 3D tolerance zone with interdependence of lubrication, blankholder force, and drawing distance [14]

lubrication the process will fail owing to wrinkles, whereas for reduced amounts the deep drawing part will tear off.

A portable sensor based on high-resolution infrared spectrometry has been developed for the measurement of oil film thickness on metal surfaces. This lightweight hand-held device is intended for use in rolling mills and sheet metal engineering workshops. The sensor permits the measurement of thin oil films and is useful for optimizing the thickness of oil coatings or pressforming lubricants. Figure 4.2-16 shows a schematic diagram of the analyzer, a two-part system consisting of a measurement head and data collection unit. The analytical measurement principle is based on the absorption of infrared radiation by hydrocarbons, the common constituent in all oils.

The optical measurement head includes a compact multichannel infrared analyzer, electronics, and an LCD display. The optics, mechanical parts, light source, and multichannel detector electronics are integrated into the measurement head to provide stable, high-resolution analyses in a production environment. A control unit includes a data processing unit, LCD display, keypad, and PC interface. The control unit collects measurement data and calculates oil amounts in terms of weight per unit area by comparing measured data with pre-calibration curves stored in memory.

The measurement is performed by placing the sensor head on the metal surface (Figure 4.2-17). Spacer pins at the measurement head stabilize the fixed distance between the measuring head and the surface. After triggering, the measured amount of oil is displayed on both the measurement and control units. The actual result is derived as an average value from several sub-results measured at different points on the surface.

Owing to optical differences in the surface texture of materials such as cold-rolled and hot-rolled steel, copper, and aluminium, the analyzer is calibrated for the type of surface to be measured. Calibration also eliminates effects caused by possible differences in oil quality. It is recommended that each calibration is made using the same type of surface and oil as is expected in actual measurement. The repeatability of the analyzer, which can be expressed by the standard deviation of readings in a single-point measurement, depends on the oil film thickness. In the case of cleaned cold-rolled steel the standard deviation is of the

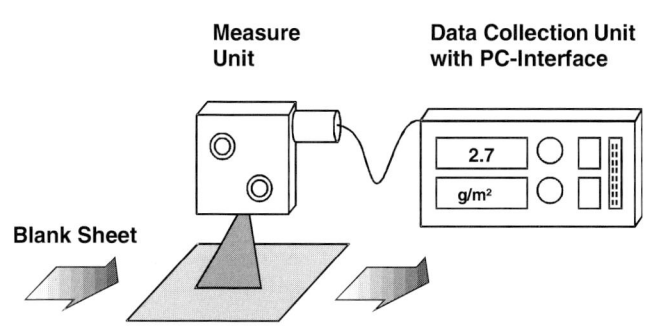

Fig. 4.2-16 Schematic diagram of the infrared analyzer

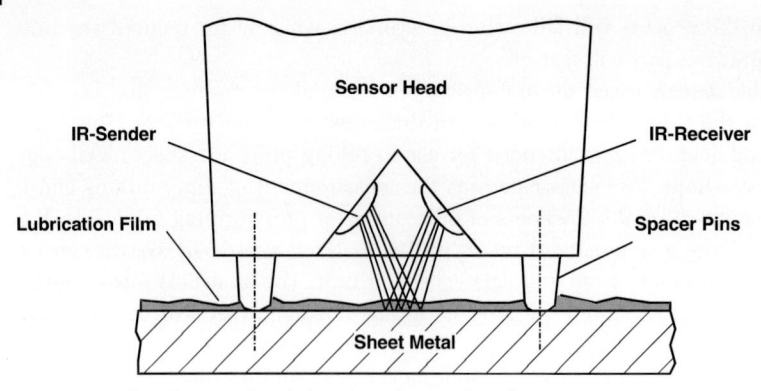

Fig. 4.2-17 Infrared sensor head placed on the metal surface

order of 1 g/m². The influence of surface textures increases the standard deviation when measurements are performed at separate points on the surface.

4.2.2.4 In-Process Control for the Deep Drawing Process

Nowadays, deep drawing processes are controlled on the basis of predetermined static values. Considering the heavy demands on the quality of deep drawn components and low production costs, it is necessary to observe process-influencing parameters. In a first step, higher process security can be obtained with the practical operation of multiple sensors located directly in the drawing process.

For process monitoring, direct process-based and time-dependent information for the characterization of the process course have to be available. This is very difficult because the forming takes place in closed tools at high forces. Therefore, it is not possible to react automatically to parameter changes which occur, eg, with the use of another coil with different forming or surface properties [15].

Flange Insertion Sensor
For the consideration of the deep drawing process, the measurement of the flange insertion offers information which contains a reliable prediction of the progress of the deep drawing process and further of the part quality. A flange insertion sensor has been developed to measure the flange insertion distance and to draw conclusions regarding stress and strain [16].

The sensor consists of an inductive position sensor with a thin metal tongue at the top. The tongue has a thickness of 0.5 mm and is brought into the gap between blankholder and die to touch the outer edge of the blank sheet from the beginning to the end of the deep drawing process (Figure 4.2-18). The flange insertion is measured over the drawing distance and can be used to detect a deviation from the tolerance field describing the non-failure area. Deviations can lead to wrinkles and cracks in the drawing part.

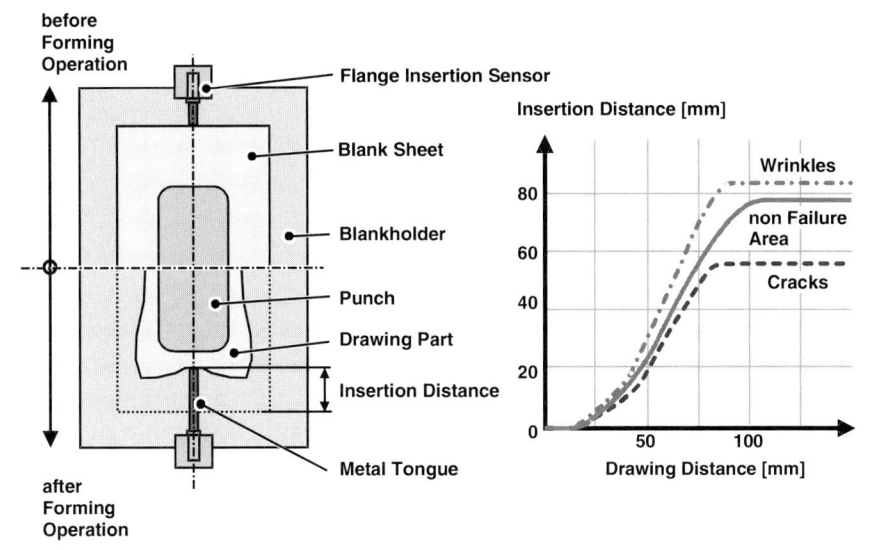

Fig. 4.2-18 Flange insertion sensor in deep drawing tool [17]

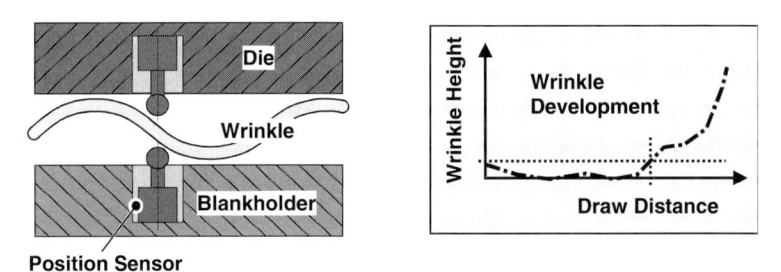

Fig. 4.2-19 Wrinkle sensor [17]

Wrinkle Sensor
With the combination of the flange insertion sensor and the wrinkle sensor, a more accurate prediction of the failure of the drawing process can be designed. Wrinkles develop at high radial tensile stress and tangential compression stress in the flange. With the acting normal pressure the blankholder avoids buckling of the material. The wrinkle sensor consists of two position sensors that detect the distance between blankholder and die (Figure 4.2-19).

The development of wrinkles can be detected by the increase in the gap between the two sensors. The wrinkle sensor is a good addition to the flange insertion sensor because it can observe the top tolerance border of the wrinkle development as a function of the flange insertion distance during the deep drawing process (Figure 4.2-18).

Roller Ball Sensor

An analysis of drawing operations shows that the material flow can be identified as a direct value for the characterization of the forming process [18]. The material flow can be defined as the dynamic local displacement of the material during the forming operation which allows direct conclusions about the failure mode cracks, necks, and wrinkles. Based on the material flow, it is possible to calculate the quality characteristics thickness and strain distribution of the drawing part. Additionally, the motion of the welding seam for tailored blanking can be investigated [19].

The material flow depends on all affecting influences of the forming process such as the tribological system, the machine, and material conditions. Thus, the material flow is the essential process providing information for deep drawing. Therefore, the practical on-line measurement of the material flow is important for the realization of a process monitoring system. The direct assessment of the material flow can be realized with a measurement concept called the roller ball sensor. With this sensor principle, it is possible to measure the direction, velocity, and distance of the material movement during the forming process. The sensor works with the same principle as a computer mouse, which detects relative motions with a ball rolling over a surface (Figure 4.2-20).

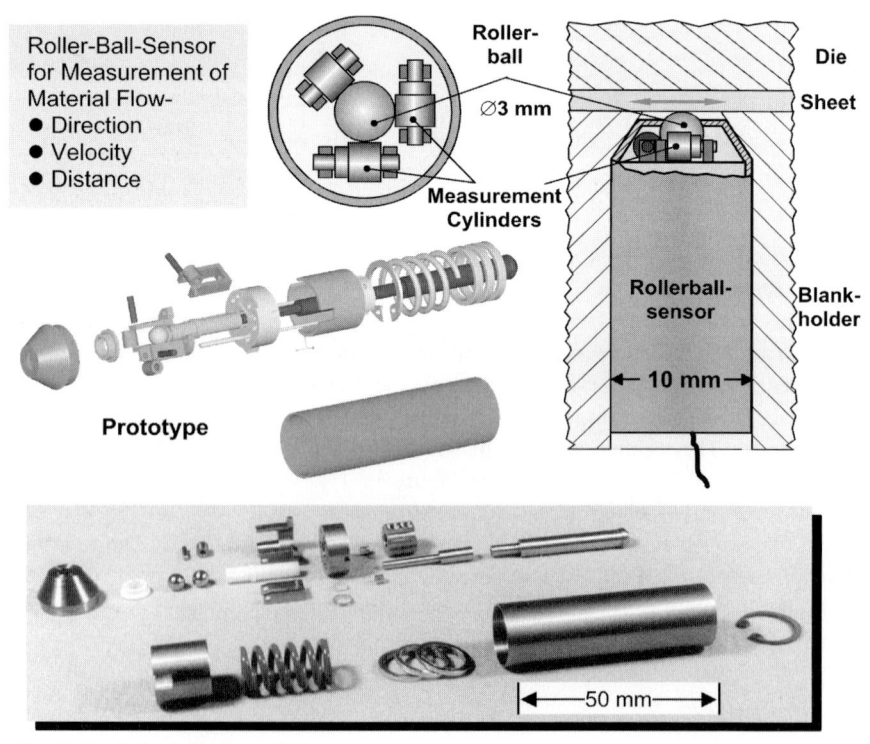

Fig. 4.2-20 Roller ball sensor [20]

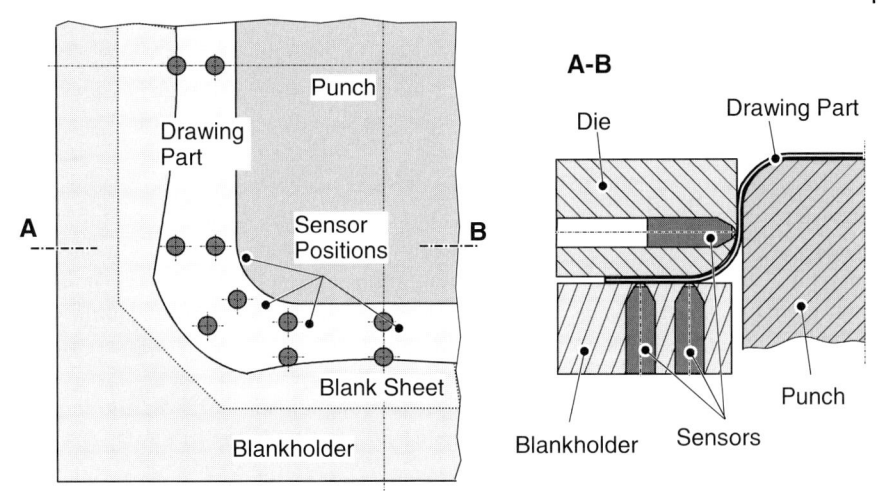

Fig. 4.2-21 Position of roller ball sensors in the deep drawing tool [20]

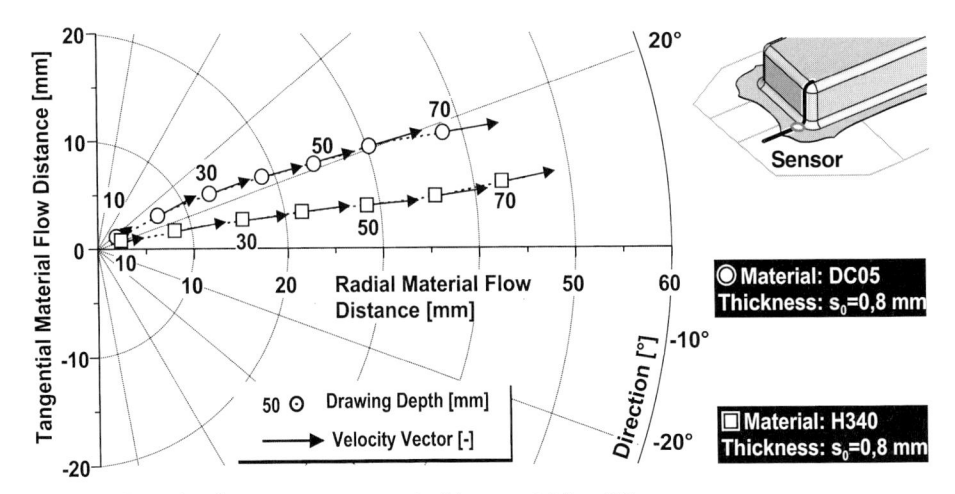

Fig. 4.2-22 Example of a measurement record of the material flow [20]

The integration of multiple sensors into the drawing tool at positions which are critical for the forming operation leads to the recording of detailed measurement data. Over the circumference of a drawing part, it is useful to place some sensors at the edges and some in the straight areas (Figure 4.2-21).

The sensors can be located in the blank holder in front of the drawing radius and also in the drawing die behind the drawing radius through bore holes with a diameter of about 3 to 6 mm into the active surfaces of the tool without causing scratches on the drawing parts. Figure 4.2-22 shows as an example a measurement record of the material flow obtained with a rectangular drawing part.

The radial scale of the diagram corresponds to the material flow distance, the abscissa to the radial distance, and the ordinate to the tangential distance. Also, a velocity vector of the material movement and the drawing depth are assigned to each material flow distance. Therefore, the curve represents the entire material movement on the sensor location during the drawing operation. A straight material flow, orthogonal to the drawing border, corresponds in the record to a straight line in the 0° direction. The continuous positive angle of the measurement curve illustrated shows that the material flows from the edge of the drawing part into the straight side. The decreasing angle means that the lateral material movement becomes smaller with increasing drawing depth. A greater difference between material flow distance and drawing depth affects in a higher plastic strain of the material, which could be detected in the sensor area.

The material flow offers further possibilities for evaluation and for monitoring of the drawing process. With the installation of several sensors in the flange, it is possible to measure the field of velocity from the material flow. This vector field is the basis for the calculation of the material deformation. The vector gradient will be derived from the field of velocity which represents the change in the velocity vector between different positions inside the field. The vector gradient can be split up into a symmetric and an antisymmetric part. The antisymmetric part is equivalent to the rigid body motion of the material and the symmetric part describes the local deformation. The tensor of strain rate is calculated and the effective strain rate according to von Mises can be determined. Numerical integration over time leads to the distribution of the effective strain in the flange and the wall of the drawing part (Figure 4.2-23). This graph shows the high and low stressed

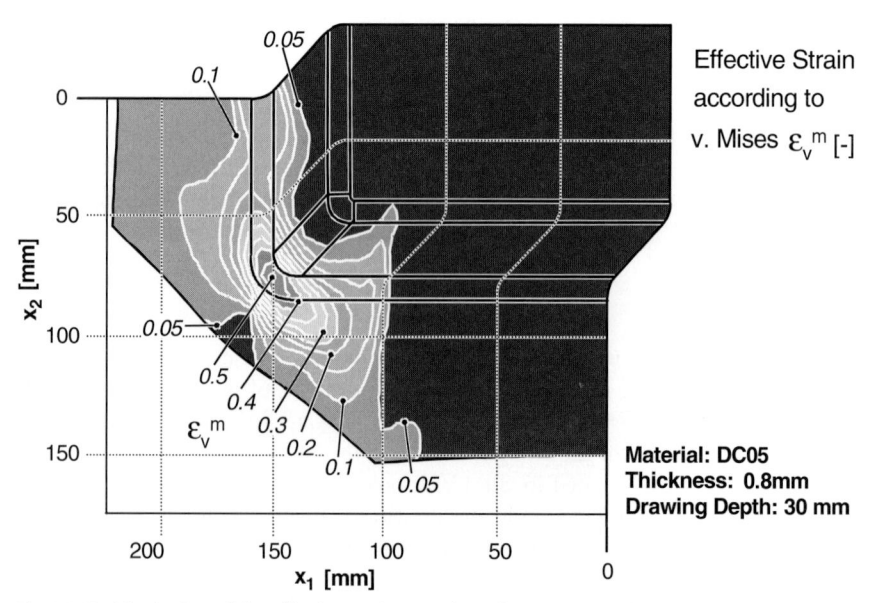

Fig. 4.2-23 Monitoring of the effective strain at a deep drawing part [20]

areas of the deep drawing part and offers the possibility of judging the failure or success of the process.

4.2.3
Sensors for the Forging Process

Whenever high strength and surface quality of massive components are required, forging parts are used. Owing to the manufacturing procedure, these parts show a more regular structure than cast components and a favorable uninterrupted fiber orientation in comparison with components manufactured in machining processes. Therefore, forging parts have the best mechanical properties.

The operating sequence of the forging process is shown in Figure 4.2-24 [21, 22]. The entire sequence can be divided into three main sections: forming (cutting, heating, forming, and clipping), heat treatment, and verification (cleaning and testing). Also, division into pre-process (cutting, heating), in-process (forming, clipping) and post-process (heat treatment, cleaning, and testing) is possible. There are processes within these main sections that do not necessarily have to take place, eg, clipping operation for precision-forged parts [23–26].

4.2.3.1 Sensors Used in Forging Processes
In order to achieve high-standard forging parts and at the same time reproducible parts in large amounts, a number of part measurements have to be taken. They have to be controlled throughout the process. According to the forging process shown in Figure 4.2-24, Figure 4.2-25 shows the necessary measurements before the deformation process (slug mass and slug temperature) and during the defor-

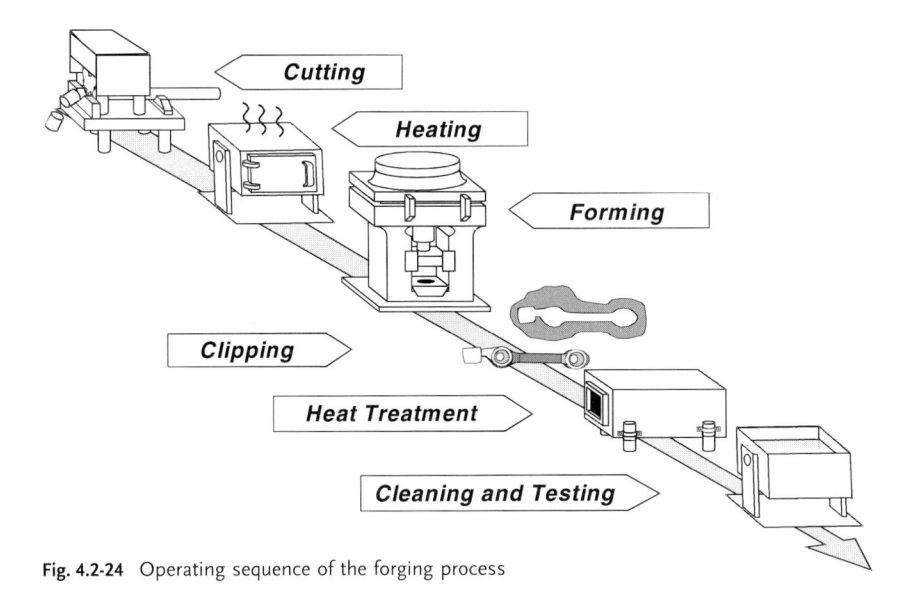

Fig. 4.2-24 Operating sequence of the forging process

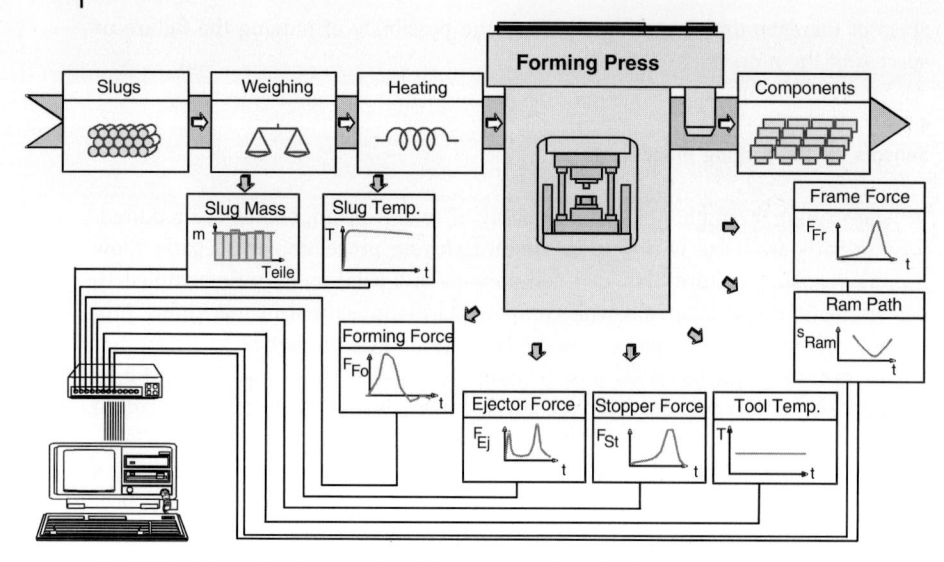

Fig. 4.2-25 Schematic diagram of useful measurements in forging processes

mation process (forming force, ejector force, stopper force, tool temperature, ram path, and frame force).

The established measurements show those process parameters by means of which a judgement about the process and the process results is possible. These characteristic process-describing values are force, temperature, and pressure. They must be measured by adequate sensor equipment.

Slug Temperature and Mass
Before the heating process, the slug mass is weighed with a highly accurate electronic scale (Figure 4.2-26, left) and the signals are transferred digitally to a measuring computer. The temperature of the heated slug is measured without contact before the forging process. For this measuring operation a quotient pyrometric system (Figure 4.2-26, center) or an infrared thermoelement (Figure 4.2-26, right) is used. Depending on the construction of these sensors, temperatures between –45 and +3000 °C can be measured with an accuracy of ±1% of the measuring result. The determination of the temperature is effected by the measurement of the optical radiation capacity depending on the temperature that is taken by a test object in the spectral region. Should the measured temperature be outside a previously defined range of tolerance, the affected slug will not be taken for the subsequent forging process.

Forces
The total load on the press, the frame force, is determined via strain gages (compare Figure 4.2-3) on the press frame. They consist of a meander-like measuring lattice in a thin carrier foil and transform strains into a modification of the electric resistance. Strain control techniques supply information about deformation

Fig. 4.2-26 Electronic scale (left) from Kilomatic GmbH, pyrometric meter (center) from Land Infrarot GmbH and infrared thermoelement (right) from ASM GmbH [27–29]

characteristics and the state of stresses in parts. They allow the realization of damagable force transducers (force measuring ring) and weighing techniques.

Common forging processes have a frame force that corresponds to the load of the tool. For the precision forging process by means of closed dies [21, 26], the frame force is the total of forming and closing force (compare Figure 4.2-32).

To measure the forming force a force sensor with the above mentioned measuring principle according to Figure 4.2-3 is employed.

The stopper force is also determined by strain gages. The strain gages are connected to a full Wheatstone bridge to compensate for thermal effects by active and passive strain gages.

The determination of the ejector force, required to eject the forged part, is carried out by measuring the pressure of the hydraulic ejector system. Therefore, a piezoelectric pressure gage is employed. When a force acts on the piezoelectric crystal, positive and negative grid points are offset. This causes a change of the amount of electricity on the crystal surface as a function of force. These piezoelectric pressure gages are shown on the left in Figure 4.2-3.

Tool Temperature

The temperature of the tools is measured with contact by means of thermoelectric couples. The most frequently used thermoelectric couples belong to type K based on NiCr-Ni. These thermoelectric couples allow the measurement of temperatures in a range from –200 to +1372 °C. They consist of two nickel cables, one including 10% chromium and the other including 5% aluminium and silver. At their ends they are joined by soldering or welding. Thus, a thermocouple has two junction points (see Figure 4.2-27).

One junction point is called the hot junction (spot mark) and designated $T_{\text{Hot-Junction}}$ and the other junction point is called the cold junction (comparison mark) and designated $T_{\text{Cold-Junction}}$. When the hot junction and cold junction are heated to different temperatures, a potential difference U_{PD} is obtained that is proportional to the temperature difference between the hot and cold junctions.

Fig. 4.2-27 Schematic diagram of the measuring principle of a thermoelectric couple

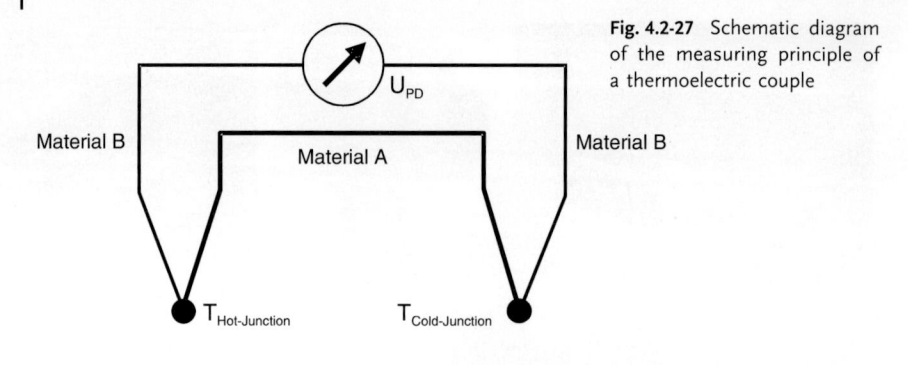

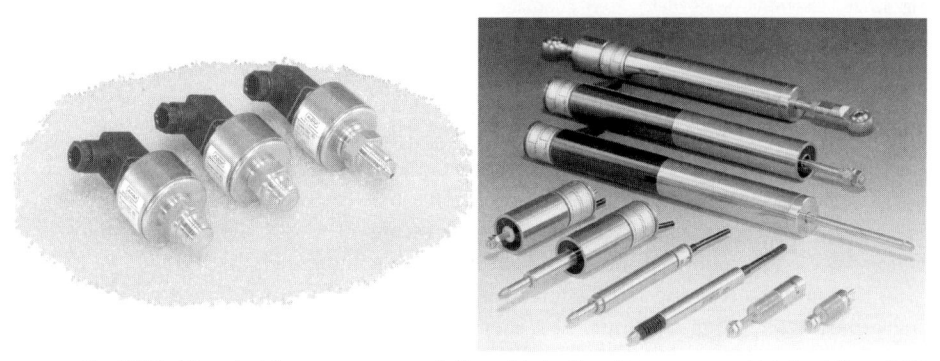

Fig. 4.2-28 Piezoelectric pressure gages (left) and inductive distance gages (right), all from ASM GmbH [29]

Ram Path
The ram path is a reference value to represent forces in diagrams. It can also be displayed as a function of time and is determined by an inductive distance gage. Inductive distance gages make use of the influence of induction depending on the distance of coil systems (AC) and caused by the displacement of iron cores (principle of solenoid plunger). A distance accuracy of 10 µm and better can be achieved. Measuring lengths are from 0.1 to several hundred mm. The Figure 4.2-28 (right) shows different constructions of inductive distance gages with different measuring lengths from 0.25 to 470 mm and a temperature stability up to 600 °C.

After having given an extensive description of the different sensors to determine the necessary measuring variables and a brief explanation of the general measuring principles for the sensors used, the following sub-sections deal with the locations for the sensors and the boundaries that have to be taken into account for mounting and measurement. Furthermore, representative measuring results will be presented and interpreted.

4.2.3.2 Sensor Application and Boundaries

Slug Temperature and Mass

For determining the slug mass and slug temperature, the sensors have to be placed in such a way that measuring the corresponding variables does not cause unnecessary delays in handling.

The slug mass is determined before heating. In this context, not the slug mass but the volume of the slug is the essential variable for the process to be monitored. The volume can be calculated from slug mass and slug density. Depending on the forging technique, variations of the volume within a range of ±0.5% can be observed. These high demands on mass accuracy have to be met for precision forging.

After heating the slugs, the slug temperature is measured without contact immediately before loading the die. For this the pyrometric meter is mounted on a plate with a swiveling ball joint and is aligned with the hot slug at an approximate distance of 1 m.

The locations for mounting the sensors which are used to measure the variables in the process, ie, forces, tool temperature, and ram path, are shown schematically in Figure 4.2-29.

Forces

To determine the frame force and stopper force, the strain gages are cemented directly on to the frame or the stopper, respectively. Two special package systems are employed for cementing. Owing to the loss of adhesion of common cements at temperatures higher than 80–120 °C, these temperatures must not be exceeded. Lower temperatures are valid for moving spot marks whereas static spot marks can withstand higher temperatures. Special heat-resistant cements for temperatures higher than 120 °C are available but the cement and also the cure technique have to meet very exact requirements.

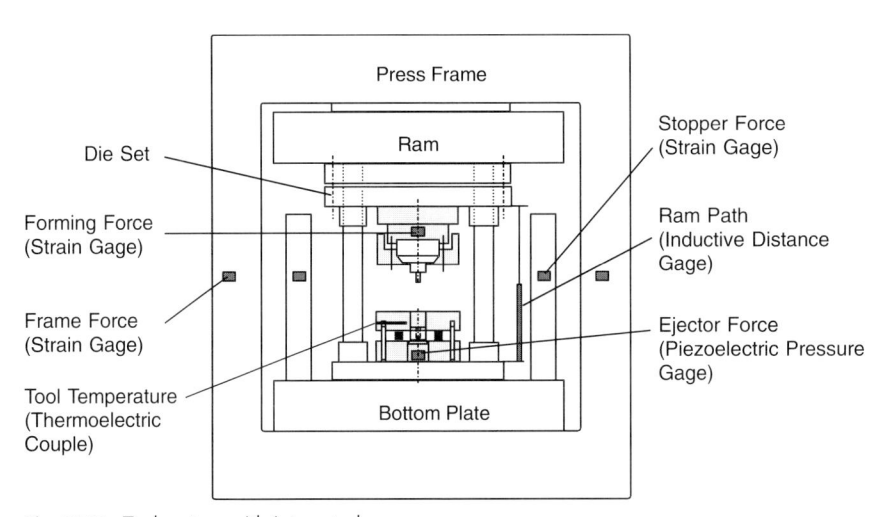

Fig. 4.2-29 Tool system with integrated sensors

4 Sensors for Process Monitoring

To determine the forming force as process-oriented as possible, the force sensor is integrated into the force flux of the forming process. On account of the high slug temperatures and the tool heating necessary for some processes, there are high temperatures in the tool system with integrated force sensors. Cooling devices are placed between the heated tool and force sensor to protect the strain gages of the force sensor from unacceptable temperatures.

All strain gages are connected to a full Wheatstone bridge to compensate for thermal effects by active and passive strain gages.

Among others, the tool system shown in Figure 4.2-30 demonstrates the force sensor integrated into the flux of the forming force and the necessary cooling device.

The mounting of the piezoelectric pressure gage can also be seen in Figure 4.2-30. This sensor belongs to the ejector of the press. The mode of mounting the sensor is specified by the press manufacturer and a change is not necessary.

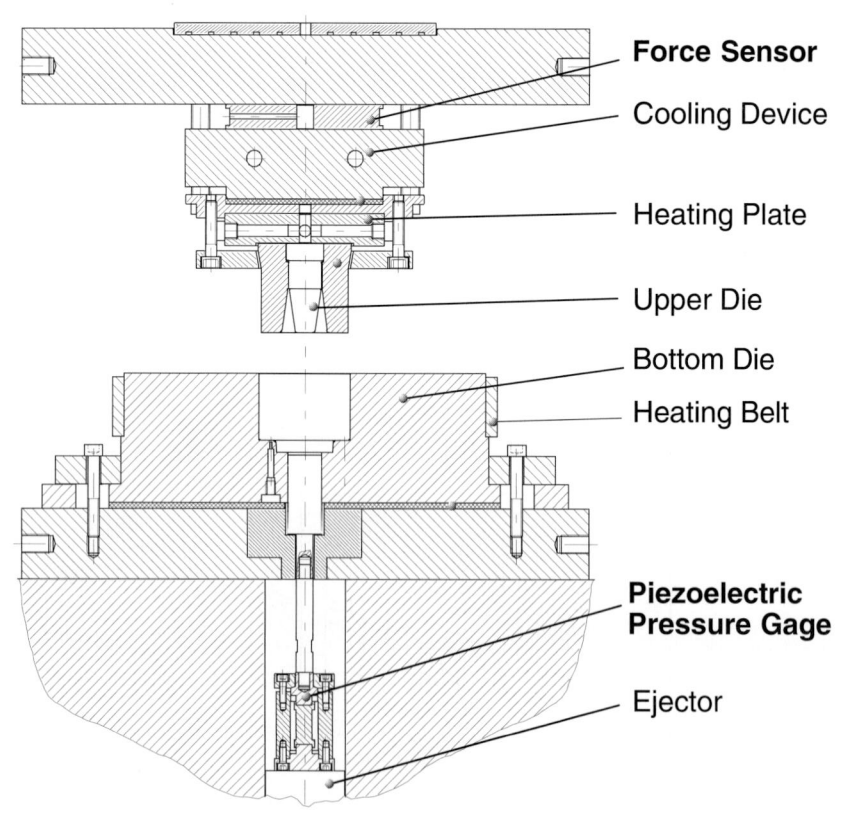

Fig. 4.2-30 Tool system for precision forging with force sensor to determine the forming force and pressure gage to determine the ejector force

Tool Temperature

Owing to high thermal and mechanical loads caused by the forming operation, the measuring system determining the temperature of forging tools has to meet very high requirements. Particularly the temperature of the effective surface during the forging operation is of great interest. Technical aspects prevent the direct measurement of this temperature. To achieve a sufficiently exact measurement of the tool surface temperature, it has to be established as close to the surface as possible.

The mounting of the thermoelectric couple into a die is shown schematically in Figure 4.2-31. The thermoelectric couple is placed as close to the surface as possible by means of a bore hole. This is executed without an unacceptable decrease in the die material strength.

Ram Path

The inductive distance gage for measuring the ram path is mounted in such a way that the solenoid plunger is connected with the upper die moving up and down. Owing to these movements, the solenoid plunger plunges into the static sucking coil of the sensor. The piezoelectric pressure gage is fixed to the press and also the inductive distance gage. The manner in which they are mounted is defined by the press manufacturer and depends on the type of press.

To avoid parasitic induction of the surroundings, only shielded cables are used. Before each measurement the force channels are rated (null balance) automatically. Hence falsification of the measuring result caused by drifting of the test amplifier or sensors can be prevented. The determined data are maintained so that a complete data record consisting of mass, temperature, force, and path can be allocated to each part (see also Figure 4.2-25).

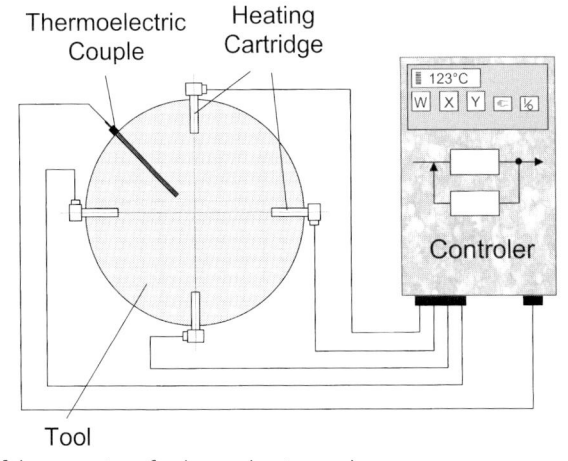

Fig. 4.2-31 Schematic diagram of the mounting of a thermoelectric couple

4.2.3.3 Typical Signals for Forces and Path

Figure 4.2-32 shows the function of a forging tool system and representative force time and ram path time courses for the precision forging of gearwheels.

The graph of process forces and ram path allows the concrete reconstruction of the entire forming process. After a ram path of 237 mm the forming operation starts when the upper die contacts the slug. This results in a simultaneous increase in forming force and frame force. Further forming of the bore hole achieved by the upper die takes place with a steady slight rise of both the forming force and frame force. Owing to the contact of the upper die and the toothed bottom die at a ram path of 250 mm, the springs become upset, which causes a considerable increase in the frame force. For an abrupt growth of the forming zone at a ram path of 253 mm a substantial increase in the forming force can be recognized. Forming the teeth of a gearwheel from a ram path of 261 mm, the two forces rise very steeply. At the bottom dead center the ram reverses its motion and the tool system and also the frame become unloaded in a very short time (10 ms).

Figure 4.2-33 shows the ejector force time courses of a forged aluminium part as an example of a new die without wear (top) and a worn-out die (bottom).

According to Figure 4.2-33, it is possible to draw conclusions from the maximum ejector force to judge the tool conditions and therefore to evaluate the surface quality of the forged parts immediately. In that way a considerable increase in the ejector force can be detected when aluminium galling on to the dies occurs. These gallings become apparent on the forged part.

A large number of components manufactured by forging, particularly forged aluminium parts, have to meet high demands regarding surface quality and physical appearance. Therefore, the measured variable ejector force can be applied to judge the quality of the forged part by using it to develop the characteristic quantity maximum ejector force.

All these variables are determined by means of suitable sensors during the entire forging process to develop characteristic quantities which allow the judgement

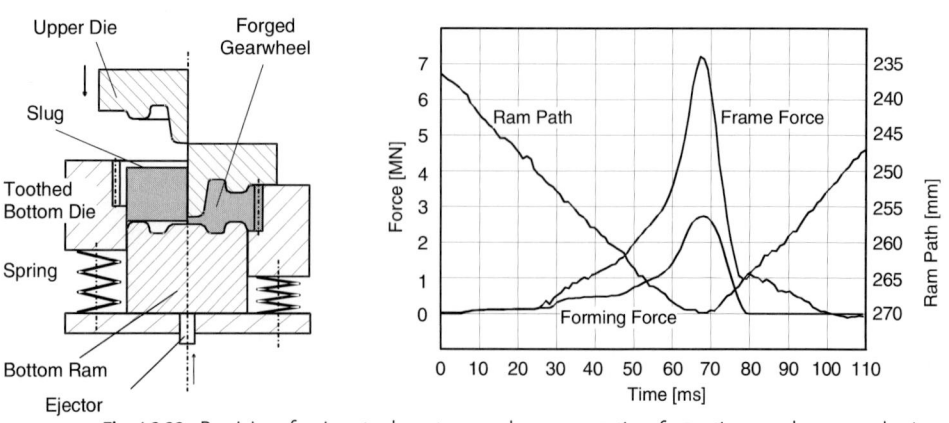

Fig. 4.2-32 Precision forging tool system and representative force time and ram path time courses

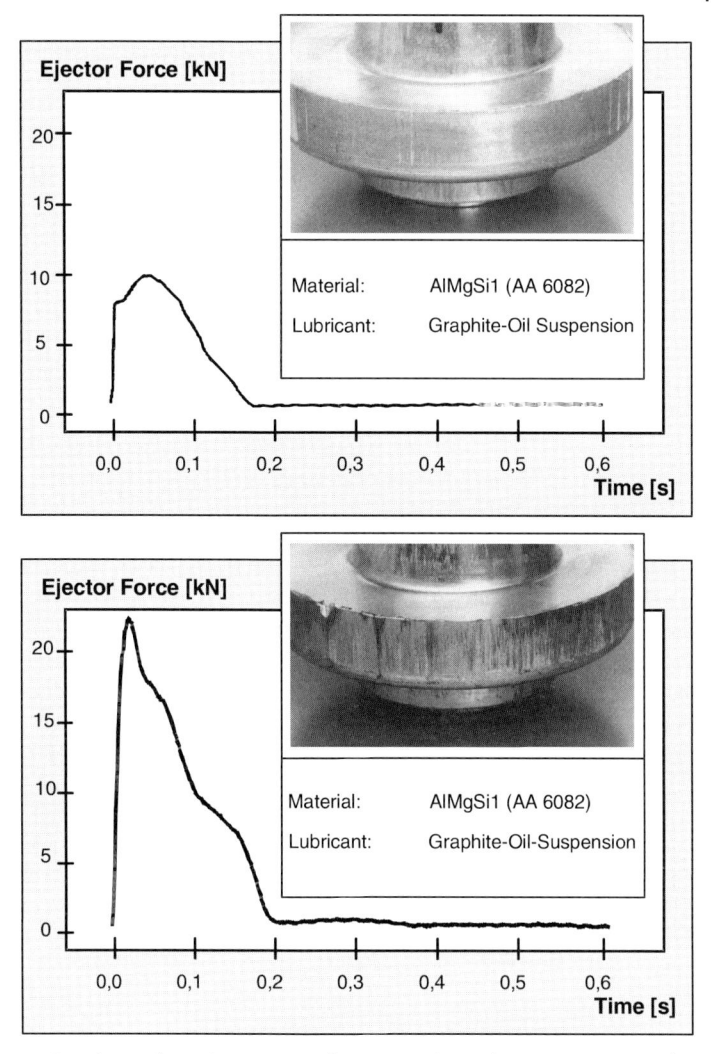

Fig. 4.2-33 Representative ejector force-time courses for a new die without wear (top) and a worn-out die (bottom)

of the process and of the quality of the forged parts immediately after the forming operation. This process-integrated quality evaluation is dealt with below.

4.2.3.4 Process Monitoring

The developed characteristic quantities of the process, eg, maximum forming force or the integrated force-path courses, are characteristic process-specific quantities for the evaluation of the entire forging process.

The mathematical evaluation of the signals is executed by a measuring and analysis program, which includes different ways of evaluating signals. The analysis of the measuring signals is carried out range by range. In this way, minimum and maximum values can be extracted from the measuring signals. Furthermore, there is the possibility of integrating or differentiating ranges of the signal. By evaluating the signals, different signals can also be linked. Such a linkage is necessary to evaluate a force path course, for example.

Figure 4.2-34 shows the manner in which to proceed when developing characteristic quantities as an example of forming force and ram path.

Today, the forging industry most frequently carries out quality inspections of geometric features for parts chosen randomly after a long cooling time. Owing to the long cooling time of forged parts, an acknowledgment of information resulting from the final inspection takes place very late. In the case of faulty forging parts, mass production leads to a large amount of scrap produced before defects in the process are recognized. To avoid this production of scrap, a process-integrated quality inspection is necessary to obtain information about the quality of the forged part immediately after the forming process. This has to be achieved by linking measured quantities during the forging process and quality features of the forged part. The advantages of an evaluation of this kind are as follows:

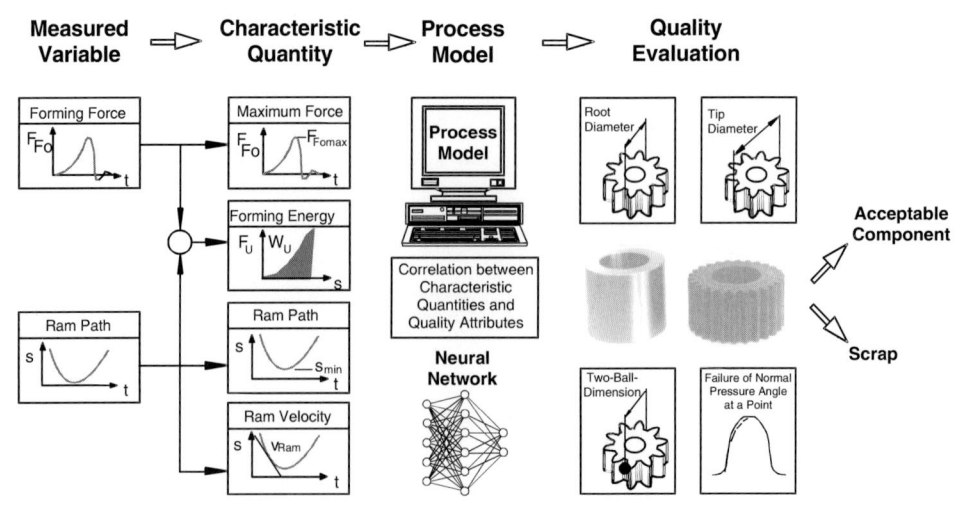

Fig. 4.2-34 Development of characteristic quantities and process model for quality evaluation

- early sorting out of scrap;
- fast detecting of failures during the process;
- reducing the scrap quota;
- 100% component inspection;
- decreasing the manual inspection work.

A correlation between characteristic quantities and quality attributes of the forged part allows the judgement of the component quality by means of a computer-based system (process model). In this way an automatic evaluation of acceptable parts and scrap can be carried out (Figure 4.2-34).

To develop a process model correlating the attributes of the part and the characteristic quantity, rule-based systems (if-then rules) are used. An additional possibility for producing this correlation is offered by intelligent systems such as artificial neural networks. A neural network-based system is adapted to its tasks not like common data processing systems by programming, but by teaching. Modifications of selected process parameters made be done very rapidly, when necessary.

An overview of developed characteristic quantities, their significance for the process, and their usage is shown in Table 2.4-1. The characteristic quantities developed from the measured variables are used to monitor the forging process and for the integrated evaluation of the parts as described above.

Tab. 4.2-1 Characteristic quantities, their significance for the process, and their usage

Characteristic quantity	Significance for the process	Usage
Slug mass	Affects the volume and shrinkage of the slug	Component evaluation Process monitoring
Slug temperature	Affects the volume and shrinkage of the slug	Component evaluation Process monitoring
Tool temperature	Affects the volume of the die cavity	Component evaluation Process monitoring
Maximum forming force time course	Describes the maximum force needed for the forming operation	Component evaluation
Time integral of the forming force time course	Describes the maximum energy needed for the forming operation	Component evaluation
Maximum stopper force	Characteristic quantity for the surplus of forming energy transmitted to the bedplate of the press	Component evaluation Process monitoring
Maximum frame force	Characteristic quantity for the stretch of the press frame	Component evaluation Process monitoring
Maximum ejector force	Characteristic quantity for the removal of the workpiece Information about wear of the die cavity and missing lubricant	Process monitoring

One must take into account that the characteristic quantities interact with each other. For example, the maximum forming force is influenced by the parameters slug volume, slug temperature, and ram velocity. However, the slug volume depends on the slug mass and slug temperature. These interactions clarify that the characteristic quantities must not be interpreted separately. The experimental determination of these inter-relationships between characteristic quantities and quality attributes has to take these interactions into account by using an appropriate experimental method.

4.2.4
References

1 Hellwig, W., *Bänder Bleche Rohre* 1 (1990) 73–78.
2 Dornfeld, D., *VDI-Ber.* 1179 (1995).
3 Allen, D.R., *SAE Pap.* March (1989) 13–19.
4 Wilhelm, R.T., *Met. Forming* May (1991).
5 Keremedjiev, G., *Met. Forming* June (1997) 75–77.
6 Terzyk, T., Oppel, F., Galle, M., *EFB-Kolloquium,* Tagungsband, T 13.
7 Cowper, S., *Sheet Met. Ind.* January (1999) 25–28.
8 Inasaki, I., Aida, S., Fukuoka, S., *JSME Int. J.* **30** (1987) 323–328.
9 Doege, E., Strache, W., in: *Proceedings of the 8th International FAIM Conference,* Portland, OR, 1998.
10 Brankamp, K., Bongartz, B., *Der Moderne Stanzbetrieb – Vom Sensormonitoring zur Geisterschicht;* Düsseldorf: VDI-Verlag, 1986.
11 Doege, E., Hütte, H., Kröff, A., Strache, W., *Blech Rohre Profile* **45**(6) (1998) 46–51.
12 Schwind, M., Engel, U., *Blech Rohre Profile* **41**(10) (1994) 686–693.
13 Raappana, T., Keränen, E., Piironen, T., Kiuru, E., presented at the 18th Biennial Congress IDDRG, Lisbon, 1994.
14 Siegert, K., in: *Beiträge zur Umformtechnik 16,* Stuttgart, 1997.
15 Thoms, V., Neugebauer, R., Klose, L., *Stahl – Eisen* **118**(6) (1998).
16 Doege, E., Schroeder, M., presented at 18th Biennial Congress IDDRG, Lisbon, 1994.
17 Straube, O., *Untersuchungen zum Aufbau einer Prozeßregelung für das Ziehen von Karosserieteilen;* Munich: Carl Hanser, 1994.
18 Doege, E., Seidel, H.-J., Griesbach, B., Mende, T., in: *Jahrestagung DFMRS, Automatische Prozeßführung, Prozeßüberwachung und Montage,* Bremen, 1999.
19 Spur, G., Thoms, V., Liewald, M., Straube, O., *Blech Rohre Profile* **41**(4) (1994).
20 Griesbach, B., *In-Prozess Stoffflussmessung zur Analyse und Führung von Tiefziehvorgängen, Dissertation;* Universität Hannover, 1999.
21 Lange, K., *Umformtechnik – Handbuch für Industrie und Wissenschaft, Band 2: Massivumformung,* Berlin: Springer, 1988.
22 Polley, W.-G., *Dissertation;* Universität Hannover, 1998.
23 Doege, E., Westerkamp, C., *Trans. N. Am. Res. Inst. SME* (1993).
24 Douglas, J.R., Kuhlamm, D., in: *Proceedings of the 4th International Precision Forging Conference,* 12–14 October 1998, Columbus, OH.
25 Janssen, St., Meiners, F., *Technica* **10** (1999) 8–14.
26 Davis, J.R., *Metals Handbook,* Desk Edition, 2nd edn; ASM International, Materials Park, O.
27 Brochure, Kilomatic GmbH, Hannover, 1999.
28 Brochure, Land Infrarot GmbH, Leverkusen, 1998.
29 Brochure, ASM GmbH – Automation, Sensoren, Meßtechnik, Unterhaching, 1999.

4.3
Cutting Processes

I. INASAKI, B. KARPUSCHEWSKI, *Keio University, Yokohama, Japan*
H. K. TÖNSHOFF, *Universität Hannover, Hannover, Germany*

4.3.1
Introduction

The mechanical removal of chips from the workpiece is called material removal. If the number of cutting edges and their macro-geometry and orientation are known, the operation is called a cutting process. These cutting processes play a major role in manufacturing because of their wide field of applications. Many different materials with a wide variety of shapes can be machined by cutting. Both roughing for high productivity and finishing to meet high precision demands can be achieved by cutting. A further distinction is made according to the number of cutting edges. Single-point cutting processes are turning as the most important method, planing and shaping. If more than one cutting edge integrated in a tool is contributing to the material removal, the process is called multi-point cutting. Milling, drilling, and broaching are the most important operations in this field. Any cutting process is possible only by applying forces to remove the chips from the workpiece. These forces may also cause deformations of the tool, the machine tool, or the workpiece, thus leading to dimensional errors on the part. The cutting energy, as a result of force application under specific speeds, is to a large extent converted to heat, which may cause thermal problems for the participating components. Mechanical and thermal loads are also responsible for a temporal change of the tool condition, leading to a change in the process output. Hence sensors are needed to monitor all the mentioned undesirable but inevitable changes of the process state to avoid any damage of equipment or machined parts.

4.3.2
Problems in Cutting and Needs for Monitoring

Major tasks, which should be attained with a monitoring system, are the detection of problems in the cutting processes and to gain information from the process condition for optimization. All cutting processes are subject to malfunctions, which lead to the production of sub-standard parts or even make it difficult to continue the process. Major problems can be related to the condition of the tool. Most critical conditions are tool breakage and the chipping of cutting edges. When these problems occur, the process should be immediately interrupted to change the tool. Therefore, the breakage and chipping of the cutting tool should be monitored and detected with high reliability. However, these failures of cutting tools made of mostly hard and brittle materials are stochastic processes, and therefore difficult to predict. Therefore, monitoring in this field is of great industrial interest [1]. The next important task is to detect the wear behavior of the tool. It deteriorates the surface quality of the machined parts and increases the cutting forces and heat generation during the process, resulting in an increase in machining errors. The tool wear is again a random process and hence the tool life is significantly scattered. Therefore, in industrial practice cutting tools are changed after a predetermined cutting time or number of machined parts, thus often wasting cutting capacity.

Formation of a built-up edge on the tool rake face, which is considered as adhesion of the workpiece material, is another serious problem in cutting processes because it also deteriorates the surface quality of the machined parts. The occurrence of this phenomenon depends on the combination of tool and workpiece materials and cutting conditions. In addition, it is affected by the supply of cutting fluids and the tool wear state. These overall tool-related problems are driving forces to develop suitable sensor systems to monitor cutting processes.

Furthermore, chatter vibrations might also occur, which can be distinguished as two types, forced vibration and self-excited vibration. Both of them will generate undesirable chatter marks on the machined surface and may even cause tool breakage. The prediction of these effects based on theoretical analysis is still difficult and thus a technique to detect any kind of chatter vibration is desirable. Other problems to mention are chip tangling and collisions due to NC errors or operator failure. Together with the ongoing trend to automate cutting processes as much as possible, all the above problems are major reasons to develop sensor systems for cutting process monitoring.

4.3.3
Sensors for Process Quantities

In any cutting operation, the removal of material is initiated by the interaction of the tool with the workpiece. Only during this contact can the resulting process quantities be measured. Their temporal and local course is determined by the effective quantities in the zone of contact, which may differ from the nominal setting quantities owing to internal or external disturbances.

The most important process quantities to be detected are forces, power consumption, and acoustic emission [1]. However, vibrations and temperatures resulting from material removal are also of interest. In the following, sensors developed to measure these different process quantities will be introduced. Figure 4.3-1 shows an overview of possible positions for sensors to determine these quantities. In a schematic set-up a portal milling machine and a lathe are equipped with different sensors for force, acoustic emission, torque, power, and vibration measurement.

4.3.3.1 Force Sensors
During material removal, the cutting edge penetrates the surface of the part to be machined owing to the relative movement between tool and workpiece. The tool applies forces to the material, which result in elastic and plastic deformations in the shear zone and which lead to shearing and cutting of material. The process

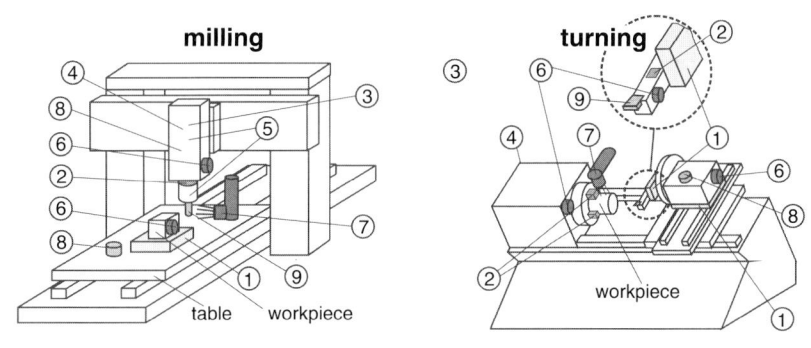

Fig. 4.3-1 Possible sensor positions to measure process quantities during cutting
① piezo-electric dynamometer platform type; ② strain gauge-based force measurement; ③ force measuring bearing; ④ power sensor; ⑤ torque sensor; ⑥ AE-sensor, surface mounted; ⑦ AE-sensor, fluid coupled; ⑧ acceleration sensor; ⑨ tool inbuilt sensor

behavior is reflected by the change in the cutting forces, hence monitoring of this quantity is highly desirable. Cutting forces have to be measured continuously. The signal evaluation can be done in different ways. Static force analysis is necessary, eg, to describe the influence of the workpiece material. Knowing the static force components, it is possible to determine the specific cutting force k_c for different materials under defined cutting edge geometry and cutting conditions [2]. They are also essential to describe the influence of different cutting parameters such as cutting speed, feed or depth of cut, and also the influence of different cutting tool materials and geometries. A more complex evaluation of the dynamic force components is applied to gain more knowledge about the current cutting tool condition. It is the purpose to detect tool chipping or breakage, the occurrence of chatter vibrations, or changes of chip breaking as fast as possible during operation to avoid any damage to the workpiece or other involved components. Different methods have been applied for further force signal processing such as frequency analysis and cepstrum analysis. Artificial intelligence techniques such as neural networks, fuzzy set theory, and combinations of the two methods have also been applied to the cutting force signals.

Whereas for the measurement of cutting forces during turning it is relatively easy to mount the tool shank on any kind of measuring system (Figure 4.3-1, right), a force measurement during milling is more complicated. Often the forces are measured with a sensor system mounted on the machine table in a stationary coordinate system (Figure 4.3-1, left). Owing to the rotation of the tool, a transformation of the force components according to the current cutting edge position is necessary. Another possibility is the simulation of a milling process by a turning operation with interrupted cut, where the milling cutting frequency is achieved by an adapted number of rotations of a workpiece with additional stripes to achieve a defined ratio of material and gaps at the circumference [3]. For larger inserted tooth cutters there is a possibility of integrating a sensor system behind one individual cutting edge.

Most of the first approaches to measure forces were based on strain gage methods. The main disadvantage of this technique is that the best sensitivity can only be achieved by applying strain gages to elements under a direct force load with reduced stiffness to generate measurable strains. Most often strain rings were used, which led to a significant weakening of the total stiffness. Owing to improvements in the sensitivity and size of strain gages, this difficulty could be reduced. In the latest applications of this method for turning a wireless transmission of the signals from the strain gages in the tool shank is realized by infrared data transfer [4]. However, for this process a different approach is also possible. Strain gages have been applied to a three-jaw chuck on a lathe for wireless force measurement during rotation of the workpiece [5]. Furthermore, an integration of strain gages in tool holders for milling with wireless data transmission has already been introduced to the market [1]. In addition to axial and radial forces, the torque can also be measured. Each tool requires to be fitted with the sensor system, which limits this approach to laboratory use.

A very reliable and accurate method is the application of piezoelectric quartz force transducers. In a dynamometer of platform type, four transducers based on this piezoelectric effect, being able to measure in three perpendicular directions, are mounted on a base plate and covered with a top plate under significant preload. These platforms are available in different sizes and are extremely stiff. They can therefore be mounted in the direct flux of force without significantly weakening the structure. Even the problem of complete protection of these sensitive transducers against coolant flow of any kind has been solved in recent years. As already mentioned, during milling or drilling a dynamometer platform is most often placed on the machine table underneath the workpiece (eg, [6]) (Figure 4.3-1, left). In turning a small dynamometer is often applied between the shank and the turret (eg, (7]) (see Section 3.3.3.1). Exemplary results of a dynamometer-based force measurement in turning and milling are shown in Figure 4.3-2. The results of hard turning reveal a linear increase in the cutting force with increasing feed for two different depths of cut [7]. In milling of high-strength steel the superior behavior of PCBN cutting tools compared with tungsten carbides and cermet is demonstrated by evaluating the maximum cutting force [6].

An installation of a piezoelectric-based dynamometer between the cross slide and the tool turret has been reported. Lee et al. performed an FEM analysis to identify the best position of the piezoelectric sensor underneath the turret housing [8]. Ziehbeil chose a special application of piezoelectric quartz force transducers in the field of fundamental research [9]. His attempt was to separate thermal and mechanical influences on the tool rake and flank face by applying adapted sensors. For the stress distribution evaluation he used a split cutting tool (Figure 4.3-3, left). The necessary force distribution on the rake and flank face was determined by four independent piezoelectric elements. With this set-up it is not directly possible to measure the normal and tangential force component on each face, because both tool parts interact due to the contact in the parting line. However, by using an adapted calibration matrix and procedure and by limiting the tests to orthogonal cutting, it was possible to determine the normal and tangential

4.3 Cutting Processes

turning

force dynamometer
cutting tool
workpiece

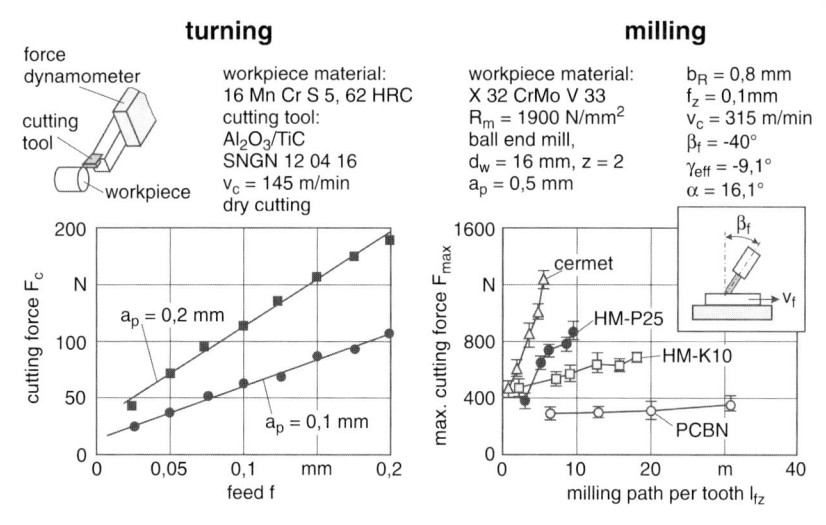

workpiece material:
16 Mn Cr S 5, 62 HRC
cutting tool:
Al_2O_3/TiC
SNGN 12 04 16
v_c = 145 m/min
dry cutting

workpiece material:
X 32 CrMo V 33
R_m = 1900 N/mm²
ball end mill,
d_w = 16 mm, z = 2
a_p = 0,5 mm

milling

b_R = 0,8 mm
f_z = 0,1mm
v_c = 315 m/min
β_f = -40°
γ_{eff} = -9,1°
α = 16,1°

Fig. 4.3-2 Dynamometer-based force measurement in turning and milling. Source: Brandt [7], Hernández [6]

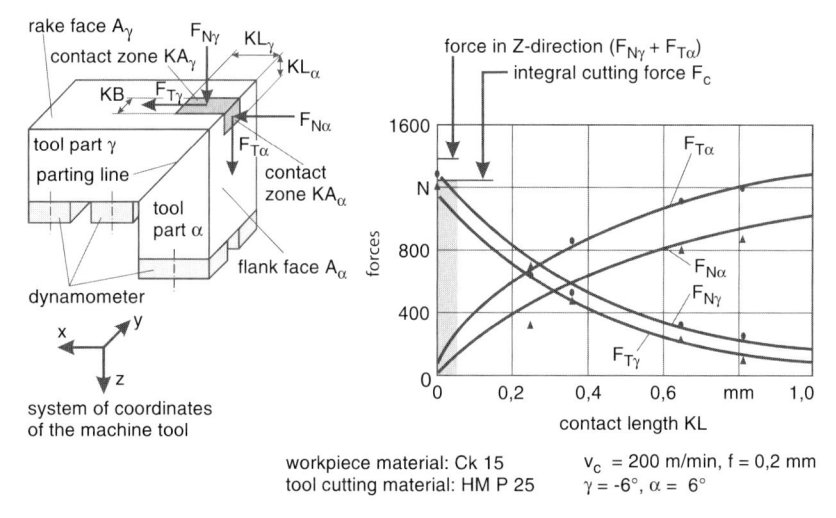

workpiece material: Ck 15
tool cutting material: HM P 25

v_c = 200 m/min, f = 0,2 mm
γ = -6°, α = 6°

Fig. 4.3-3 Piezoelectric force measurement on a split tool. Source: Ziehbeil [9]

force components $F_{N\gamma}$ and $F_{T\gamma}$ on the rake face and $F_{N\alpha}$ and $F_{T\alpha}$ on the flank face. Figure 4.3-3 (right) shows the result of measurements with different parting line positions. For comparison, the integral cutting F_c is also shown; the results of the split tool are nearly identical. These forces together with the corresponding cutting lengths were used to calculate the stress distribution. The most complex dynamometer development so far is a rotating system for milling applications. It consists of four quartz components for the measurement of forces and torque.

Also, four miniature charge amplifiers are integrated in the rotating system and the transmission of data is realized via telemetry. This system is especially attractive for five-axis milling, where the force transformation from a stationary platform-type dynamometer is extremely complex.

Direct force measurement using stationary dynamometers can be regarded as state of the art. They are widely used in fundamental research, but their application in industrial production is very limited for basically two reasons. First, these systems are only available at very high cost, and second, no overload protection is available, leading to severe damage of the dynamometer in the case of any operator or machine error [1]. For this reason, platform- and ring-type sensors based on quartz transducers or strain gages have been implemented in shunt with the process forces [1, 10]. They are mounted either behind the spindle flange of milling machines or at the turret interface on lathes. These sensors are overload protected, because they are only subjected to a small part of the load. Although commercially available, these sensors still do not work reliably owing to their sensitivity to many disturbing factors such as coolant supply or thermal expansion of components.

Force measuring bearings have also been introduced, either with strain gages at circumferential grooves of the bearing ring or in an additional bushing [1]. Owing to the necessary filtering of the obtained signal to eliminate the ball contact frequency, they are not able to measure high-frequency signals. Furthermore, the rigidity of the spindle is reduced, which limits this method to a very few cases.

Another method for force monitoring became possible with the introduction of spindles with active magnetic bearings. By evaluating the power demand of the stationary magnets at the circumference of the rotor to keep it in a desired position with constant gaps from the different magnets, the cutting forces can be determined without further equipment [11]. These spindles are very attractive, especially for high-speed cutting, because they allow rotational speeds of more than $100\,000$ min^{-1}. However, the high cost of this spindle type limits their application to a very few cases at present.

Force dowel pins or extension sensors detect the cutting force indirectly if they are correctly applied to force-carrying components. However, the effort to find the most suitable fitting position and the poor sensitivity limit the application of these sensors in many cases to tool breakage detection during roughing processes [1]. Husen [12] used dowel pins for strain measurement in the housing of a multi-spindle drilling head. It was possible to detect individual tool breakage on eight different spindles by applying only one sensor [12].

Summarizing the available sensor solutions for direct force measurement, it can be said that piezoelectric transducers can be regarded as the most suitable but most expensive solution. The application of strain gages is also very popular, and sufficient sensitivity can be achieved without severe weakening of the total stiffness. Solutions integrated in the tool or tool holder are complex and expensive, which limits their application to laboratory use.

4.3.3.2 Torque Sensors

The measurement of torque is most suitable for drilling and milling processes. Several different principles can be applied. One attempt was to integrate two preloaded piezoelectric quartz elements in the main machine spindle [13]. However, the high effort and the additional required space are limiting factors. A spindle-integrated system incorporating a torsional elastic coupling or two toothed discs or pulleys was also introduced, but the practical use is again limited for the above-mentioned reasons [1].

A brief explanation of sensors integrated in the tool holder was given in the previous section. Either rotating systems with piezoelectric transducers or with strain gages are also able to measure torque. A complex sensor based on strain gages for torque and thrust also incorporating thermocouples for temperature measurements was introduced [14]. Furthermore, a special piezoelectric dynamometer for torque measurements is available, which operates stationary and has to be placed underneath the workpiece on the machine table. It is used in fundamental investigations for drilling processes. A different approach for torque measurement has been published [15]. The supervision of the main spindle rotational speed by using a pulse generator in the spindle motor was proposed. By investigating the fluctuation pattern of the signal during one revolution and applying a vector comparison algorithm, it was possible to determine tool breakage and chatter vibrations.

Two other techniques are based on magnetic effects and will be explained below.

The first sensor uses the magnetostrictive effect [12]. The permeability of ferromagnetic materials changes under mechanical load. Changes due to torque load on the shaft of a drill can be detected by applying an adapted system of coils. One excitation coil and four receiving coils are integrated in a miniature sensor system, which is able to measure on drills with a diameter of 2.0 mm or more. The measuring distance is 0.5 mm (Figure 4.3-4, left). Figure 4.3-4 (right) shows an exemplary result of one drilling operation. The results reveal that by analyzing the torque sensor signal in the time domain it is possible to detect process disturbances. Transient torque peaks in an earlier state (c) indicate the occurrence of continuous chatter in state (d) due to reduced cutting ability. These torque peaks are related to the drilling depth, tool type, and wear state regarding their form and distribution. Typical frequencies were found between 200 and 600 Hz. Monitoring of the lifetime of a drill is therefore possible. With the sensor indication the drill can be removed from the machine tool before tool breakage or workpiece damage occurs. Owing to the small size of the sensor with a diameter of 5 mm, integration in almost any machine tool environment is possible. Parallel monitoring of different drills in a multi-spindle head may also be considered, although Husen [12] has developed a special solution based on strain dowel pins for this application.

The second solution is based on magnetic films, which are deposited on the tool shank [16] (Figure 4.3-5). Torque of the shaft due to mechanical load will lead to a change in the permeability of the films. The films are magnetized with the

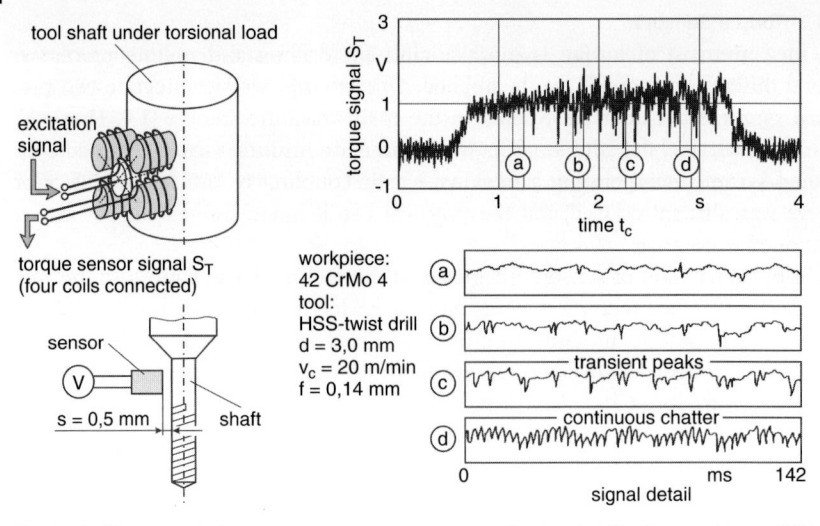

Fig. 4.3-4 Magnetostrictive torque measurement on small twisted drills. Source: Husen [12]

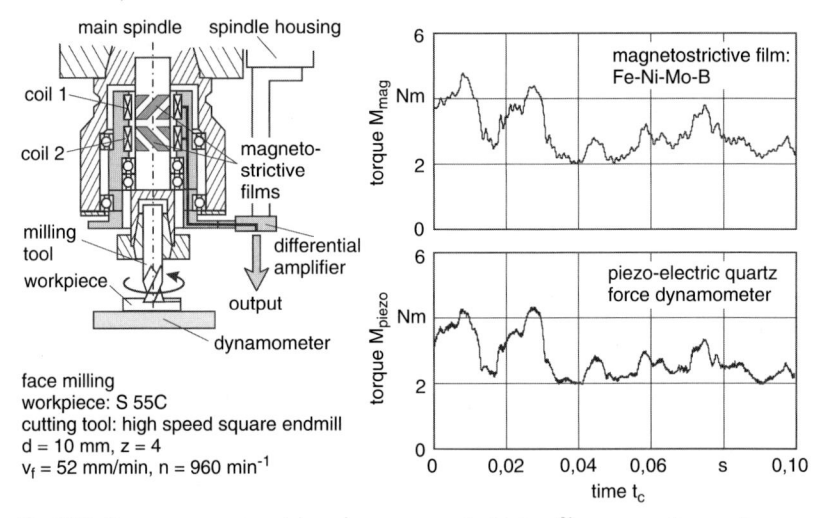

Fig. 4.3-5 Torque measurement based on a magnetostrictive film sensor. Source: Aoyama et al. [16]

surrounding circular coils. Owing to the different orientations of the upper and lower film, this sensor system is very sensitive to the torque load on the shaft by using an adapted bridge circuit. The material for the film Fe-Ni-Mo-B was chosen because of its high sensitivity and low hysteresis loss. Figure 4.3-5 (right) shows the results of a milling test. The signal of the magnetostrictive film sensor is compared with the measurement of a stationary piezoelectric dynamometer, placed underneath the workpiece. The face milling experiments demonstrate the sensitivity of the magnetostrictive sensor and the suitable dynamic characteristics. Even experiments with spindle revolutions of 3500 min^{-1} could be performed successfully [16]. The major disadvantage is the high effort to install the system with the need to modify the spindle end. Also, the necessity for additional bearings limits the possible maximum spindle rotation.

To summarize the available solutions for torque measurement, it can be said that expensive piezoelectric or strain gage-based systems are available which offer the necessary functionality. For laboratory use other complex systems have shown suitable performance. The most promising low-cost version for industrial use seems to be the non-contact magnetostrictive sensor with five coils, because this solution does not need any major changes to the machine set-up.

4.3.3.3 Power Sensors

The measurement of power consumption of a spindle drive can be regarded as technically simple. Depending on the type of system used, current, voltage, and/or phase shift can be detected. The sensors are not even located in the workspace of the machine tool and therefore have no negative impact on the process. Also, the amount of investment is very moderate, thus making this sensor type attractive for industrial application. It is even possible to gain information about the actual power demand from the drives from the machine tool control without additional sensors. However, the sensitivity of this measuring quantity is limited, because the power required for cutting is only a portion of the total consumption (see also Section 3.3.3.2). Most often power monitoring is used to prevent overload of the spindle and to detect collisions. Nevertheless, attempts have been made to use the motor current of the feed drive in milling to determine process conditions and tool breakage. Using permanent magnet synchronous AC servo motors for direct drive of the feed axis the dynamic changes of the current can be determined. By applying special algorithms, which include the average cutting force residuals and the force vibration of each cutter, a successful determination of tool breakage from the current measurement is possible. Further developments in the field of dynamic drive systems in combination with the latest machine tool controls will further increase the importance of this monitoring strategy, even without additional sensors.

4.3.3.4 Temperature Sensors

As already explained in Section 3.3, every cutting process generates a significant thermal impact on the workpiece material. The measurement of the temperature distribution in the cutting zone is therefore of great importance for the fundamental understanding of tool wear and workpiece surface integrity. A distinction in measuring systems based on heat conduction or heat radiation can be made [2]. The most popular systems are shown in Figure 4.3-6. The systems based on heat conduction use the thermoelectric effect. Direct methods are the single-tool and the twin-tool methods.

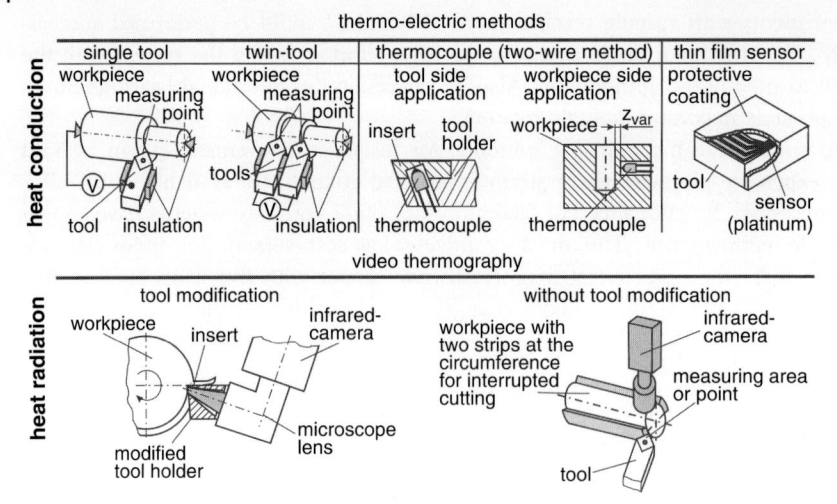

Fig. 4.3-6 Temperature-measuring systems in cutting

The effect is based on the fact that workpiece and tool material form a thermocouple in the heated contact zone. A second contact point has to remain at a defined temperature to determine the average temperature in the cutting zone by measurement of the thermal voltage after calibration. The major problem for the single-tool method is the insulation of the relevant components and the calibration. Any kind of temperature distribution is not detectable. The method is furthermore restricted to electrically conductive materials. Another method is to integrate a thermocouple in the tool or the workpiece. In case of a single-wire method the conductive tool material or the contacting chip will serve as the second element of the thermocouple. If several thermocouples or different measurement positions are applied, a temperature distribution can be determined. A method for the evaluation of the temperature distribution in the cutting zone is based on thin-film sensors [9]. The ohmic resistance of pure metals such as platinum changes with variation in temperature, while a pressure influence can be neglected. A layer of 12 platinum sensors with a thickness of 0.2 μm and a width of 25 μm at a distance of 0.1 mm to each other was evaporated on an Al_2O_3 cutting tool and protected by an additional 2 μm coating of Al_2O_3 on top (Figure 4.3-7, left). The results reveal that it is possible to determine the local temperatures at the rake face even at a cutting speed of 800 m/min (Figure 4.3-7, right). The measured temperatures are slightly higher than the melting point of the machined aluminium alloy at normal pressure, but melting of the chip bottom surface was not observed. The melting temperature of the material is shifted towards higher values because of the high mechanical load in the zone of contact. The pressure in the corresponding area has been determined to be in the region of 500 MPa [9]. This sensor development helped considerably in understanding the fundamentals of cutting and in calibrating simulation programs [17]. Unfortunately, it is not possible to machine harder materials than the chosen aluminium alloy, because

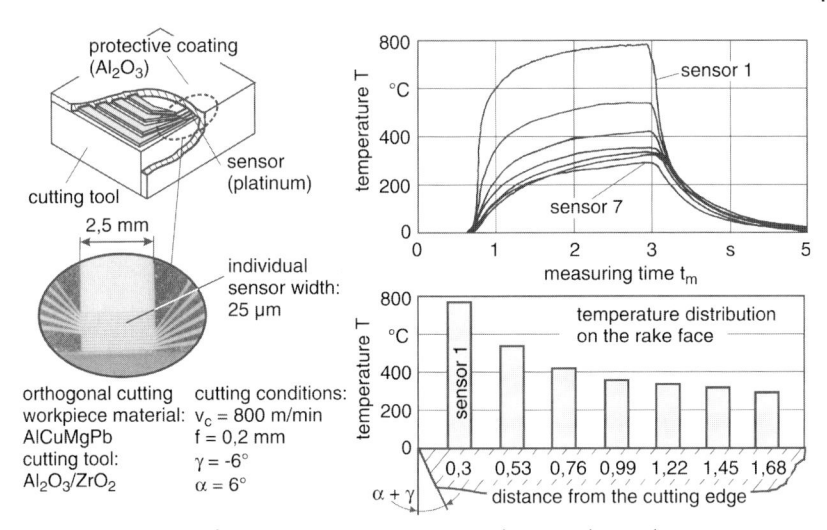

Fig. 4.3-7 Thin-film sensor for temperature measurement during orthogonal cutting. Source: Ziehbeil [9]

the protective layer is exposed to fast wear if steel materials are cut. The basic idea is still very attractive and further developments are currently under way [18].

Temperature measurement methods based on heat radiation use infrared films, video-thermography, or total radiation pyrometry [2]. With the first two methods the temperature distribution can be determined by analyzing the degree of exposure of the film and the tube to the heat radiation. The latter is used to collect the total radiation of the measuring area with the aid of a lens system and focus it to an indicator. As a major advantage of these methods, modification of tool or workpiece is often not necessary, only optical access to the measuring area has to be guaranteed. Nevertheless, solutions with modified tools are also in use as shown in Figure 4.3-6 (bottom left) (see Section 3.3.3.3). The emission coefficients of the investigated materials are temperature dependent, and easier calibration is possible if the measurement is restricted to a single spot [3]. In Figure 4.3-8 results of measurements with an infrared camera using the single-spot method are shown. An interrupted cut comparable to milling is achieved by applying a workpiece with two strips at the circumference. The temperatures for different ceramic cutting tools increase with increasing cutting speed and with a change of the workpiece material from cast iron to steel. The same tendency was found with increasing feed.

Summarizing, it can be said that all systems are limited to application in the laboratory because of their complexity and the often necessary modification of components. The developed solutions have significantly supported the fundamental understanding of heat transfer in the cutting zone. However, the industrial use of any sensor as a means of process monitoring is not available.

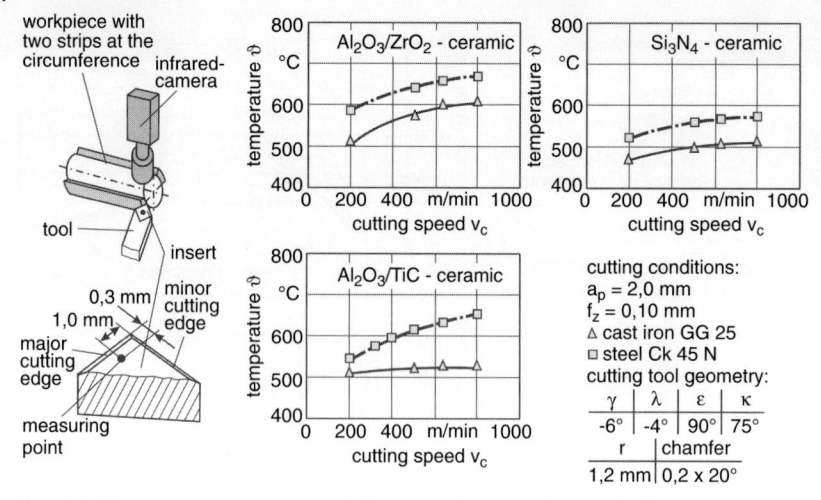

Fig. 4.3-8 Temperature measurement in interrupted cutting. Source: Denkena [3]

4.3.3.5 Vibration Sensors

The measurement of vibrations in machine structures can be done in two different ways. On the one hand, acceleration sensors are used which consist of a seismic mass and a spring-damping system connected with a displacement pickup. Often the piezoelectric effect of quartz is used to register the movement of the mass. The second solution is based on a relative displacement pick-up between two elements of a vibrating structure, eg, between the spindle and the housing on a milling machine [19]. The frequency range of these sensors is adapted to the type of phenomenon to be registered. Their application field is seen in the frequency range well below 150 kHz [1]. In other work the frequency range of vibration sensors was limited to 15 kHz [20]. In any case, these characteristics qualify vibration sensors also for tool condition monitoring. Acceleration sensors fulfil the demands of reliability and robustness, because they are designed for the use in rough environments. They can be easily applied to a machine tool component and do not need mounting very close to the zone of contact, because the frequencies to be detected do not suffer severe attenuation or distortion such as high frequency acoustic emission (AE) signals [21]. A pure mechanical sensor coupling is applied; small air gaps do not have a relevant influence. The terminology concerning vibration measurement is not clearly defined, and terms such as low-frequency acoustic emission [1] and ultrasonic vibration [21] are in use. In different publications the suitability of vibration sensors for cutting process monitoring was stated (eg, [19, 21]). Figure 4.3-9 shows a typical result of vibration analysis in turning [22]. The average amplitude spectra of a tool life cycle reveal clear differences between a new and a worn ceramic cutting insert. The vibration signal of a new tool is composed of low-frequency natural vibration modes of the lathe, the chip segmentation frequency at 34 kHz, and vibrations exceeding 50 kHz induced by friction and deformation. A change of the vibration pattern is visible after the

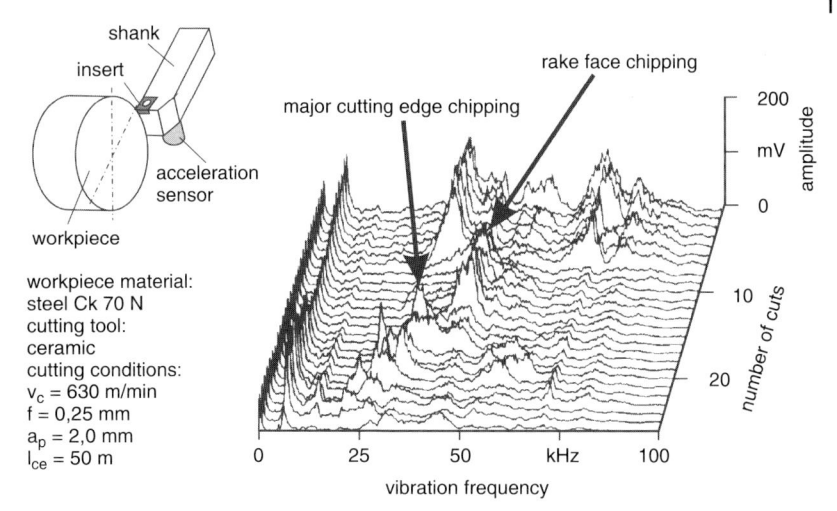

Fig. 4.3-9 Vibration amplitude spectrum change during turning of steel. Source: Warnecke and Bähre [22]

eighth cut owing to rake face chipping. The resulting increase in the rake angle shifts the chip segmentation frequency towards 45 kHz and reduces the amplitude in the frequency range 60–90 kHz. After cut No. 14 chipping at the major cutting edge is detected, leading to a quasi-chamfered edge geometry. The chip segmentation frequency decreases to 33 kHz. In the following cuts this segmentation frequency alternates owing to continuing chipping of the cutting tool. Finally, tool breakage occurs at the 26th cut.

This result proves the efficiency of vibration sensors in cutting processes. Owing to the mentioned advantages and the relatively low investment, they are very popular as monitoring devices. If frequencies above 100 kHz are to be investigated, the sensor system has to be changed to an AE sensor.

4.3.3.6 Acoustic Emission Sensors

AE sensors must be regarded as the most popular monitoring equipment in cutting processes over the last 20 years, despite force measurement. A large number of publications have dealt with the application and signal processing of AE systems. In a survey conducted in 1994, more than one fourth of 539 listed publications dealt with AE techniques in cutting processes [23]. The sources of AE signals have already been explained in Section 3.3. Basically two types of AE sensors have to be distinguished, wideband sensors and resonance systems. In the first case the sensor does not have a seismic mass to reduce an unwanted sensitivity in the low-frequency range. The sensitive element, most often piezoelectric based, is mechanically damped, sometimes by applying a relatively large damping mass. These sensors can be used up to the MHz range. The resonance type sensor has still a seismic mass and is in principle of the same type as an acceleration sensor.

The frequency range is limited owing to the design. An usual upper threshold is 250 kHz. If the measuring range is shifted to low frequencies in the region of 20 kHz, there is no clear difference from a vibration analysis (see Section 4.3.3.5). The coupling conditions are important; the signal transmission can be significantly improved by using grease in the gap between component surface and sensor. The sensor signals need further processing such as filtering, amplifying, and rectifying until the desired quantities can be deduced. Methods of artificial intelligence such as neural networks and fuzzy logic systems have also been applied to AE signals.

In most of the investigations on turning, the AE sensor is mounted on the tool shank, which is very close to the signal origin. For industrial application with the need for fast tool changes, this solution has some limitations. AE sensors are basically used to determine tool breakage and wear behavior. The first phenomenon will lead to a significant increase in AE energy. Many authors have used this clear signal for monitoring [1].

More complicated and challenging is the detection of tool wear using AE sensors because of two different effects. Increasing width of flank wear land increases the contact area between tool and workpiece and also leads to a temperature rise. On the one hand the energy of the friction-related acoustic waves increases, and on the other the shear strength of the workpiece material, shear angle, and contact length also change [10]. Hence a variation of the AE signal is likely to occur. In some publications an increase in bursts was reported, and rising root mean square values due to wear have also been published (eg, [24, 25]). Representative results of root mean square values during turning of steel with coated and uncoated tungsten carbide tools are shown in Figure 4.3-10 [25]. For uncoated tools a relatively low signal increase is visible over the very short tool life, while coated tools are much more suitable for this cutting operation and gen-

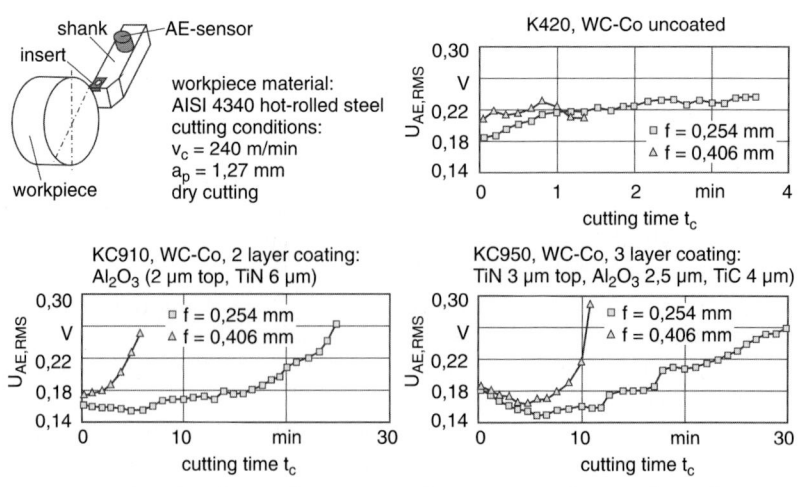

Fig. 4.3-10 AE-based tool wear determination during turning. Source: Cho and Komvopoulos [25]

erate a significant increase in the AE signal after the coating is worn. The initial problem of the r.m.s. value depends on the type of surface layer. The three-layer tool has a rough TiN top coating. During wear propagation of this layer, the signal decreases; after reaching the Al_2O_3 intermediate layer, an increase until total coating failure can be observed.

In [23], the peak position of the amplitude distribution curve of the AE signal, called AE mode, was proposed for improved identification of tool wear, because the influence of randomly appearing bursts is eliminated. Figure 4.3-11 shows representative results of this quantity and their determination procedure. The AE sensor was integrated in a modified tool shank and totally protected against coolant and other process residues. The AE mode values also show a clear increase with continuing flank wear. The nonlinear behavior is explained by the superposition of the effects of flank wear and crater wear. Whereas the former generates a clear AE signal increase due to the enlarged contact area, the latter leads to an increased effective rake angle, which reduces the AE activity. Finally, the flank wear dominates the signal with a further increase until large chipping of the tool occurs.

These two types of wear often occur at the same time. In [26], it was reported that the development of crater wear could totally compensate a further increase in the AE signal. In [27], even a decrease in the AE signal with increasing wear was observed. The choice of the frequency range for the AE analysis also has a major influence. As shown in Section 3.3.3.4 for surface integrity monitoring in hard turning it is possible to identify a narrow bandwidth of the AE signal for specific correlation purposes, where the signal also decreases with proceeding wear.

AE sensors have also been tested in a wide variety of milling experiments. The measuring task is more difficult because of the permanent change of chip thick-

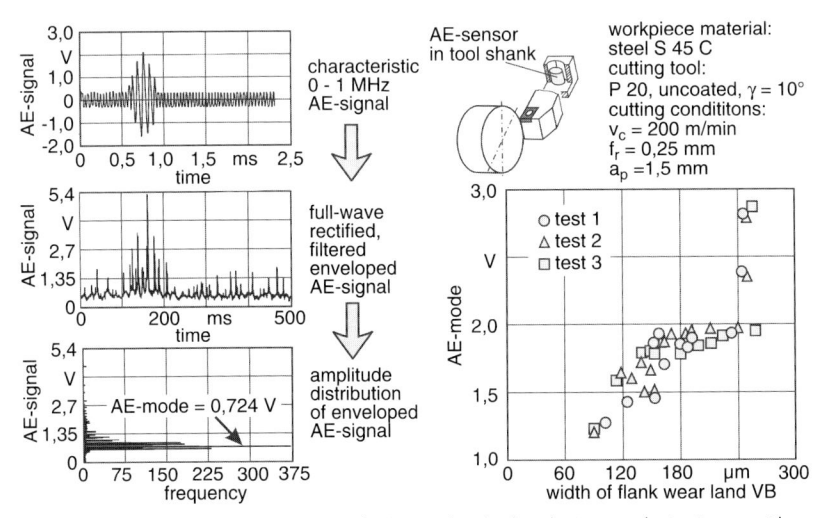

Fig. 4.3-11 Wear determination in turning with AE amplitude distribution analysis. Source: Blum and Inasaki [23]

ness and the pulses from the entrance and exit of the single cutting edges. Tool breakage leads to the same burst type increase of the AE activity as mentioned in turning. Different solutions for a suitable AE sensor position have been chosen. A mounting at the workpiece, at the rotating tool, under the insert, or by applying a special device at the spindle top for transmitting the signal through a magnetic liquid have been published. All these solutions have specific limitations for industrial use. In [10], a practical approach to establish an AE system for milling was made. Different sensors and mounting positions have been evaluated.

Figure 4.3-12 shows the result of the most suitable solution for tool breakage detection. A wideband AE sensor in the frequency range 100–500 kHz was mounted on the X-table of a horizontal milling machine, not influencing either the process or tool or workpiece change. The strategy is based on the application of dynamic thresholds. The problem is to separate tool breakage from disturbance signals. The upper part of Figure 4.3-12 shows the $U_{AE,RMS}$ value together with the dynamic threshold. One real tool breakage and several other peaks are apparent. Two criteria are used to distinguish between these signals. The duration of exceeding the threshold is significantly larger for the tool breakage (6.2 ms) (Figure 4.3-12, middle) than for the disturbance peak (1.9 ms) (Figure 4.3-12, right). The shape of the pulse is described by signal differentiation (Figure 4.3-12, bottom). The disturbance peak shows only one oscillation, whereas the breakage signal oscillates over a longer period. The proposed tool breakage monitoring system has to check both criteria, exceeding time and differential signal shape, to trigger an alarm signal [10].

Again, the determination of wear influence on the AE signal in milling is more complicated. In [28], results of experiments on a vertical milling machine during single- and multi-tooth face milling of steel are presented. The influence of different input quantities was evaluated. Wear could only be determined during single-

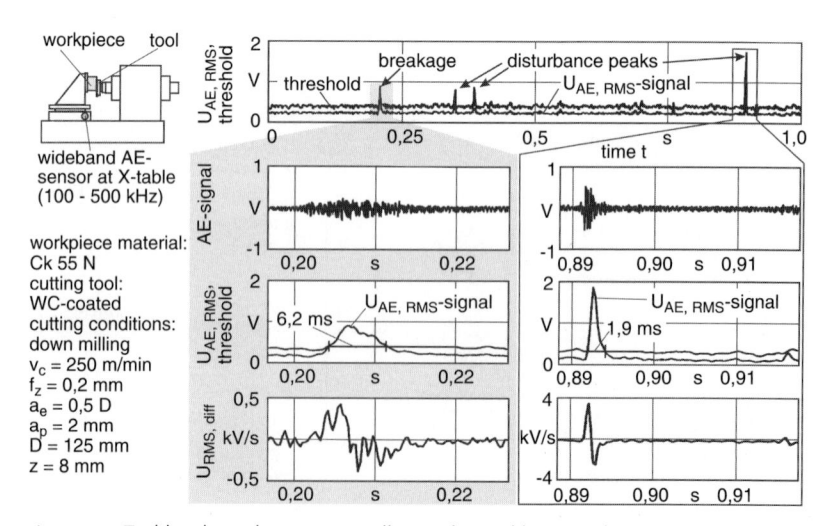

Fig. 4.3-12 Tool breakage detection in milling with suitable AE analysis. Source: Ketteler [10]

tooth milling, where a fluctuating increase in the root mean square value was found with proceeding wear. A contrary tendency of wear and the AE root mean square value were reported in [29] for cutting speeds above 220 m/min. The same decrease of $U_{AE,RMS}$ with rising tool wear was reported in [30] during face milling of steel at a cutting speed of 300 m/min, using a special type of fiber-optic interferometer for acoustic emission measurement. Kettler used a resonance-type sensor in the lower frequency range of 20–100 kHz for the determination of tool wear [10]. In Figure 4.3-13, results of his investigation are shown. The sensor is mounted on the pallet of the workpiece. The major problem for this measurement is the occurrence of additional, not directly wear-related AE sources such as chip contact, chip breakage, or burr formation. By analyzing the AE signal in the lower frequency range of the resonance-type sensor, it is possible to suppress these influences and to identify single teeth of the tool. The root mean square value $U_{AE,RMS}$ increases with proceeding wear over the cutting length. Teeth 7 and 8 have major cutting edge chipping at the highest cutting length of 45 m, whereas the maximum width of flank wear land for tooth 1 is 0.4 mm at this stage. In this investigation, the average of the root mean square value was found to be most suitable for wear characterization.

The discussed results reveal that the AE signal also has good potential for milling process monitoring. Although tool breakage detection is relatively easy to achieve, further work has to be done for wear determination, especially for processes with coated cutting inserts because of their complex wear behavior. The contradictions of AE activity and wear behavior are not yet fully understood. Further research is necessary until robust systems for industrial use are available.

It should be mentioned that AE sensors have also been applied to many other processes with geometrically well-defined cutting edges. Publications on drilling [31], reaming [32], deburring [33], tapping [34], and planing [35] are available. The

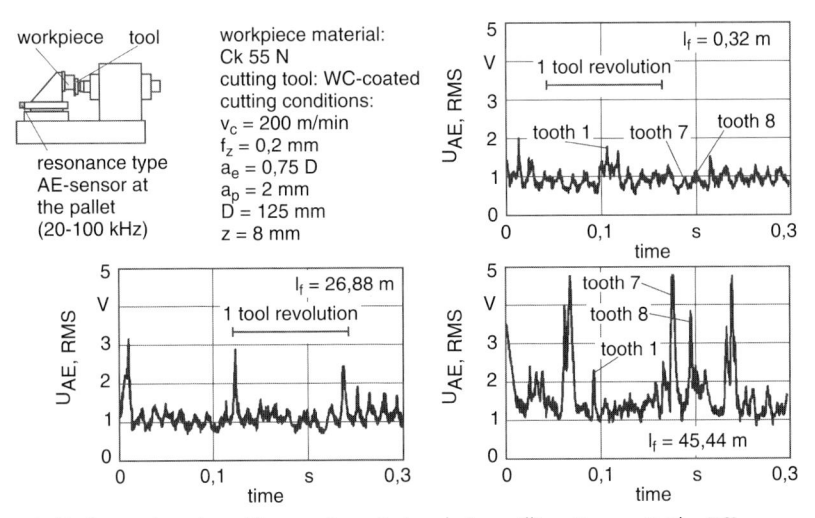

Fig. 4.3-13 Tool wear detection with acoustic emission during milling. Source: Kettler [10]

basic principles of sensor application, signal processing, and evaluation strategies can be compared with the mentioned systems.

Summarizing the potential of AE, it can be clearly stated that these sensors offer a wide range of interesting applications. Almost any phenomenon can be detected, at least in the laboratory. Future should be concentrated on systems applicable to the rough environment of cutting processes without hindering the process or changing procedures and on robust and fast algorithms to allow on-line decisions and to avoid false diagnosis.

4.3.4
Tool Sensors

All sensor systems which are directly related to the tool try to obtain information about the geometric features of the cutting edge. As already explained, every tool is assumed to wear owing to the occurrence of thermal, mechanical, and in certain cases also chemical load. Based on these loads, typical different mechanisms of wear are present such as abrasive, adhesive, diffusive, and oxidation wear, as well as crack formation and breakage [2]. These mechanisms result in different wear forms, the most important ones being found on the rake and flank face, such as width of flank wear land VB or crater depth. These quantities can be determined during laboratory investigations by removing the tool from the machine at specific intervals to be checked by microscope or stylus measurements. The purpose of any tool sensor is to determine these or similar quantities in the working space of the machine tool; the optimum solution would be a direct measurement during cutting.

In Figure 4.3-14, an overview of the most important methods is given. The solutions presented comprise sensor systems for both stationary objects and rotating

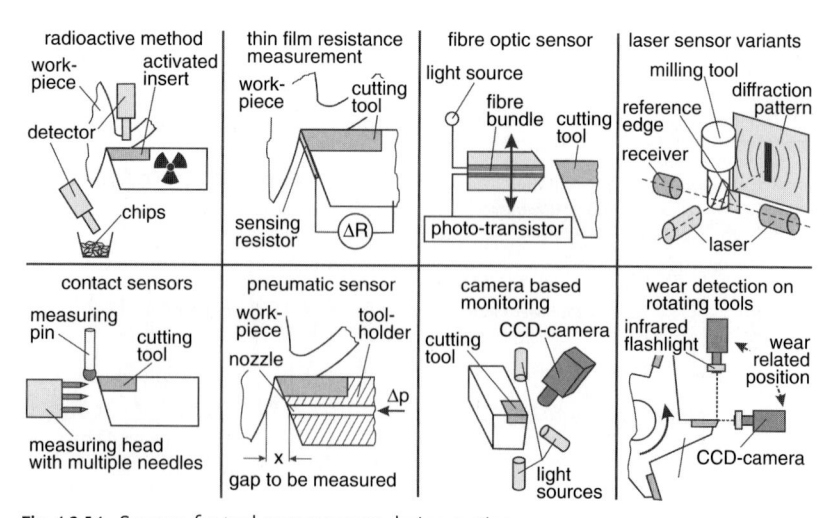

Fig. 4.3-14 Sensors for tool measurement during cutting

tools. In the early days of cutting research one of the favorite methods was to use radioactive cutting tools for tool-life testing. The tool was activated by irradiation. Worn tool material, which was transferred to the chips, was monitored by collecting the chips and measuring their radioactivity or by determination of the remaining radioactivity of the tool with a suitable detector. Although at least in later applications of this method the radioactivity was very small, this method could not achieve practical importance owing to the dangers with any kind of radioactivity, the necessity for irradiation facilities, and the limited sensitivity under specific conditions. Attempts were also made to use chemical analysis for fast tool-life testing by pickling the chips after 1 min cutting to analyze the residues with respect to tool material tracing [36]. However, the chemical efforts incorporating a calorimetric analysis were not practical.

Contact sensors of different kinds have reached practical application because of their low investment, robust design, and relatively easy integration and signal evaluation. Measuring pins with wide geometric variations are available and have been used in lathes and milling machines. In most cases the purpose is to determine the displacement of the cutting edge in the face direction, because any change in this dimension directly influences the geometric accuracy of the machined part. By measuring its cutting edge displacement, the depth setting of the tool can be adjusted to maintain the desired workpiece diameter. A contacting measurement of the width of flank wear land in the machine tool was also tried by using a measuring head with several contacting needles (Figure 4.3-14, bottom left). Although the resolution was sufficient with 10 µm, systems of this type could not achieve wider acceptance mainly because of the time and space demands in the machine tool.

The use of thin-film sensors was proposed to measure the wear of the flank face (Figure 4.3-14, top middle-left) [37]. The material was printed on the flank face, insulated by a heat-resistant paint and exposed to continuing wear. The length decrease of the thin film due to the increase in the width of flank wear land was registered via resistance measurement. This interesting idea of thin-film sensors on tool surfaces was further developed to measure temperatures directly (see Section 4.3.3.4).

Pneumatic sensors are used to determine the displacement of the cutting edge in the face direction by measuring the gap between the tool holder and the generated workpiece surface (Figure 4.3-14, bottom middle-left). The sensor is based on the nozzle-bounce plate principle. The nozzle is integrated in the tool holder and the pressure change in dependence of the gap width is measured. The set-up is simple, the measuring range was around 400 µm, and an accuracy of 2–3 µm could be achieved. However, variations in temperature, pressure, and surface quality have some influence on the reliability and sensitivity of the sensor.

Apart from these mentioned sensor systems, most solutions are based on optical measurement. One of the first attempts was a sensor based on fiber optics to measure the width of flank wear land (Figure 4.3-14, top middle-right). One half of the fiber bundle illuminates the flank face of the tool and the reflected light is registered by the other half. The displacement of the cutting edge has to be taken into account, because the light intensity decreases with larger distances. An accuracy of 10 µm was reported.

The consequent next step was to use a charge coupled device (CCD) camera for measuring the cutting edge geometry [37, 38] (Figure 4.3-14, bottom middle-right). The features of the camera itself are most often very similar. The sensitivity of the approach is directly related to the quality of the lightning. Fiber optics with additional equipment to diffuse the light or a laser light source with a diffraction grating is one of the applied methods. With additional movement of the camera and the lightning, separate analyses of flank wear and crater wear are possible. If the system is integrated in the working space of the machine tool, it is, of course, only possible to measure the cutting edge geometry in auxiliary times. The measuring time, space demand, and protection efforts to cover the system during cutting operations under coolant supply are still limiting factors for wider application. Outside the machine tool these systems have already had a considerable impact. A complete 3D analysis of a cutting insert is possible by applying projected fringes to the tool surface and using a CCD camera with adapted software [38]. The most complex optical system for cutting tools is shown schematically in Figure 4.3-15. The measuring machine is built to determine automatically all relevant geometric data on rotational cutting tools such as milling cutters. The machine has a four-axis CNC control to focus the optical system on every part of the tool of interest. With a combination of two CCD cameras, one working in the direct illumination mode and the other with transillumination, it is possible to determine the whole complex geometry of the tool. A very important feature is the lightning system with a special light-emitting diode (LED) array, which allows control of the light intensity of different diode sections individually. The machine was originally designed as a measuring machine in combination with a six-axis CNC tool grinding machine to generate automatically the NC program of the grinding operation after tool measurement especially for re-grinding of worn tools.

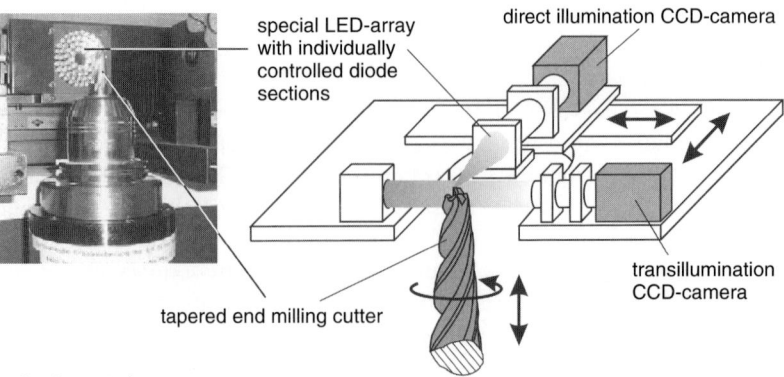

- 4 axis control
- repeatability of diameter measurement: ± 1 µm
- repeatability of length measurement: ± 2 µm

Fig. 4.3-15 Optical measurement of the total geometry of rotary tools. Source: Walter-Vialog AG, Hannover

However, the very precise determination of tool wear of any kind with this system is a clear motivation to extend the application field. Integration in the working space of the machine tool is not possible, but an implementation in the tool magazine seems to be very promising. Initial studies have already been made; with a further decrease in the system costs by reducing the accuracy demands of currently 2 µm to more practical values, a realistic return of investment can be achieved in a very short time.

For milling operations, a measurement of the rotating tool would be the best achievable solution. Several attempts have already been made, all based on optical methods. In [39] it was proposed to use the diffraction pattern of laser light, which is sent to a small slit between the rotating tool and a reference edge (Figure 4.3-14, top right). Flank wear states between 0.2 and 0.4 mm could be distinguished by analyzing the fringe pattern, but the major drawback is the extreme importance of the exact and repeatable positioning of the tool at the reference cutting edge, because this directly influences the result.

A solution which has already been introduced to practical application is the use of a laser light barrier in the working space of a milling machine tool. The purpose of this method is to determine the envelope curve of the rotating tool in order to adapt the process parameters to the individual length or diameter and to identify tool breakage. In Figure 4.3-16 some results of investigations in a state-of-the-art milling machine tool are shown.

It can be seen that the accuracy of the measurement is directly related to the approach speed. Every pass through the light barrier generates a trigger signal for the control of the machine. The second and third approaches are made with significantly reduced speed. The length and diameter determination is done by reading the machine tool axis data after passing through the light barrier. Hence the

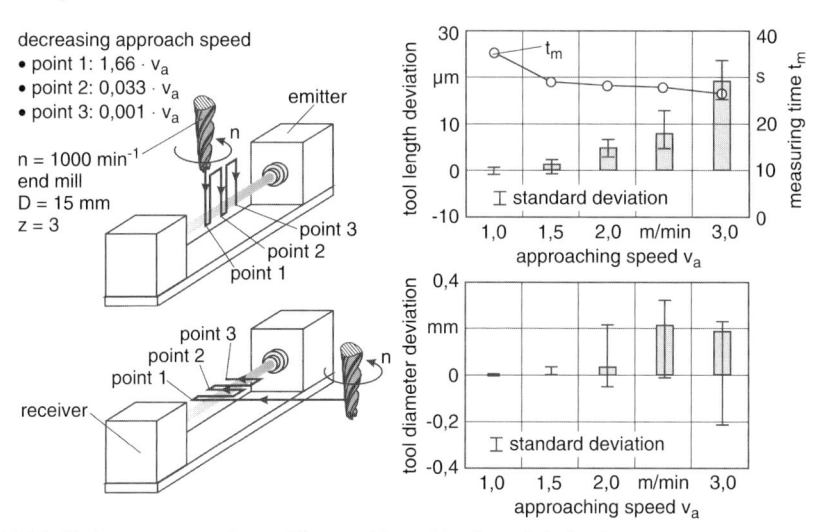

Fig. 4.3-16 Tool measurement in a milling machine with a laser light barrier

precision of the measurement is dependent on the machine tool accuracy and the processing speed of the trigger signal. In the investigated case with an unknown tool and a measuring cycle with three transits through the light barrier, a measuring time of approximately 30 s is necessary; an increase in the approach speed has only a negative influence on the accuracy and will not reduce the time considerably. The achievable accuracy is sufficient for most practical applications and is comparable to that of external tool pre-set techniques. Mechanical shutters during cutting can protect the system. The importance of this sensor type will increase further with the capability of modern milling machine controls to effect automated radius compensation of the NC program with decreasing tool diameter due to wear or re-grinding before re-coating.

A different approach was chosen in [40]. Based on a laser sensor, both displacement and intensity techniques are applied to determine the wear of a milling cutter.

The whole set-up is mounted on the main spindle and able to measure even during tool rotation (Figure 4.3-17). On the right, results of displacement and intensity signals for three different tool wear states are presented. The chipping of the tool and the length of flank wear could be determined to an accuracy of 40 µm. Even the first attempts to measure during face milling under coolant are reported. Applying a special compensation method could drastically reduce the influence of coolant and chips on the measurement. Furthermore, a 3D reconstruction of the tool based on the laser sensor signals is possible. This method seems to have a high potential for further development. Because of the massive reduction of flexibility and increased danger of collision, application on the main spindle is possible only for laboratory investigations. However, a sensor application in the workspace of the machine tool for intermittent fast measurement will be much more acceptable.

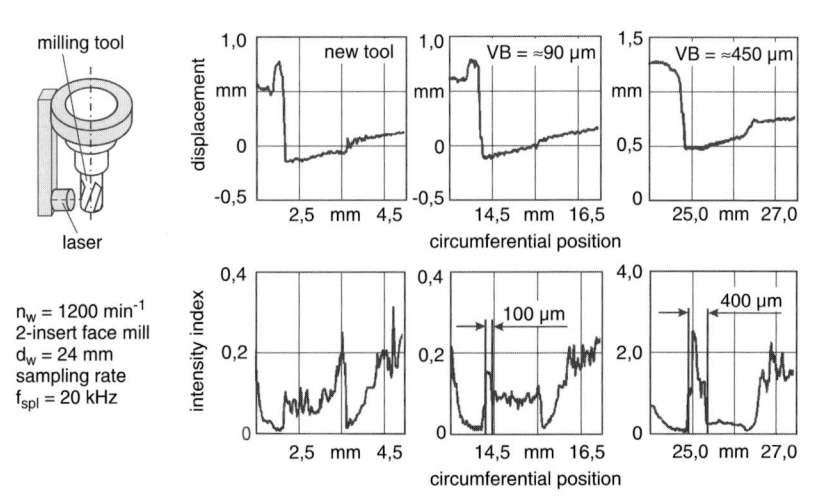

Fig. 4.3-17 Laser sensor for wear determination during rotation of milling tools. Source: Ryabov et al. [40]

The last optical method to be mentioned is based on a CCD camera with a special circular infrared flashlight [41] (Figure 4.3-14, bottom right). With this equipment it is possible to measure flank wear of individual inserts during rotation of the milling cutter at 750 min^{-1}. By shifting the camera to a top view position of the rake face, crater wear should also be detectable. The idea of increasing the measuring speed by using a high-speed flashlight has further potential, but it is expected to remain a technique for laboratory investigations.

Summarizing the discussed sensor systems for tool measurement, it can be said that contacting pins and laser barrier techniques have already reached industrial maturity. The application of any kind of complex optical sensors such as CCD cameras will remain a problem owing to the rough conditions in the working space of a machine tool, but future integration in the tool magazine seems to be very promising even under industrial conditions.

4.3.5
Workpiece Sensors

The measurement of the workpiece quality during cutting processes is in most cases related to geometric features. Sensors for physical properties to determine the surface integrity state are very rare and have already been discussed in Section 3.3. The geometric quantities can be further divided into macro- and micro-geometry (see also Section 4.4.5). The numbers of sensor systems and applications and published results are not as large as for abrasive processes. The main reason is the reduced accuracy demand on cutting operations compared with abrasive processes, which are mostly applied for finishing. Nevertheless, different sensors have been used during turning and milling and will be discussed. The sensors

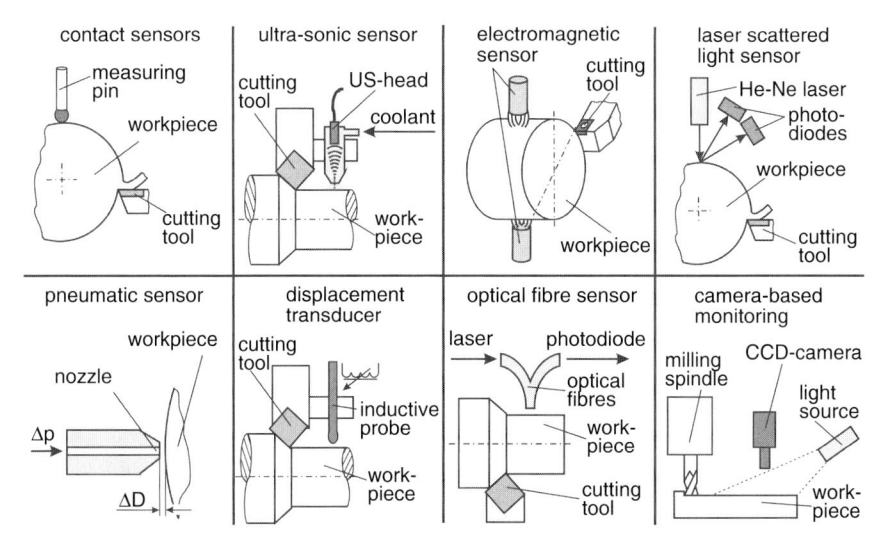

Fig. 4.3-18 Workpiece sensors applied to cutting processes

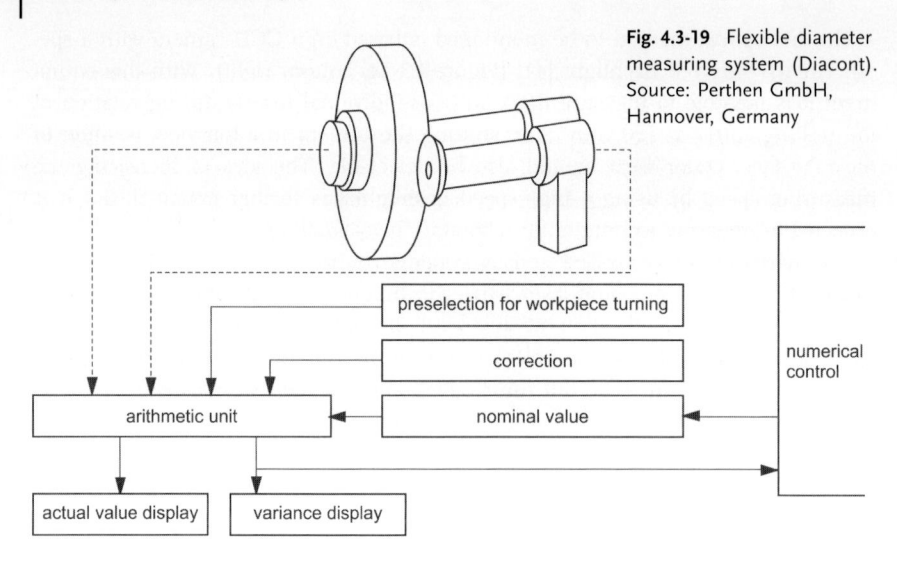

Fig. 4.3-19 Flexible diameter measuring system (Diacont). Source: Perthen GmbH, Hannover, Germany

used for macro-geometric quantities are mostly of the contact kind (Figure 4.3-18, top left), mainly applied before or after the machining process.

Pin-shaped sensors are applied to both rotational and for prismatic parts. Usually, these devices use the measuring and positioning system of a numerical machine. Therefore, they do not work independently. Figure 4.3-19 shows a contact sensor which measures absolutely.

It returns the diameter measurement to the determination of the circumference by comparing the number of rotations of a contact roller with the known diameter and the number of rotations of the workpiece to be measured. This sensor can be applied during operation on outer and inner diameters and its application is therefore very flexible. Problems are the unavoidable slip, the elastic deformations of the contacting bodies, and their dependence of the workpiece roughness. This means that the measuring accuracy is limited. Nevertheless, the sensor is successfully used in practice for the measurement of large diameters.

Quality control after machining and an initial measurement of the workpiece geometric orientation in the working space of a machine tool before cutting are possible. During turning an in-process measurement is also used; in this case a size deviation of the workpiece due to displacement of the cutting edge will be recognized in order to conduct a further cut or to remove the tool. Pneumatic sensors are also in use. The principle is the same as already explained in the previous section. The only difference is that the air supply is not passed through the tool holder. An independent sensor is mounted on the opposite side of the cutting tool. Again, it is the purpose to measure size deviations and to register cutting edge displacement.

The same geometric information can also be achieved by applying an ultrasonic sensor to determine the distance between the cutting tool and the workpiece (Figure 4.3-18, top middle-left). The signal coupling is done via coolant, which flows around the ultrasonic head to the workpiece surface. This technique can be used for cutting speeds up to 200 m/min; a further increase will lead to breakdown of the fluid coupling. The idea of ultrasonic surface testing was taken up again and extended from the determination of macro-geometric quantities to surface roughness measurement. First a post-process measurement with the workpiece placed in an immersion tank was applied. Further development allowed on-line measurement via coolant coupling in the chosen milling machine [42]. The sensor was mounted on the spindle housing and the measurement was performed at the chosen table speed. This technique is not negatively influenced by coolant as in all optical methods, remaining chips on the surface being washed away by the coolant flow. However, the major problem is to keep the coolant clean and bubble-free, because any change in the coupling conditions will have a significant influence on the measurement result. Nevertheless, these initial results seem to be promising and further research is under way to improve the system performance and reliability.

Displacement transducers are one further possibility to determine a change in the tool-workpiece distance during turning operations due to cutting edge displacement or chatter vibrations (Figure 4.3-18, bottom middle-left). Both in hard turning and in precision machining, inductive-based displacement sensors have been used to determine the vibrations during turning.

Electromagnetic sensors on conductive materials can monitor the change in the workpiece diameter (Figure 4.3-18, top middle-right). In [43] a sensor based on this principle was applied during turning to deduce tool wear from the diameter change. The two sensors were set up in a differential mode to compensate vibrations and deflections. The gap between workpiece and sensor should be in the range 0.5–2.2 mm for optimum sensitivity. Diameter changes of 7 µm could be detected at a cutting speed of 170 m/min, but it was not possible to distinguish between nose wear and flank wear.

The following techniques are all related to optical methods. The use of fiber optics is one of the favorite methods for guiding appropriate lightning to the workpiece surface and for registering the reflected light (Figure 4.3-18, bottom middle-right). In [44], a sensor based on fiber optics was used to measure the surface profiles of milled workpieces. By evaluating the waviness of the machined surfaces, the development of tool wear could be estimated. In [45] the reflectance of a turned surface was measured in-process with a fiber-optic sensor. With additional use of a neural network, the tool wear state could be deduced from the optical results.

Figure 4.3-18 (top right) shows a set-up with an He-Ne laser as light source for roughness measurement during turning. The inclining laser beam can be split to give information in the feed direction of the cut. Results from in-process application revealed that it is possible to determine the peak-to-valley height of the workpiece during turning with acceptable accuracy, but chipping of the tool and chatter vibrations increased the deviations from a post-process reference measurement.

Further solutions are based on the application of cameras. Either in turning or in milling CCD cameras have been applied to determine the topographical state of the machined surface. While the significance of optical parameters was first investigated in the laboratory, the cameras were then moved into the working space of the machine tool. The major problem with this approach is that increasing tool wear does not always lead to a deterioration of the surface quality. A severely worn tool can produce a better surface quality than a new tool. Of course, a correlation of optical parameters with stylus roughness quantities could be found. At least the standard deviation of the gray level distribution could indicate increasing wear.

In [46], the CCD camera was integrated in a milling machine to measure the generated surface on-line (Figure 4.3-18, bottom-right). As a special feature, an area-based fractal approach was chosen to describe the roughness state. After a necessary calibration a good correlation between the fractal dimension and the standardized roughness quantities could be achieved. Limiting factors are the restriction to dry machining and the finite speed of the frame grabber. However, also a fast post-process area measurement of the roughness distribution is of great interest for industrial applications.

The last optical method to be mentioned is the use of high-speed cameras. This technique, also known as micro-cinematography, has been used in fundamental investigations to visualize the chip formation (eg, [47]). With major technical effort incorporating a hollow spindle for the camera access and an etched workpiece pressed to a quartz glass window, it is possible to monitor continuously the material behavior in the shear zone during orthogonal cutting. The created movies contributed substantially to a better understanding of the procedures during the tool-workpiece interaction.

Summarizing the sensor solutions for workpiece measurement, it can be said that currently only sensors based on workpiece contact have found wide acceptance in industry. All other mentioned solutions have either specific limitations or were definitely developed for fundamental research. Optical methods will always have problems with coolant and process residues during on-line measurement, but offer a wide range of fast and increasingly reliable solutions for post-process measurement in the machine tool surrounding. Ultrasonic sensors do not suffer from these problems in the same way and may therefore gain more importance in the future.

4.3.6
Chip Control Sensors

The type of chip removal directly influences the result of any cutting operation. Thus monitoring of the chips and their breaking and removal has been the subject of many investigations. Although the chips are part of the workpiece material and monitoring is possible only during the process, sensors for this purpose will be mentioned separately in this section. Figure 4.3-20 gives an overview of the most popular solutions for chip monitoring. Many researchers try to use an already installed sensor system also for this special purpose. Thus piezoelectric dynamometers and AE sensors are employed for chip breakage detection (Figure 4.3-20, left). The sensor mounting and signal processing do not differ significantly from those in the previous mentioned applications.

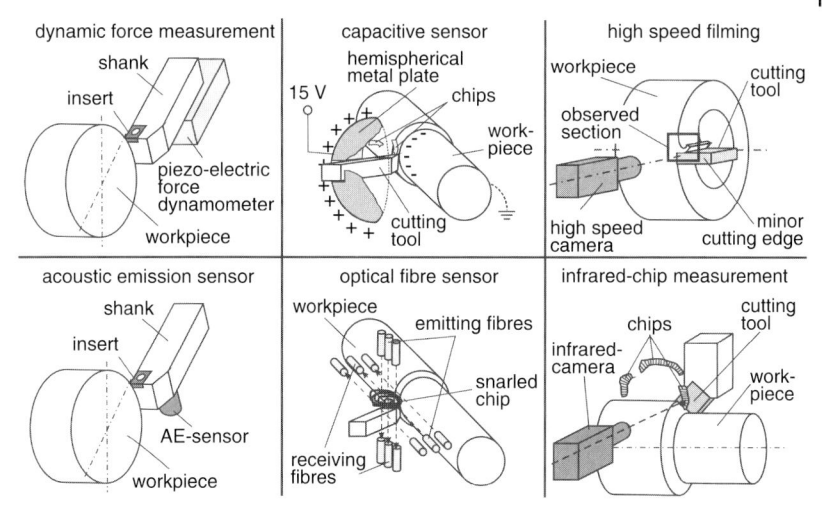

Fig. 4.3-20 Sensors for chip monitoring during cutting processes

During force measurement, the strategy is based on the evaluation of the dynamic components of the force signal (eg, [48]). However, this method has specific frequency limitations. Especially for high-speed cutting with high chip lamination frequencies, force-based chip monitoring is not possible. As explained, AE sensors offer a suitable range of frequency sensitivity and are thus used for chip-form detection. However, with the availability of AE signals from the process, most authors concentrate on directly correlating them with tool wear.

A different approach is based on the use of a capacitive sensor [49]. Venuvinod and Djordjevich formed a capacitor from a hemispherical metal plate and the corresponding part of the metallic workpiece (Figure 4.3-20, top middle). A fixed potential of 15 V is applied on the plate and the machine is grounded. The tool itself has to be isolated from the plate to prevent continuous discharge. During cutting, every chip is electrically charged and likely to collide with the plate. This contact will generate a small discharge and only applying an additional charge can restore the initial potential. This is done by electric current, and the registered current peak is related to the size of the contacting chip. Geometric information such as chip thickness cannot be derived; the major task is to register continuous chips or chip entanglement, which would lead to a total discharge of the capacitor [49].

Using optical fibers can also solve this problem. With a rectangular arrangement of emitting and receiving fibers it is possible to detect the occurrence of chip entanglement [49] (Figure 4.3-20, bottom middle). A distinction between flying broken chips, which will also cause an interruption in the light transmission, and snarled chips is made by time considerations. The distance between emitter and receiver can amount to 30 cm, but still the installation in the working space of a machine tool does not seem to be possible for practical applications.

Figure 4.3-20 (top right) shows an often-used set-up for fundamental investigations concerning chip formation. A high-speed camera is applied to monitor the

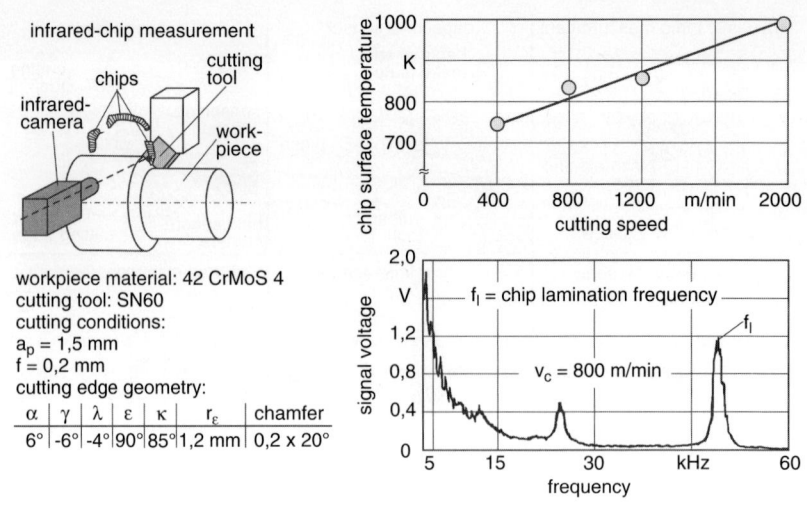

Fig. 4.3-21 Chip temperature monitoring during turning with an infrared camera. Source: Winkler [47]

cutting process, which is often restricted to orthogonal cutting (eg, [48]). Single pictures of the observed section reveal the mechanisms of chip formation. High-speed cameras are also suitable for monitoring turning operations with interrupted cut on workpieces with stripes as a simulation of milling [10].

A technique based on heat radiation is also suitable to obtain information about the chips (Figure 4.3-20, bottom right). In [47] an infrared camera was used to determine the temperature on the chip surface at different cutting speeds and to gain access to the chip lamination frequency by directly computing the temperature signal in the time domain (Figure 4.3-21). The results reveal a rise of the chip surface temperature with increasing cutting speed, because the heat generated has no time to penetrate into either the workpiece or tool. It is concentrated in the shear zone and removed with the chips. By applying a special signal processing it is possible to determine the chip lamination frequency, which is around 48 kHz for the chosen conditions at a cutting speed of 800 m/min.

Further, it should be mentioned that ultrasonic techniques and thermocouples have also been used to obtain information about the chip contact length and chip formation.

All the presented sensor solutions except AE systems are not really suitable for industrial application. Their purpose is to increase the fundamental knowledge about chip formation and to help toolmakers in the design of new cutting tool geometries. With the fast-developing computer software and hardware they can also contribute to calibrate simulation programs for chip forming by quantifying the effect of different calculated design developments.

4.3.7
Adaptive Control Systems

With the introduction of computer-based machine tool control units, the rapid development of adaptive control (AC) systems also took place. The major purpose of any AC system is to maintain the desired workpiece quality over a long period of manufacturing time and to increase productivity. Pre-set parameters are transmitted to the CNC machine tool control and transferred to setting quantities for the process such as feed, speeds, etc. The process is exposed to additional disturbing quantities, which may influence the output. With the application of sensor systems for the monitoring of both output quantities and process quantities, it is possible to adjust the setting quantities to any change during operation of the machine tool. This measurement of a quantity deduced from the machining operation to be held on a defined level in a control loop is called adaptive control (Figure 4.3-22).

The approach to operate cutting processes in a desired condition by controlling machining parameters according to the measurement of process or output quantities can be further divided into the two following cases:

- Adaptive control constraint (ACC). The chosen process quantity is not allowed to exceed a *fixed limit* value. Even under the influence of disturbing factors such as changing workpiece material characteristics or tool wear, this limit will be approached as close as possible without exceeding the given maximum value of the setting quantity (eg, the feed). Usually the regulating quantity in such a system is a process quantity.
- Adaptive control optimization (ACO). Based on a chosen strategy the process is conducted to reach a *desired optimum value*. This optimum has to be defined, eg, it can be either the achieved workpiece quality or the machining time. This type of system can be operated with both process and output quantities. The

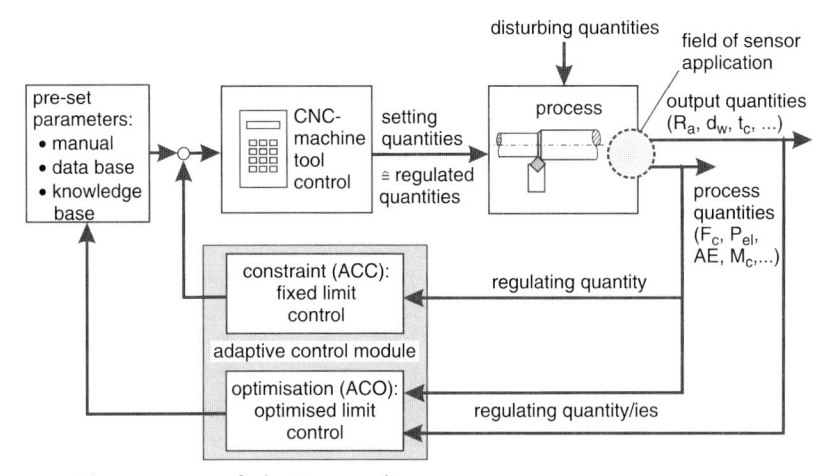

Fig. 4.3-22 Schematic set-up of adaptive control systems

most complex ACO systems not only regulate the setting quantities through the CNC unit, but will also modify the pre-set parameters if necessary. In principle, an ACO system should provide a better performance than an ACC system.

To visualize the effect of AC application for cutting processes, the advantages are shown schematically in Figure 4.3-23.

In conventional machining without AC use, the feed has to be fixed on the basis of the most difficult operation to be expected on the corresponding workpiece. The application of AC allows the setting quantities to be modified according to the actual cutting conditions. Thus in the case shown, changing depth of cut, different material properties, air gaps, and tool wear will lead to an adjustment of the feed either for turning or milling operations. From this very general approach it is evident that the major application field for AC systems in cutting is the roughing process. A change of the feed influences the surface topography and corresponding roughness values and might not be allowed in several cases. Any modification is only possible, of course, if the change in cutting conditions is detected with suitable sensors. In addition to the distinction between ACC and ACO systems, there is also a difference between geometric and technological AC systems. The first group is based on the measurement of basically macro-geometric quantities of the workpiece, whereas the latter uses process quantities to achieve the best possible productivity. All sensors which have been tested for AC systems in cutting have already been explained in previous sections.

In geometric AC systems for turning pneumatic sensors, inductive displacement probes and optical sensors were applied to measure the workpiece diameter (eg, [50, 51]). Technologically oriented AC systems are often based on cutting force measurement either by strain gages or piezoelectric dynamometers. Further-

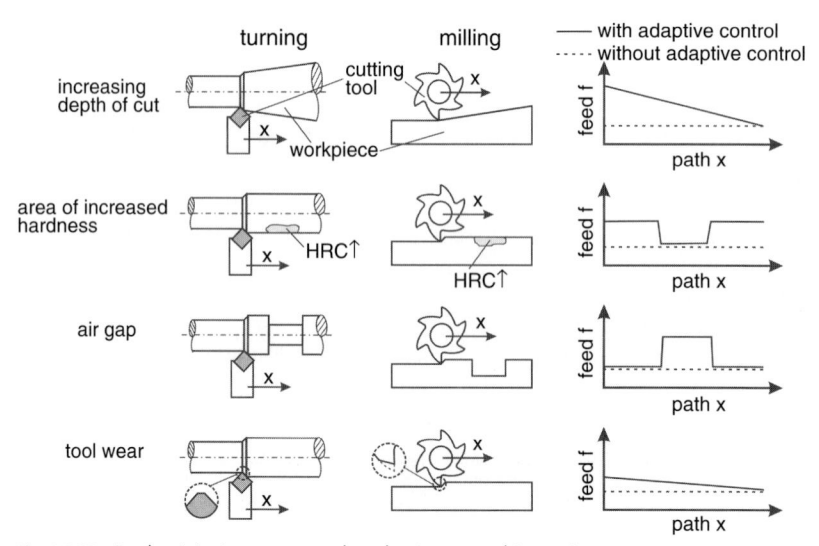

Fig. 4.3-23 Productivity improvement by adaptive control in cutting processes

more, chatter detection by force, torque, or acceleration measurement was included in AC systems. Systems for wear detection have also been tested based on pneumatic, ultrasonic or thin-film resistance sensors, but could not achieve practical importance. Motor power or current monitoring is another opportunity to establish an AC system, but special attention has to be paid to the influence of temperature. The main features of these AC systems for turning incorporate a strategy to allow idle paths at rapid feed to reduce the overall machining time, automated distribution of cutting paths, and tool life monitoring [50]. With the introduction of CAD-based machine tool programming, some of the described features lost their importance. Nowadays the determination of tool wear and breakage and chatter detection are the main monitoring tasks of sensors. Their signals are fed back to the control unit of the lathe in the sense of an AC system to reach the desired process condition and workpiece quality.

The development of AC systems for milling processes is characterized by the higher complexity of the process with rotating tools consisting of multiple cutting edges and changing chip thickness. Developments also started more than 30 years ago. Systems based on strain gages in the spindle for torque monitoring, spindle deflection measurement by displacement sensors, vibration sensors, motor power consumption or force measurement, as well as AE sensors have been applied to AC systems. In most cases the feed or feed speed is the regulated quantity; during chatter detection, a reduction of the depth of cut can also be chosen. Furthermore, in some cases the rotational speed of the spindle was adjusted. The basic features of the first-developed AC systems for milling incorporated a contact detection procedure to increase the feed during idle paths, a suppression of chatter [50], and an automated cutting path distribution [50, 51]. As already explained for turning, also for milling some of the described functions of an AC system have lost their significance. Still, tool wear and breakage recognition and chatter detection remain important topics also during milling and the described sensor solutions are being further developed together with suitable signal-processing strategies.

4.3.8
Intelligent Systems for Cutting Processes

There is no clear definition of an intelligent system. Many authors have used this term to describe an unattended machining process, where the tool cuts the workpiece while the process is monitored and controlled by the aid of suitable sensors (eg, [52]). Also, 'intelligent tools' have been presented (eg, [18]), which consist of a specific sensor as an integral part of the tool design. Furthermore, authors sometimes refer to 'intelligent machining operations', which means a model-based cutting simulation for pre-process cutting parameter optimization, followed by an adaptive controlled machining operation [53]. In the previous sections different sensor solutions and typical measurement results have been explained, where every sensor was treated as isolated. However, in many investigations the authors have chosen a multiple sensor approach to solve the desired task of monitoring the whole cutting process. Just to mention a few possibilities, force measurement was often combined with acceleration measurement or power or acoustic emission monitoring. The individual task of the single sensor is still the same, the major challenge is to combine the different signals in a system to obtain as much information as possible about the current process conditions. This parallel consideration of several sensors signals in a computer-based system with the aim of adjusting the process conditions on-line in case of determined problems might be regarded as an intelligent system in the context of this survey. One of the most important factors to promote this idea is the development of open-architecture control units for CNC machine tools. The different sensors necessary will also have a modular design to allow easy and fast integration in the machine tool control [52]. Owing to the complexity of cutting processes, techniques of artificial intelligence such as neural networks and fuzzy logic are attracting increasing attention and are already in use (eg, [35, 54]). The effort to train these artificial intelligence systems is still very great, but because of the rapid progress in computer hardware and software, further time reductions for some tasks are expected. Still one major problem is to understand clearly and define the role and monitoring task of every chosen sensor. As reported in [54], the parallel signal processing of different sensors in a neural network to estimate drill wear does not necessarily improve the monitoring accuracy. Rather, a deterioration of the correct estimation occurred, because the sensors for power, thrust force, and torque gave only redundant information for the specific task of drill wear detection. This result reveals the definite necessity to choose initially the right sensor, mounting position, and signal processing strategy before an intelligent system for cutting processes can be established. In this context, the previous sections should provide some useful information.

4.3.9
References

1 BYRNE, G., DORNFELD, D.A, INASAKI, I., KETTELER, G., KÖNIG, W., TETI, R., Ann. CIRP 44 (1995) 541–567.
2 TÖNSHOFF, H.K., Spanen – Grundlagen; Berlin: Springer, 1995.
3 DENKENA, B., Dr.-Ing. Dissertation; Universität Hannover, 1992.
4 SANTOCHI, M., DINI, G., TANUSSI, G., BEGHINI, M., Ann. CIRP 46 (1997) 49–52.
5 SPUR, G., STELZER, C., VDI-Z. 134 (12) (1992) 50–55.
6 HERNÁNDEZ CAMACHO, J., Dissertation; Universität Hannover, 1991.
7 BRANDT, D., Dissertation; Universität Hannover, 1995.
8 LEE, J.M., CHOI, D.K., CHU, C.N., Ann. CIRP 43 (1994) 81–84.
9 ZIEHBEIL, F., Dissertation; Universität Hannover, 1995.
10 KETTELER, G., Dissertation; RWTH Aachen, 1997.
11 TÖNSHOFF, H.K., KARPUSCHEWSKI, B., LAPP, C., ANDRAE, P., in: Proceedings of the International Seminar on Improving Machine Tool Performance, San Sebastian (Spain), 6–8 July 1998, Vol. I, pp. 65–76.
12 HUSEN, H., Dissertation; Universität Hannover, 1994.
13 WECK, M., FÜRBASS J.-P., Ind.-Anzeiger 109 (37) (1987) 10–15.
14 NAGAO, T., HATAMURA, Y., Ann. CIRP 37 (1988) 79–82.
15 TAKATA, S., NAKAJIMA, T., AHN, J.H., SATA, T., Ann. CIRP 36 (1987) 49–51.

16 AOYAMA, H., OHZEKI, H., MASHINE, A., TAKASHITA, J., *VDI-Ber.* **1179** (1995) 319–333.
17 CERETTI, E., KARPUSCHEWSKI, B., WINKLER, J., in: *AMST '99: Advanced Manufacturing Systems and Technology, 5th International Conference CISM, Udine* 6, 3–4 June 1999, No. 406, pp. 145–154.
18 KLOCKE, F., REHSE, M., *Prod. Eng.* **4** (1997) 65–68.
19 TLUSTY, J., TARNG, Y.S., *Ann. CIRP* **37** (1988) 45–51.
20 VARMA, A., KLINE, W.A., *VDI-Ber.* **1179** (1995) 223–233.
21 HAYASHI, S.R., THOMAS, C.E., WILDES, D.G., *Ann. CIRP* **37** (1988) 61–64.
22 WARNECKE, G., BÄHRE, D., *Prod. Eng.* **1** (2) (1994) 13–18.
23 TETI, R., in: *Proceedings of the CIRP Workshop on Tool Condition Monitoring, Paris*, 1994, Vol. III, pp. 98–137.
24 Blum, T., Inasaki, I., *J. Eng. Ind. Trans. ASME* **112** (1990) 203–211.
25 CHO, S.-S., KOMVOPOULOS, K., *J. Manuf. Sci. Eng.* **119** (1997) 238–246.
26 NAERHEIM, Y., LAN, M.-S., in: *Proceedings of the 16th NAMRC, SME, Champaign*, May 1988, pp. 240–244.
27 EMEL, E., KANNATEY-ASIBU, E., *J. Eng. Ind.* **110** (1988) 137–145.
28 DIEI, E.N., DORNFELD, D.A., *J. Eng. Ind.* May (1987) 92–99.
29 SCHMENK, M.J., *Am. Soc. Mech. Eng. PED* **14** (1984) 95–106.
30 CAROLAN, T.A., HAND, D.P., BARTON, J.S., JONES, J.D.C., WILKINSON, P., *J. Manuf. Sci. Eng.* **118** (1996) 428–433.
31 INASAKI, I., in: *Proceedings of the Machine Tool Design and Research Conference*, 1985, pp. 245–250.
32 MATHEWS, P.G., SHUNMUGAM, M.S., *J. Mater. Process. Technol.* **62** (1996) 81–86.
33 DORNFELD, D.A., LISIEWICZ, V., *Ann. CIRP* **41** (1992) 93–96.
35 SCHMITTE, F.-J., PETUELLI, G., BLUM, G., *VDI-Ber.* **1255** (1996) 253–258.
35 NEUMANN, U., *Dissertation*; TU Chemnitz, 1998.
36 UEHARA, K., *Ann. CIRP* **22** (1973) 23–24.
37 GIUSTI, F., SANTOCHI, M., in: *Proceedings of the 20th MTDR Conference, Birmingham (UK)*, September 1979, pp. 351–360
38 LEOPOLD, J., *J. Mater. Process. Technol.* **61** (1996) 34–38.
39 FAN, Y., DU, R., *J. Manuf. Sci. Eng.* **118** (1996) 664–667.
40 RYABOV, O., MORI, K., KASASHIMA, N., *Ann. CIRP* **45** (1996) 97–100.
41 WEULE, H., SPATH, D., WEIS, W., *Prod. Eng.* **2** (1) (1994) 95–98.
42 COKER, S.A., SHIN, Y.C., *Int. J. Machine Tools Manuf.* **36** (1996) 411–422.
43 EL GOMAYEL, J.I., BREGGER, K.D., *J. Eng. Ind. Trans. ASME* **108** (1986) 44–47.
44 WILKINSON, P., REUBEN, R.L., CAROLAN, T., HAND, D., BARTON, J.S., JONES, J.D.C., in: *Cond. Monitoring, International Conference, Swansea* 1994, pp. 592–610.
45 CHOUDHURY, S.K., JAIN, V.K., RAMA RAO, C.V.V., *Int. J. Machine Tools Manuf.* **39** (1999) 489–504.
46 ZHANG, G., GOPALAKRISHNAN, S., *Int. J. Machine Tools Manuf.* **36** (1996) 1137–1150.
47 WINKLER, H., *Dissertation*, Universität Hannover, 1983.
48 JAWAHIR, I.S., LUTTERFELD, C.A., *Ann. CIRP* **42** (1993) 659–693.
49 VENUVINOD, P.K., DJORDJEVICH, A., *Ann. CIRP* **45** (1996) 83–86.
50 STÖCKMANN, P., et al., *Ind.-Anzeiger* **99** (1974) 1696–1710.
51 WECK, M., *Werkzeugmaschinen, Band 3 – Automatisierung und Steuerungstechnik*; Düsseldorf: VDI-Verlag, 1978.
52 ALTINTAS, Y., MUNASINGHE, W.K., *J. Manuf. Sci. Eng.* **118** (1996) 514–521.
53 TAKATA, S., *Ann. CIRP* **42** (1993) 531–534.
54 NOORI-KHAJAVI, A., KOMANDURI, R., *Ann. CIRP* **42** (1993) 71–74.

4.4
Abrasive Processes
I. INASAKI, B. KARPUSCHEWSKI, *Keio University, Yokohama, Japan*

4.4.1
Introduction

Abrasive processes, which are mostly applied for achieving high accuracy and high quality of mechanical, electrical, and optical parts, can be divided into two categories, fixed abrasive processes and loose abrasive processes. In fixed abrasive processes, grinding wheels or honing stones are used as tools and the abrasives are held together with bonding material, also providing sufficient pores for chip removal. In loose abrasive processes, the individual grains are not fixed and are usually supplied together with a carrier medium. Among various types of abrasive processes, grinding is most widely applied in the industry. This is due to the fact that the modern grinding technology can meet the demands of not only high-precision machining but also a high material removal rate. In addition, some difficult to cut materials, such as engineering ceramics, can only be machined with abrasive processes.

4.4.2
Problems in Abrasive Processes and Need for Monitoring

The behavior of any abrasive process is very dependent on the tool performance. The grinding wheel should be properly selected and conditioned to meet the requirements on the parts. In addition, its performance may change significantly during the grinding process, which makes it difficult to predict the process behavior in advance. Conditioning of the grinding wheel is necessary before the grinding process is started. It becomes necessary also after the wheel has finished its life to restore the wheel configuration and the surface topography to the initial state. This peripheral process needs sufficient sensor systems to minimize the auxiliary machining time, to assure the desired topography, and to keep the amount of wasted abrasive material during conditioning to a minimum.

Sensor systems for a grinding process should also be capable of detecting any unexpected malfunctions in the process with high reliability so that the production of sub-standard parts can be minimized. Some major problems in the grinding process are chatter vibration, grinding burning, and surface roughness deterioration. These problems have to be identified in order to maintain the desired workpiece quality.

In addition to problem detection, another important task of the monitoring system is to provide useful information for optimizing the grinding process in terms of the total grinding time or the total grinding cost. Optimization of the process will be achieved if the degradation of the process behavior can be followed with the monitoring system. The information obtained with any sensor system during the grinding process can be also used for establishing databases as part of intelligent systems.

4.4.3
Sensors for Process Quantities

As for all manufacturing processes, it is most desirable to measure the quantities of interest as directly and as close to their origin as possible. Every abrasive process is determined by a large number of input quantities, which may all have an influence on the process quantities and the resulting quantities. Brinksmeier proposed a systematic approach to distinguish between different types of quantities to describe a manufacturing process precisely [1]. The hardware components used such as machine tool, workpiece, tools, type of coolant, etc., are described as system quantities. The settings are further separated into primary and secondary quantities; the former comprise all relevant input variables of the control which describe the movement between tool and workpiece whereas the latter do not have an influence on the relative motion for material removal, such as dressing conditions or coolant flow rate. In addition, disturbing quantities also have to be taken into account, often leading to severe problems concerning the demand for constant high quality of the manufactured product. All these input quantities have an effect on the process itself, hence the mechanical and thermal system transfer behavior is influenced. Owing to the interaction of tool and workpiece, the material removal is initiated and the zone of contact is generated. Only during this interaction process can quantities be detected. The measurement of these by use of adequate sensors is the subject of this section.

The most common sensors to be used in either industrial or research environments are force, power, and acoustic emission (AE) sensors [2]. Figure 4.4-1 shows the set-up for the most popular integration of sensor systems in either surface or outer diameter grinding.

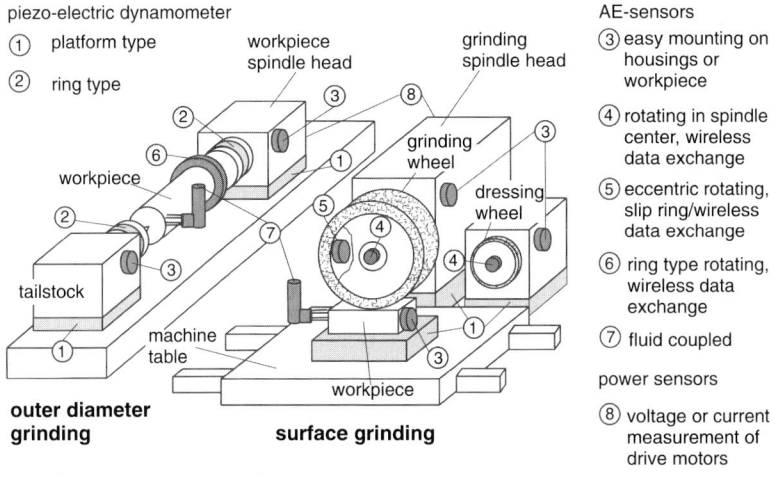

Fig. 4.4-1 Possible positions of force, AE, and power sensors in grinding

4.4.3.1 Force Sensors

The first attempts to measure grinding forces go back to the early 1950s and were based on strain gages. Although the system performed well to achieve substantial data on grinding, the most important disadvantage of this approach was the significant reduction in the total stiffness during grinding. Hence research was done to develop alternative systems. With the introduction of piezoelectric quartz force transducers, a satisfactory solution was found. In Figure 4.4-1, different locations for these platforms in grinding are shown. In surface grinding most often the platform is mounted on the machine table to carry the workpiece. In inner (ID) or outer (OD) diameter grinding this solution is not available owing to the rotation of the workpiece. In this case either the whole grinding spindle head is mounted on a platform or the workpiece spindle head and sometimes also the tailstock are put on a platform.

Figure 4.4-2 shows an example of a force measurement with the grinding spindle head on a platform during ID plunge grinding. In this case the results are used to investigate the influence of different coolant supply systems while grinding case hardened steel. The force measurements make it clear that it is not possible to grind without coolant using the chosen grinding wheel owing to wheel loading and high normal and tangential forces. However, it is also seen that there is a high potential for minimum quantity lubrication (MQL) with very constant force levels over the registered related material removal [3]. For OD grinding it is also possible to use ring-type piezoelectric dynamometers. With each ring again all three perpendicular force components can be measured; they are mounted under preload behind the non-rotating center points. To complete possible mounting positions of dynamometers in grinding machines, the dressing forces can also be monitored by the use of piezoelectric dynamometers, eg, the spindle head of rotating dressers can be mounted on a platform. Besides these general solutions, many special set-ups have been used for non-conventional grinding processes

workpiece:
16 MnCr 5, 62 HRC
d_w = 40 mm, v_{ft} = 1 m/s
Q'_w = 1 mm³/mms
grinding wheel:
5SG100LVS (MK-Al$_2$O$_3$)

d_s = 30 mm, v_c = 40 m/s
dressing conditions:
diamond cup wheel D301
U_d = 20, q_d = 0,6
a_{ed} = 3 µm, a_{pd} = 1 mm

conventional cooling:
mineral oil, Q = 11 l/min
MQL (minimum quantity lubrication):
ester, Q = 0, 4 ml/min

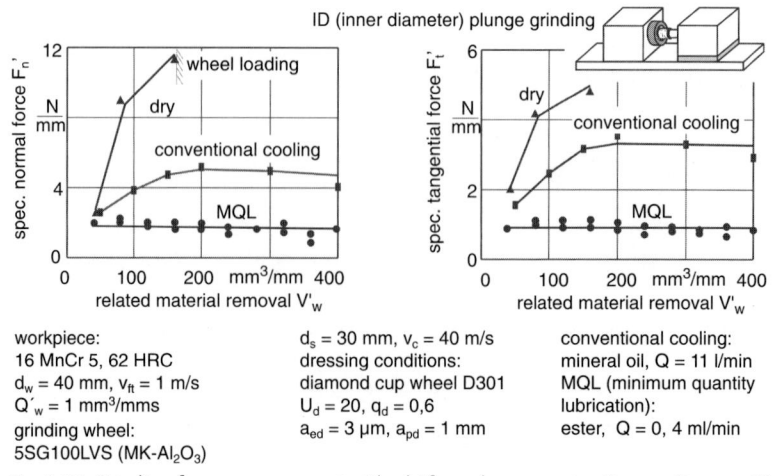

Fig. 4.4-2 Grinding force measurement with platform dynamometer. Source: Brunner [3]

such as ID cut-off grinding of silicon wafers and ID grinding of long small bores with rod-shaped tools.

As already stated for cutting, also for abrasive processes the application of dynamometers can be regarded as state of the art. The problems of high investment and missing overload protection are also valid.

However, wire strain gages are also still in use. For example, the force measurement in a face grinding process of inserts is not possible with a piezoelectric system owing to limited space. In this case an integration of wire strain gages with a telemetric wireless data exchange was successfully applied [4].

4.4.3.2 Power Measurement

As explained for cutting in Section 4.3.3.3, the measurement of power consumption of a spindle drive can be regarded as technically simple. Also for abrasive processes the evidence is definitely limited. The amount of power used for the material removal process is always only a fraction of the total power consumption. Nevertheless, power monitoring of the main spindle is widely used in industrial applications by defining specific thresholds to avoid any overload of the whole machine tool due to bearing wear or any errors from operators or automatic handling systems. However, there are also attempts to use the power signal of the main spindle in combination with the power consumption of the workpiece spindle to avoid grinding burn. This approach is further discussed in Section 3.3.

4.4.3.3 Acceleration Sensors

In Section 4.3.3 the difficulty of separating acceleration sensors from AE sensors has already been mentioned. In abrasive processes the major application for acceleration sensors is related to balancing systems for grinding wheels. Especially large grinding wheels without a metal core may have significant imbalance at the circumference. With the aid of acceleration sensors the vibrations generated by this imbalance are monitored during the rotation of the grinding wheel at cutting speed. Different systems are in use to compensate this imbalance, eg, hydro compensators using coolant to fill different chambers in the flange or mechanical balancing heads, which move small weights to specific positions. Although these systems are generally activated at the beginning of a shift, they are able to monitor the change of the balance state during grinding and can continuously compensate the imbalance.

4.4.3.4 Acoustic Emission Systems

Systems based on AE must be regarded as very attractive for abrasive processes. An introduction to the AE technique and a brief explanation of the physical background is given in Section 3.3.3.4. Figure 4.4-1 shows the possible mounting positions for AE sensors on different components of a grinding machine. Either the spindle drive units, the tool and grinding wheel, or the workpiece can be equipped with a sensor. In addition, fluid-coupled sensors are also in use without any direct mechanical contact to one of the mentioned components. As pointed out before, the time domain course of the root mean square value $U_{AE,RMS}$ is one of the most important quantities for characterizing the process state. In Figure 4.4-3 as an example the correlation between the surface roughness of a ground workpiece and the root mean square value of the AE signal is shown [5].

A three-step OD plunge grinding process with a conventional corundum grinding wheel was monitored. It is obvious that for a dressing overlap of $U_d=2$ the generated coarse grinding wheel topography is leading to a high initial surface roughness of $R_z=5$ μm. Owing to continuous wear of the grains, the roughness even increases during the material removal. For the finer dressing overlap of $U_d=10$ a smaller initial roughness with a significant increase can be seen for the first parts followed by a decreasing tendency. This tendency of the surface roughness is also represented by the AE signal. Higher dressing overlaps lead to more cutting edges, thus resulting in a higher AE activity. The sensitivity of the fine finishing AE signal is higher, because the final roughness is mainly determined in this process step. Meyen [5] has shown in many other tests that monitoring of the grinding process with AE is possible.

In recent years, research has been conducted on high-resolution measurement of single cutting edges in grinding. The root mean square value must be regarded as an average statistical quantity, usually often low-pass filtered and thus not really suitable to reveal short transient effects such as single grit contacts. Webster et al. observed burst-type AE signals of single grits in spark-in and spark-out stages of different grinding operations by analyzing the raw AE signals with a special high-speed massive storage data acquisition system [6].

In addition to these time-domain analyses, the AE signal can also be investigated in the frequency domain. Different effects such as wear or chatter vibration have different influences on the frequency spectrum, so it should be possible to

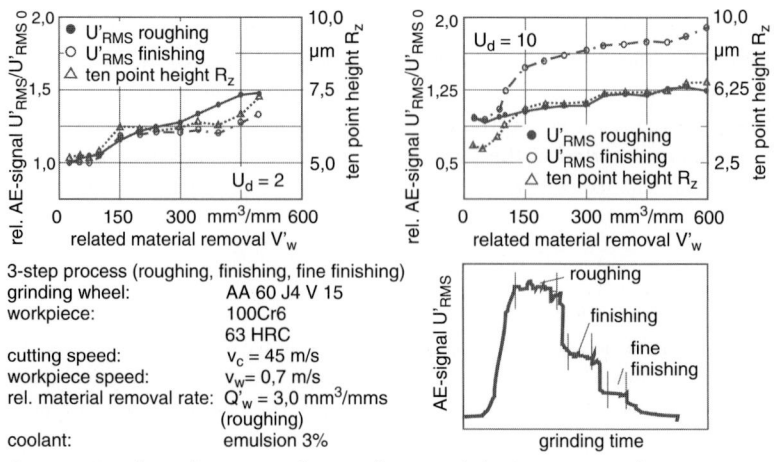

Fig. 4.4-3 Correlation between surface roughness and the AE r.m.s. signal. Source: Meyen [5]

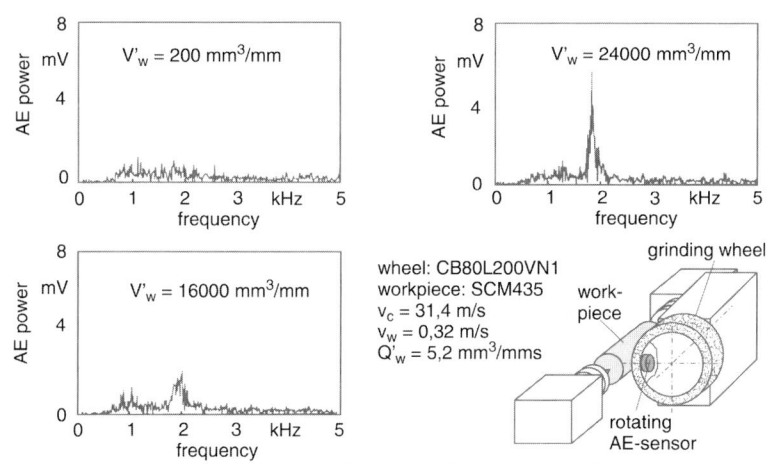

Fig. 4.4-4 Acoustic emission frequency analysis for chatter detection in grinding. Source: Wakuda et al. [7]

use the frequency analysis to separate these effects. Figure 4.4-4 shows the results of a frequency analysis of the AE signal in OD plunge grinding with a vitreous bond CBN grinding wheel [7]. As a special feature the AE-sensor is mounted the grinding wheel core and transfers the signals via a slip ring to the evaluation computer, so both grinding and dressing operations can be monitored. The results reveal that no significant peak can be seen after dressing and first grinding tests. Only after a long grinding time do specific frequency components emerge from the spectrum which show a constantly rising power during the continuation of the test. The detected frequency is identical with the chatter frequency, which could be determined by additional measurements. The AE-signals were used as input data for a neural network to identify automatically the occurrence of any chatter vibrations in grinding [7].

Owing to the general advantages of AE sensors and their variety, almost any process with bond abrasives has already been investigated with the use of AE. Surface grinding, ID and OD grinding, centerless grinding, flexible disk grinding, gear profile grinding, ID cut-off grinding of silicon wafers, honing, and grinding with bond abrasives on tape or film type substrates have all been subjects of AE research.

4.4.3.5 Temperature Sensors

In any abrasive process, mechanical, thermal, and even chemical effects are usually superimposed in the zone of contact. Grinding in any variation generates a significant amount of heat, which may cause a deterioration of the dimensional accuracy of the workpiece, an undesirable change in the surface integrity state, or increased wear of the tool. In Section 3.3.3.3 some sensors for temperature measurement have already been explained. Figure 4.4-5 shows the most popular temperature mea-

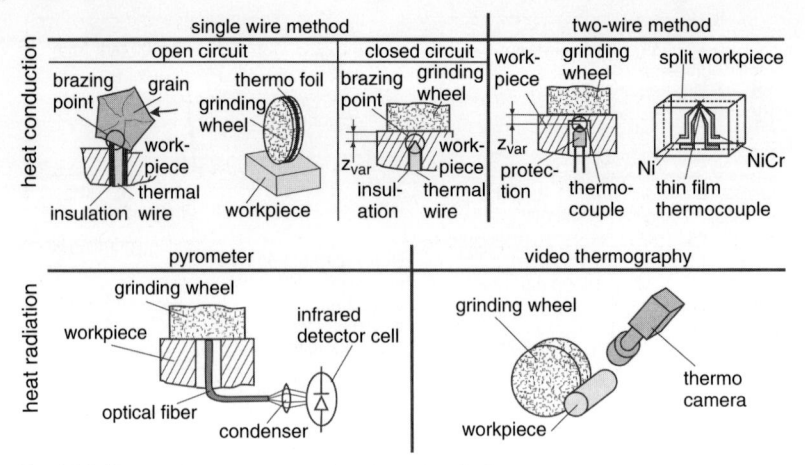

Fig. 4.4-5 Temperature measurement systems in grinding

surement devices. The preferred method for temperature measurement in grinding is the use of thermocouples. The second metal in a thermocouple can be the workpiece material itself; this set-up is called the single-wire method.

A further distinction is made according to the type of insulation. Permanent insulation of the thin wire or foil against the workpiece by use of sheet mica is known as open circuit. The insulation is interrupted by the individual abrasive grains, hence measurements can be repeated or process conditions varied until the wire is worn or damaged. Many workers (eg, [8]) have used this set-up. Also the grinding wheel can be equipped with the thin wire or a thermo foil, if the insulation properties of abrasive and bond material are adequate. In the closed-circuit type, permanent contact of the thermal wire and the workpiece by welding or brazing is achieved. The most important advantage of this method is the possibility of measuring temperatures at different distances from the zone of contact until the thermocouple is finally exposed to the surface. For the single-wire method it is necessary to calibrate the thermocouple for each different workpiece material. This disadvantage is overcome by the use of standardized thermocouples, where the two different materials are assembled in a ready-for-use system with sufficient protection. A large variety of sizes and material combinations are available for a wide range of technical purposes. With this two-wire method it is again possible to measure the temperatures at different distances from the zone of contact. This approach can be regarded as most popular for temperature measurement in grinding. A special variation of this two-wire method is the use of thin-film thermocouples [8, 9], (see also Section 3.3.3.3). The advantage of this method is an extremely small contact point to resolve temperatures in a very small area and the possibility of measuring a temperature profile for every single test depending on the number of evaporated thermocouples in simultaneous use.

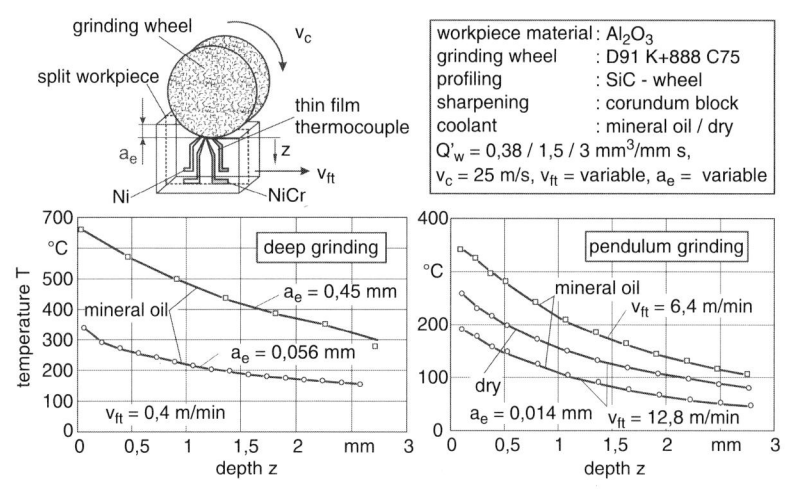

Fig. 4.4-6 Grinding temperature measurement with thin-film thermocouples. Source: Lierse [9]

In Figure 4.4-6, temperature measurements during grinding of Al_2O_3 ceramic with a resin-bonded diamond grinding wheel using these thin-film thermocouples are shown [9]. Obviously the setting quantities have a significant influence on the generation of heat in the zone of contact. Especially the heat penetration time is of major importance. In deep grinding with a very low tangential feed speed, high temperatures are registered, whereas higher tangential feed speeds in pendulum grinding lead to a significant temperature reduction. As expected, the avoidance of coolant leads to higher temperatures compared with the use of mineral oil.

However, in any case for either single- or two-wire methods the major disadvantage is the great effort needed to carry out these measurements. Owing to the necessity to install the thermocouple as close as possible to the zone of contact, it is always a technique where either the grinding wheel or workpiece have to be specially prepared. Hence all these methods are only used in fundamental research; industrial use for monitoring is not possible owing to the partial destruction of major components.

In addition to these heat conduction-based methods, the second group of usable techniques is related to heat radiation. Infrared radiation techniques have been used to investigate the temperature of grinding wheel and chips. By the use of a special infrared radiation pyrometer, with the radiation transmitted through an optical fiber, it is even possible to measure the temperature of working grains of the grinding wheel just after cutting [10]. Also the use of coolant was possible and could be evaluated. In any case, these radiation-based systems need careful calibration, taking into account the properties of the material to be investigated, the optical fiber characteristics, and the sensitivity of the detector cell. However, again, for most of the investigations preparation of the workpiece is necessary, as shown in Figure 4.4-5 (bottom left).

The second heat radiation-based method is thermography. For this type of measurement, the use of coolants is always a severe problem, because the initial radiation generated in the zone of contact is significantly reduced in the mist or direct flow of the coolant until it is detected in the camera. Thus the major application of this technique was limited to dry machining. Brunner was able to use a high-speed video thermography system for OD grinding of steel to investigate the potential of dry or MQL grinding [3].

All the mentioned temperature sensors can also be distinguished with regard to their measurement area. Video thermography is a technique to obtain average information about the conditions in the contact zone. For this reason it might be called macroscopic temperature measurement. Pyrometers can either give average information, but as Ueda and others have shown, single-grain measurements can also be conducted depending on the diameter of the optical fibers. Concerning the use of thermocouples, the situation is more difficult. Standard thermocouples and the closed-circuit single-wire method are used to measure at a specific distance from the zone of contact. Thus the average temperature at this point can be detected; the measurement spot might be extremely small, especially in the case of evaporated thin-film thermocouples. This might be called microscopic temperature measurement, but single-grain contact detection is not possible. The open-circuit method with the thin thermal wire, which is exposed to the surface, is the only real microscopic temperature measurement technique, because in this case single grains generate the signal. However, the response time of this system is significant, so it must be established critically whether all single contacts can be registered.

4.4.4
Sensors for the Grinding Wheel

The grinding wheel state is of substantial importance for the achievable result. The tool condition can be described by the characteristics of the grains. Wear can lead to flattening, breakage, and even pullout of whole grains. Moreover, the number of cutting edges and the ratio of active to passive grains are of importance. Also the bond of the grinding wheel is subject to wear.

Owing to its hardness and composition, it influences significantly the described variations of the grains. In any case, wheel loading generates negative effects due to insufficient chip removal and coolant supply. All these effects can be summarized as grinding wheel topography, which changes during the tool life between two dressing cycles. As a resulting effect, the size of the grinding wheel and its diameter are reduced. In most cases dressing cycles have to be carried out without any information about the actual wheel wear. Commonly, grinding wheels are dressed without reaching their end of tool life in order to prevent workpiece damage, eg, workpiece burn. Figure 4.4-7 gives an overview of different geometric quality features concerning the tool life of grinding wheels. As a rule the different types of wheel wear are divided into macroscopic and microscopic features. Many attempts have been made to describe the surface topography of a grinding wheel and to correlate the quantities

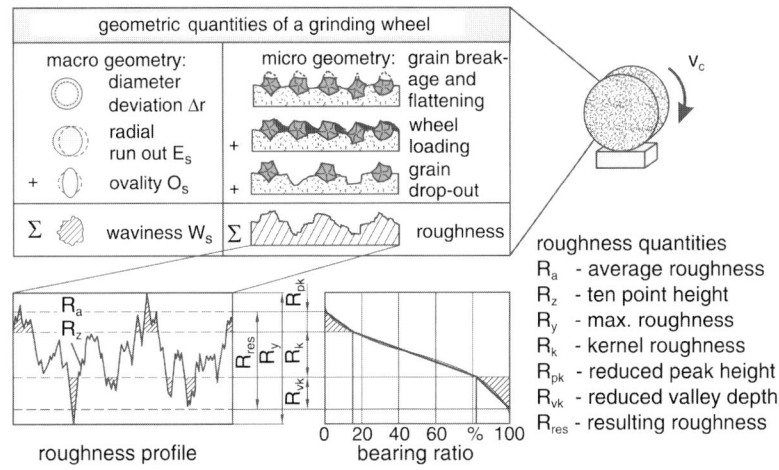

Fig. 4.4-7 Geometric quantities of a grinding wheel

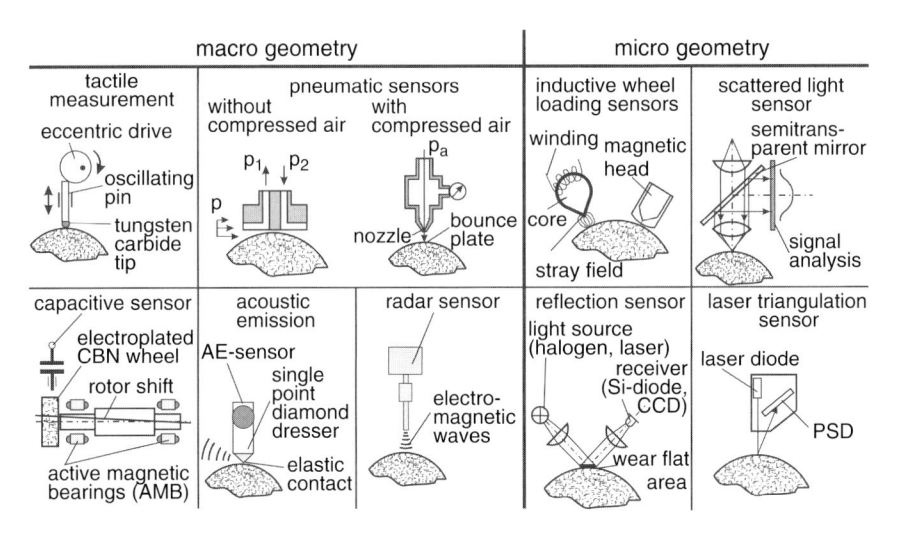

Fig. 4.4-8 Sensors for grinding wheel topography measurement

with the result on the workpiece. All methods that need a stationary object in a laboratory surrounding, which means that the grinding wheel is not rotating and even dismounted, will not be discussed. Attention is focused on dynamic methods, which are capable of being used in the grinding machine during the rotation of the tool. If only the number of active cutting edges is of interest, some already introduced techniques can be used. Either piezoelectric dynamometers or thermocouple methods have been used to determine the number of active cutting edges.

In Figure 4.4-8 other methods are introduced that are suitable for dynamic measurement of the grinding wheel. Most of the systems are not able to detect all micro- and macro-geometric quantities, and can only be used for special purposes.

4.4.4.1 Sensors for Macro-geometric Quantities

The majority of sensors are capable of measuring the macro-geometric features. Any kind of mechanical contact of a sensor with the rotating tool causes serious problems, because the abrasives always tend to grind the material of the touching element. Only by realizing short touching pulses with small touching forces and by using a very hard tip material such as tungsten carbide is it possible to achieve satisfactory results. For instance, such a system with an eccentric drive to realize the oscillation of the pin to measure the radial wheel wear at cutting speeds of up to 35 m/s has been used [11]. However, coolant supply and corundum grinding wheels with a porous vitreous bond caused severe problems. In any case, these tactile-based methods on rotating tools are only suitable for macro-geometric measurement and are limited to a few studies.

Another group of sensors for the measurement of grinding wheels is based on pneumatic systems. Although this method is in principle also not able to detect the micro-geometric features of a grinding wheel owing to the nozzle diameter of 1 mm or more, they are important for determining the macro-geometry. Systems with compressed air supply and those without have to be distinguished. The latter are characterized by measurement of the airflow around the rotating grinding wheel. The results obtained reveal a dependence of the airflow on the distance of the sensor from the surface, on the circumferential speed, and to a small extent on the topography of the grinding wheel. The method with a compressed air supply is based on the nozzle-bounce plate principle, with the grinding wheel being the bounce plate. These systems are capable of measuring the distance changes and radial wear with a resolution of 0.2 µm. Especially this feature and the comparatively easy set-up and moderate costs are the main reasons why pneumatic sensors have already found acceptance in industrial application.

Another possibility of registering the macro-geometry of a grinding wheel has been reported [12]. In high-speed ID grinding with CBN wheels, a spindle with active magnetic bearings (AMBs) was used to achieve the necessary circumferential speed of 200 m/s with small-diameter wheels. These spindles have the opportunity to shift the rotor from rotation around the geometric center axis to the main axis of inertia to compensate any imbalance. Especially if electroplated CBN wheels are used without the possibility of dressing, it is necessary to use balancing planes. To measure the runout of these grinding wheels on the abrasive layer at very high circumferential speeds, capacitive sensors have shown the best performance.

The AE signal can also be used to determine the macro-geometry of the grinding wheel. In [13] a system was proposed consisting of a single-point diamond dresser equipped with an AE sensor to detect exactly the position of the grinding wheel surface. Because AE signals can be obtained without contact of the dresser and the wheel due to turbulence, in total three different contact conditions can be distinguished, non-contact, elastic contact, and brittle contact. It is proposed to use the AE level of the elastic contact range to monitor the exact position of the grinding wheel. The only disadvantage is the current limitation to a single-point dresser. To overcome this demerit, an extension to rotating dressing tools is the subject of current research [14].

Another principle used to determine radial wheel wear is based on a miniature radar sensor [15]. The usual radar technique is known from speed and traffic control applications with a maximum accuracy in the centimeter range. The sensor used for grinding works on an interferometric principle. With an emitting frequency of 94 GHz and a wavelength of $\lambda = 3.18$ mm, this sensor has a measuring range of 1 mm and a resolution of 1 µm. The main advantages are the robustness against any dust, mist, or coolant particles and the possibility of measuring on any solid surface. The sensor has been used in surface grinding of turbine blades with continuous dressing (CD). A control loop was established to detect and control the radial wear of the grinding wheel taking into account the infeed of the dressing wheel.

4.4.4.2 **Sensors for Micro-geometric Quantities**
In addition to these systems for macro-geometric features, other sensors are able to give information about the micro-geometry. The loading of a grinding wheel with conductive metallic particles as a special type of micro-geometric wear can be detected by using sensors based on inductive phenomena. The sensor consists of a high permeability core and a winding. It is positioned at a short distance from the surface. The metallic particles generate a change in the impedance, which can be further processed to determine the state of wheel loading. A conventional magnetic tape recorder head may also be used to detect the presence and relative size of ferrous particles in the surface layer of a grinding wheel. As only this special type of wear in grinding of metallic materials can be detected, these sensors have not achieved practical application.

The mentioned limitations of all the so-far introduced techniques turn the attention towards optical methods. These seem to be very promising because of their frequency range and independence of the surface material. A scattered light sensor was used to determine the reflected light from the grinding wheel surface by using charge-coupled device (CCD) arrays. The first attempts at an optical-based measurement of the topography at cutting speeds were reported in [16]. An opto-electronic sensor with a fast Si photodiode as receiver and either a xenon vapor lamp or halogen light source was used to measure the pulses of reflected light on so-called wear flat areas. Tests have shown that the number of pulses changes during grinding, hence a possible monitoring of the wear state was proposed. However, hardware limitations, especially problems with the light sources, did not lead to further success at that time. Gotou and Touge took up the same principle again [17], keeping the Si photodiode as receiver but this time using a laser source with 670 nm wavelength and a personal computer for control. Grinding wheels in wet-type grinding at 30 m/s could be measured. Again, it was stated that the wear flat areas are registered by the output signal and that these areas change during grinding.

The optical method with the highest technical level so far is based on laser triangulation. The measurement principle and results of micro-geometric characterization of the grinding wheel surface to determine surface integrity changes are

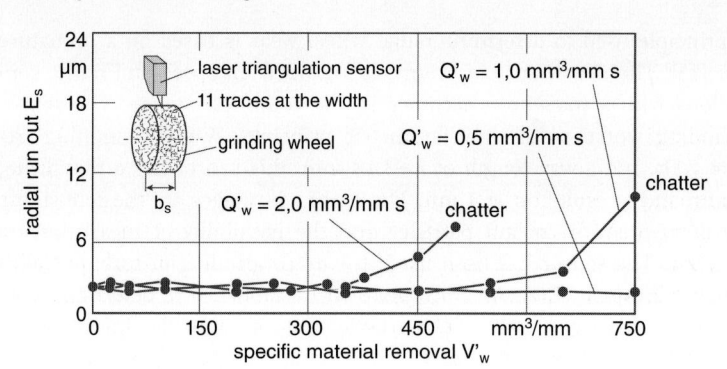

Fig. 4.4-9 Optical macro-geometric grinding wheel topography measurement

grinding conditions:
grinding wheel: EKW 80K5 V62
q = 60
v_c = 30 m/s = v_{mea}
spark out time: 10 s
multi point diamond dresser
workpiece: 100 Cr 6
ball bearing steel

given in Section 3.3.4. In the following, results of macro-geometric measurements will be presented.

As mentioned, no practical limitations exist for the determination of macro-geometric quantities such as radial runout, and the maximum surface speed may even exceed 300 m/s [18]. Figure 4.4-9 shows the result of an investigation of OD plunge grinding of ball-bearing steel with a corundum grinding wheel. Using three different material removal rates, the change of the radial runout as a function of the material removal at 30 m/s was plotted. For the smallest material removal rate no change is detectable from the initial value after dressing. However, for increasing material removal rates of Q'_w=1.0 and 2.0 mm³/mm s the radial runout rises after a specific material removal. In the latter cases the increasing radial runout leads to chatter vibrations with visible marks on the workpiece surface. Obviously the system is capable of detecting significant macro-geometric changes due to wear of the grinding wheel. The limitations of the system regarding the micro-geometric characterization have been discussed in Section 3.3.4.

The examples presented of grinding wheel sensors reveal that the majority of systems are related to macro-geometric features. However, many attempts have been made to establish especially optical systems for the measurement of micro-geometric quantities. The overall limitation for these techniques will always be the rough conditions in the working space of a grinding machine with coolant and process residues in the direct contact with the object to be measured. In many cases it is therefore preferable to measure directly the manufactured workpiece itself.

4.4.5
Workpiece Sensors

Two essential quality aspects determine the result of an abrasive process on the workpiece. On the one hand the geometric quality demands have to be fulfilled. These are dimension, shape, and waviness as essential macro-geometric quantities. The roughness condition is the main micro-geometric quantity. However, increasing attention is also paid to the surface integrity state of a ground workpiece owing to its significant influence on the functional behavior. The physical properties are characterized by the change in hardness and residual stresses on the surface and in sub-surface layers, by changes in the structure, and the likely occurrence of cracks (see Section 3.3). All geometric quantities can be determined by using laboratory reference measuring devices. For macro-geometric properties any kind of contact system can be used, eg, 3D coordinate measuring machines, contour stylus instruments, or gages. Roughness measurement is usually performed with stylus instruments giving standardized values, but optical systems are also applied in some cases. Methods to determine physical quality characteristics are mentioned in Section 3.3.

4.4.5.1 Contact-based Workpiece Sensors for Macro-geometry

The determination of macro-geometric properties of workpieces during manufacture is the most common application of sensors in abrasive processes, especially grinding. For decades contact sensors have been in use to determine the dimensional changes of workpieces during manufacturing. A wide variety of in-process gages for all kinds of operation are available. In ID or OD grinding the measuring systems can either be comparator or absolute measuring heads, with the capability of automatic adjustment to different part diameters. The contact tips are usually made of tungsten carbide, combining the advantages of wear resistance, moderate costs, and adequate frictional behavior. The repeatability is in the region of 0.1 µm [19]. Internal diameters can be gaged starting from 3 mm. If constant access to the dimension of interest during grinding is possible, these gages are often used as signal sources for adaptive control (AC) systems (see also Section 4.4-8). The conventional technique for measuring round parts rotating around their rotational axis can be regarded state of the art. The majority of automatically operating grinding machines are equipped with these systems. In the survey of contact sensors for workpiece macro-geometry in Figure 4.4-10 (top left), a more complex measurement set-up is shown. Owing to the development of new drives and control systems for grinding machines, continuous path-controlled grinding of crankshafts has now become possible [20]. The crankshaft is clamped only once in the main axis of the journals. For machining the pins the grinding wheel moves back and forth during rotation of the crankshaft around the main axis to generate a cylindrical surface on the pin. An in-process measurement device for the pin diameter has to follow this movement. A first prototype system was installed in a crankshaft grinding machine. The gage is mounted on the grinding wheel head and

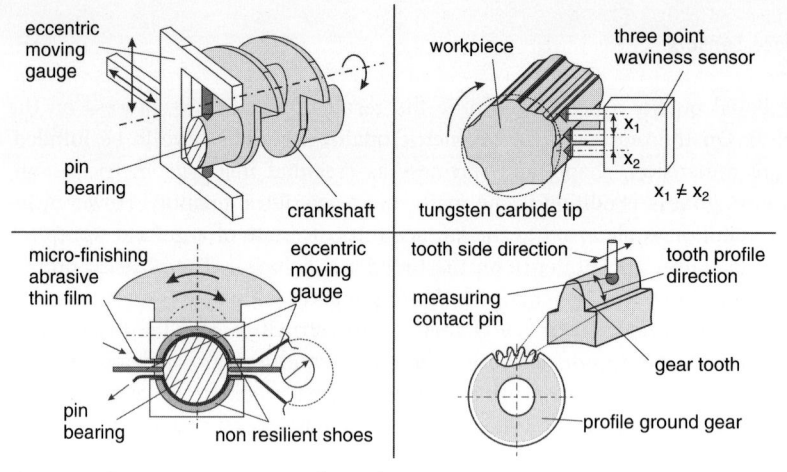

Fig. 4.4-10 Contact sensor systems for workpiece macro-geometry

moves back and forth together with the grinding wheel. A swivel joint effects the height balancing. The same problem of measurement during eccentric movement of a crankshaft pin occurs in the micro-finishing process (Figure 4.4-10, bottom left). This last process in the production chain using single-layer abrasives bond to a thin plastic belt is applied to give the pins and journals the desired final micro-geometry concerning roughness, bearing ratio, and crowning. The abrasive film is automatically indexed before each cycle and pressed by hard, non-resilient, and exactly formed shoes to the workpiece surface at specific controlled pressure. The crankshaft rotates and oscillates for a specific cycle and drags single arms with shoes, belt supply, and measuring gage for each pin and journal at the same time. Size control is realized by a moving gage with contacting pins, which allow stopping of the micro-finishing process on each bearing individually, when the final dimension is reached. A machine tool with this in-process moving size control sensor system is already available.

The detection of waviness on the circumference of rotating symmetrical parts during grinding is more complex owing to the demand for a significantly higher scanning frequency. Foth has developed a system with three contacting pins at non-constant distances to detect the development of waviness on workpieces during grinding as a result of, eg, regenerative chatter [19] (Figure 4.4-10, top right). Only by using this set-up was it possible to identify the real workpiece shape, taking into account the vibration of the workpiece center during rotation. The signal of the waviness sensor was fed back to the control unit of the machine tool. If increasing waviness was determined during grinding, the speed ratio between the rotating grinding wheel and workpiece was changed to suppress regenerative effects. Although the system performance was satisfactory and could meet all industrial demands concerning robustness, it was only used to confirm theoretical simulations. The knowledge gained can be directly applied to grinding machine controls to avoid regenerative chatter, hence waviness sensors are not really needed.

The last example of contact-based macro-geometric measurement in a machine tool is related to gear grinding (Figure 4.4-10, bottom right). Especially for manufacturing of small bath sizes or single components of high value, it is essential to fulfil the 'first part good part' philosophy. For these reasons several gear grinding machine tool builders have decided to integrate an intelligent measuring head in their machines to be able to measure the characteristic quantities of a gear, eg, flank modification, pitch, or root fillet. Usually a measurement is done after rough grinding, before the grinding wheel is changed or redressed for the finish operation. Sometimes also the initial state before grinding is checked to compensate for large deviations resulting from distortions due to heat treatment. Of course, the measurement can only be done if the manufacturing process is interrupted. However, the main advantage is still a significant saving of time. Any removal of the part from the grinding machine tool for checking on an additional gear measuring machine will take a longer time. Also the problem of precision losses due to rechucking is not valid, because the workpiece is rough machined, measured, and finished in the same set-up. These arguments are generally true for any kind of high-value parts with small bath sizes and complex grinding operations. Hence it is not surprising that also in the field of aircraft engine manufacturing new radial grinding machines are equipped with the same kind of touch probe system in the working space. Geometric quality data are acquired on the machine tool before the next grinding operation in the same chuck position is started [21]. Nevertheless, the use of a measuring head in a complex gear or turbine blade grinding machine is not a pure sensor application. The measurement is only possible in auxiliary process time, but between succeeding process steps. It must be stated as a borderline case, but should be included because of the high technical level and industrial relevance.

4.4.5.2 Contact-based Workpiece Sensors for Micro-geometry

The determination of micro-geometric quantities on a moving workpiece by using contact sensor systems is a challenging task. Permanent contact of any stylus with the surface is not possible, because the dynamic demands are much too high. Only intermittent contacts can be used to generate a signal, which should be proportional to the roughness. Saljé has introduced a sensor based on a damped mass spring element [22]. The surface of the fast-moving workpiece stimulates self-oscillations of the sensing element, which are correlated with the roughness.

The system was improved and modified in the following years, and a set-up with parallel springs was successfully applied to the honing process [23], (Figure 4.4-11). The sensor was integrated in the honing tool and the pre-amplified signal was transmitted to the evaluation unit via slip ring contact. Figure 4.4-11 shows the result of the calibration of this sensor system with conventional stylus roughness measurements. A linear correlation in the range of interest of $R_z = 2\text{--}20$ μm was found.

Rotating roughness sensors for OD grinding have also been tested, but different limitations concerning diameter and width of the workpiece did not allow

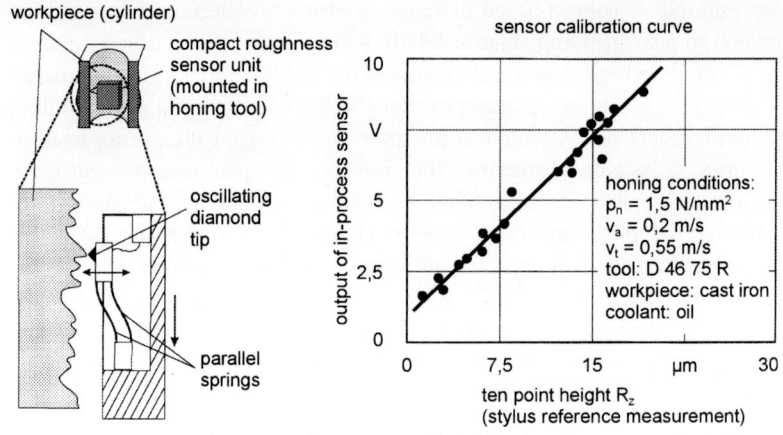

Fig. 4.4-11 Contact workpiece roughness sensor for ID honing. Source: von See [22]

practical application. A second important problem is related to the measuring direction of these in-process sensors. Whether the sensor was combined with the tool in the case of honing or fed towards the workpiece by auxiliary systems, the measuring direction was always in the direction of the abrasive process. Any stylus-type reference measurement is usually done perpendicular to the grinding or honing direction. In the parallel direction the diamond tip is likely to stay in just one groove and then suddenly jump out to the next one. Hence a parallel measurement does not give substantial information on the roughness state and is usually avoided. Although attempts have been made with additional axial feed of the sensor to generate a scroll-type movement on the surface [22], the idea of contacting the surface for roughness measurement did not lead to industrial success.

4.4.5.3 Contact-based Workpiece Sensors for Surface Integrity
The range of contact sensors on workpieces is completed with systems related to surface integrity measurement. A description of the available techniques is given in Section 3.3.5.

4.4.5.4 Non-contact-based Workpiece Sensors
All the mentioned restrictions of contact sensor systems on the workpiece surface gave a significant push to develop non-contact sensors. As for grinding wheels, again optical systems seem to have a high potential. In Figure 4.4-12 different optical systems and two other non-contacting sensor principles are introduced.

As a very fast optical system for measuring macro-geometric quantities, a laser-scanner is shown. The scanner transmitter contains primarily the beam-emitting He-Ne laser, a rotating polygonal mirror and a collimating lens for paralleling the diffused laser beam. The set-up of the scanner receiver contains a collective lens and a photodiode. The electronic evaluation unit counts the time when the photo-

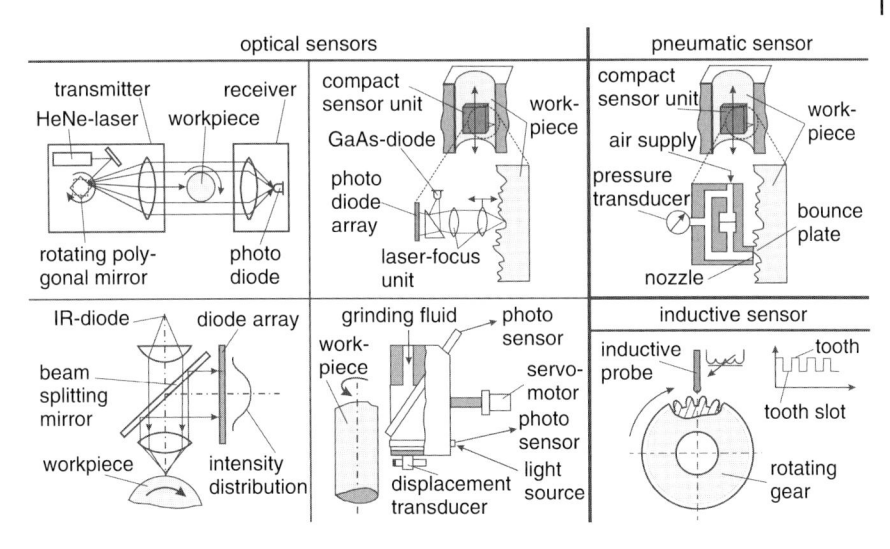

Fig. 4.4-12 Non-contact sensor systems for workpiece quality characterization

diode is covered by the shadow of the object. The diameter is a function of the speed of the polygonal mirror and the time during which the laser beam does not reach the covered photodiode. Conicity can be evaluated by an axial shifting of the workpiece. In principle this optical measurement cannot be performed during the application of coolant. During grinding the system can be protected by air barriers and mechanical shutters. Laser-scanners were first installed in grinding machines to measure the thermal displacement of machine tool components or to determine the profile accuracy of the dressed grinding wheel. For a detailed workpiece characterization, a set-up with a laser-scanner outside of the working space of the grinding machine was preferred. In [24] the layout and realization of a flexible measurement cell incorporating a laser-scanner for the determination of macro-geometric properties was introduced. The system is able to measure automatically the desired quantities within the grinding time, and the information can be fed back to the grinding machine control unit.

For the determination of macro- and micro-geometric quantities a different optical system has to be applied. The basis of a scattered light sensor for the measurement of both roughness and waviness is the angular deflection of nearly normal incident rays. The set-up of a scattered light sensor is shown in Figure 4.4-12 (bottom left). A beam-splitting mirror guides the reflected light to an array of diodes. This array is able to record the distribution only in one optical flat. The alignment of the sensor is therefore of essential importance. To obtain information about the circumferential waviness and roughness, the array has to be perpendicular to the rotation axis of the workpiece. The transverse roughness according to the stylus testing is measurable with a 90° rotation of the sensor. A commercially available system was introduced in the 1980s [25] and used in a wide range of tests. The optical roughness measurement quantity of this system is

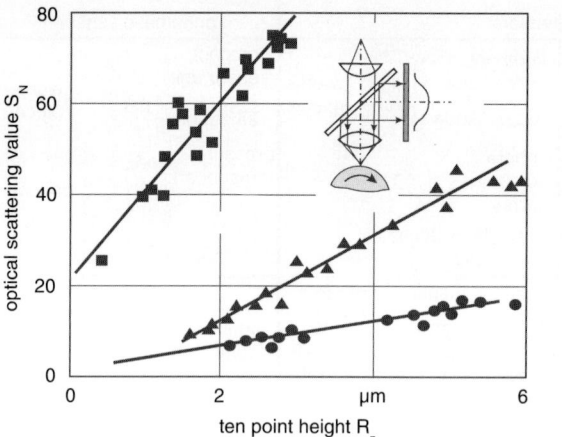

Fig. 4.4-13 Different correlation curves for an optical scattered light sensor. Source: König und Klumpen [27], von See [23], present authors

called the scattering value, S_N, and is deduced from the intensity distribution. In different tests the scattered light sensor was directly mounted in the working space of the grinding machine to measure the workpiece roughness [26]. A compressed air barrier protected the optical system. In all investigations it was tried to establish a correlation between optical and stylus roughness measurements. It is possible to obtain such a close relationship while grinding or honing with constant process parameters [23, 26, 27] (Figure 4.4-13). This restriction is indispensable, because a change of input variables such as dressing conditions or tool specification may lead to workpieces with the same stylus roughness values R_a or R_z but different optical scattering values S_N. If a quantitative roughness characterization referring to stylus values is demanded, a time-consuming calibration will be always necessary. As shown in Figure 4.4-13, the measuring direction also has to be clearly defined to achieve the desired correlation. A second limitation is seen in the sensitivity of the system. The scattered light sensor is able to determine differences in high-quality surfaces, but for roughness states of ten-point height $R_z > 5.0$ µm the scattering value S_N reaches its saturation with decreasing accuracy already starting at $R_z = 3.0$ µm [23]. Hence some relevant grinding or honing operations cannot be supervised by this sensor system.

In addition to the installation in the grinding machine, such a scattered light sensor was also integrated in the mentioned flexible measurement cell [24]. This set-up of optical systems outside the grinding machine but integrated in the close-to-machine control loop seems to be superior to the tests under coolant supply in the grinding machine. After all the mentioned investigations, no industrial application of a scattered light sensor in the working space of a grinding machine tool has been reported.

A different optical sensor is based on a laser diode [28] (Figure 4.4-12, top middle). The sensor is equipped with a gallium arsenide diode, which is commonly used in a CD player. With a lens system the beam is focused on the surface and

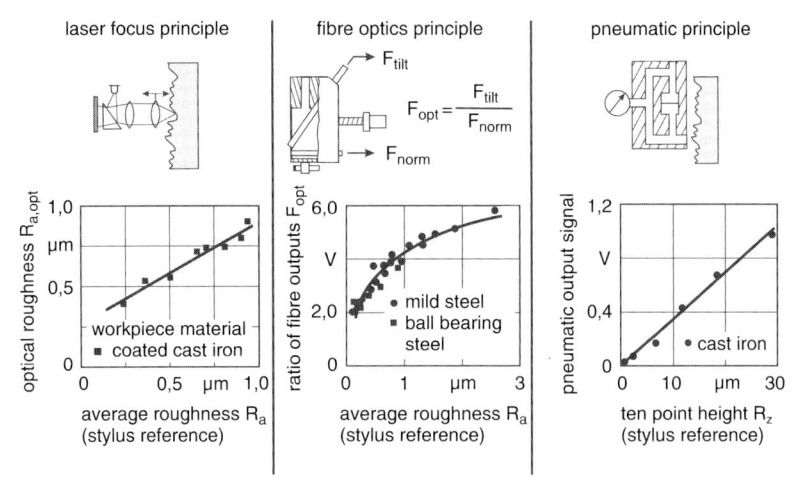

Fig. 4.4-14 Correlation curves for different workpiece roughness sensors. Source: Westkämper [28], present authors

the reflected light is registered on an array of four photodiodes. This system can be used as an autofocus system, with the signal from the four diodes, the focus lens is moved until the best position for minimum diameter is reached. The correlation of the obtained optical average roughness $R_{a,\,opt}$ with the stylus reference measurement is shown in Figure 4.4-14 (left). An almost linear dependence of the two different roughness quantities could be found. However, the system is much too slow to be used for any in-process measurement. By using the focus-error signal of the four diodes without moving the lens, it is possible to increase the measurement speed significantly. Another optical approach for in-process roughness measurement is based on the use of optical fiber sensors [29]. The workpiece surface is illuminated through fiber optics and the intensity of the reflected light is detected and evaluated (Figure 4.4-12, middle bottom).

The latter set-up was chosen to increase the sensitivity of the sensor system. The photo-sensor in the normal direction will register less intensity, whereas the inclined photo-sensor will detect more intensity with larger light scattering due to increased roughness. The ratio of the two photo-sensors is related to roughness changes (Figure 4.4-14, middle). A second advantage of the set-up with two fiber optics despite the increased sensitivity is the achieved independence of the workpiece material. Coolant flows around the whole sensor head to make measurement possible during grinding. It is essential to keep the coolant as clean as possible during operation, because the reflection conditions are definitely influenced by the filtering state of the fluid. This is the major drawback of the sensor system, because the coolant quality is not likely to be stable in production. In addition to these mentioned systems, some other optical techniques for on-line measurement of surface topography have been proposed, eg, speckle patterns. Although the measurement speed may allow the installation of these systems in a production line operation, their use as sensors in the machine tool working space is not realistic.

In summary, owing to all the problems related to coolant supply, it must be stated that these conditions do not allow the use of optical systems during grinding or honing as reliable and robust industrial sensors. Only optical sensor applications measuring with interruptions of coolant supply either in the working space of the machine tool or in the direct surrounding have gained importance in industrial production.

In addition to optical sensors, two other principles are also used for non-contact workpiece characterization. A pneumatic sensor as shown in Figure 4.4-12 (top right) was designed and used for the measurement of honed cylinders [30]. The measurement is based on the already mentioned nozzle-bounce plate principle. A correlation with stylus measurements is possible (Figure 4.4-14, right). The main advantages of this system are the small size, the robustness against impurities and coolant, and the fact that an area and not a trace is evaluated. Hence in principle no movement of the sensor during measurement is necessary.

The last system to be introduced as a non-contact workpiece sensor is based on an inductive sensor. The sensor is used in gear grinding machines to identify the exact position of tooth and tooth slot at the circumference of the pre-machined and usually heat-treated gear (Figure 4.4-12, bottom right). The gear rotates at high speed and the signal obtained is evaluated in the control unit of the grinding machine. This signal is used to index the gear in relation to the grinding wheel to define the precise position to start grinding and to avoid any damage to a single tooth. It is also possible to detect errors in tooth spacing. Gears with unacceptable distortions after heat treatment can be identified and rejected to avoid overload of the grinding wheel, especially when using CBN as abrasive.

4.4.6
Sensors for Peripheral Systems

Primary motion between the tool and workpiece characterizes the grinding process, but as already stated at the beginning of Section 4.4 also supporting processes and systems are of major importance. In this section basically the monitoring of the conditioning process and the coolant supply will be discussed.

4.4.6.1 Sensors for Monitoring of the Conditioning Process

The condition of the tool i.e. the grinding wheel is a very decisive factor for the achievable result during the process. Hence the grinding wheel has to be prepared for the desired purpose by using a suitable conditioning technology. The major problem in any conditioning operation is the possible difference between nominal and real conditioning infeed. There are four main reasons for these deviations. The unknown radial grinding wheel wear after removal of a specific workpiece material volume must be regarded as a significant factor. Also the changing relative position of grinding wheel and conditioning tool due to thermal expansion of machine components is relevant. As a third reason, infeed errors related to friction of the guide-ways or control accuracy have to be considered, although their influence is declining in modern grinding machines. The last reason to mention is the wear of the conditioning tool, which of course depends on the individual type. Especially for rotating dressers only after regular use for several weeks can the first wear effects be registered.

Owing to the great importance of the grinding wheel topography, the monitoring of the conditioning operation has been the subject for research for decades. Already in the early 1980s it was first tried to use an AE-based system for the monitoring of the dressing operation [31]. At that time the work was concentrated on the dressing of conventional grinding wheels with a static single-point diamond dresser. The AE sensor was mounted on the dresser and connected to an evaluation unit. It was possible to detect first contact of the dresser and the grinding wheel and the AE intensity could be used to determine the real dressing infeed as a function of dressing feed rate and grinding wheel speed. The dressing feed speed could also be identified by the AE signal [32]. In addition, it was stated that the AE signal reacts significantly faster to the first contact of dressing tool and grinding wheel compared with the monitoring of the spindle power. Further improvements also allowed information to be obtained about the actual profile of the single-point diamond dresser. The limitation to straight cylindrical profiles was overcome by Meyen, who developed a system capable of detecting dressing errors on any complex grinding wheel profile [5] (Figure 4.4-15).

The strategy comprises the determination of a sliding average value with static and dynamic thresholds for every single dressing stroke. The different geometry elements are identified and the currently measured AE signal is compared with the reference curve, which has to be defined in advance. With the calculation of further statistical quantities such as standard deviation or mean signal inclination, it is possible to identify the typical dressing errors in the case of the thresholds being exceeded.

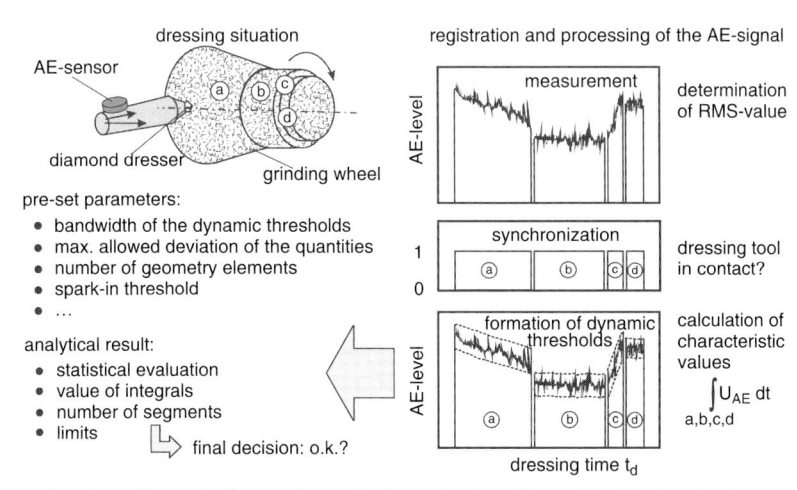

Fig. 4.4-15 Dressing diagnosis for random grinding wheel profiles with AE signals. Source: Meyen [5]

As a consequent next step, AE systems were tested for conditioning operations of superabrasives such as CBN (eg, [7, 33]). The high hardness and wear resistance of these grinding wheels require a different conditioning strategy and monitoring accuracy compared with conventional abrasives. The conditioning intervals due to the superior wear resistance can amount to several hours. The dressing infeed should be limited to a range between 0.5 and 5 µm instead between 20 and 100 µm for conventional wheels in order to save wheel costs. Especially for vitreous bonded CBN grinding wheels it was proposed to use very small dressing infeeds more frequently in order to avoid additional sharpening. This strategy, known as 'touch dressing', revealed the strong demand to establish a reliable contact detection and monitoring system for dressing of superabrasives. In most cases rotating dressing tools are used. The schematic set-up of a conditioning system with a rotary cup wheel, which is often used on internal grinding machines, is shown in Figure 4.4-16.

The conditioning cycle consists of four stages: fast approach, contact detection, defined infeed, and new initiation. The setting parameters such as number of strokes, dressing infeed, and dressing feed rate are stored in the control unit of the grinding machine. The fast approach is done with the NC axis of the machine. The dressing tool is moved to the last-stored dressing position with an additional safety distance of, eg, 20 µm to avoid any undesired contact. Then the dressing sub-program is started with the chosen infeed and feed rate of the dressing tool for each stroke. In this state, a reference signal for the chosen contact detection system can be recorded. In addition to AE techniques, other methods have also been tested. Heuer additionally investigated the possibility of using either the required power of the dressing tool spindle or a piezoelectric force measurement for monitoring [33]. The latter technique was possible, because a piezoelectric actuator was installed as a high-precision positioning system for the infeed of the dressing tool. With this additional equipment infeed extents of 0.25 µm could be

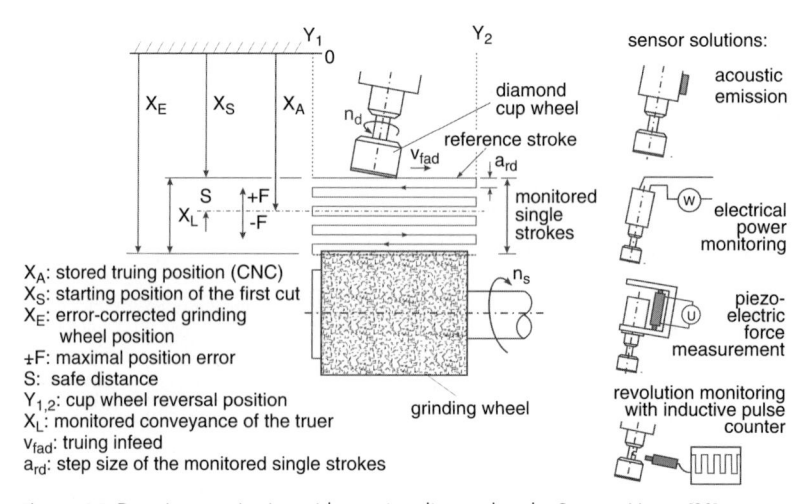

Fig. 4.4-16 Dressing monitoring with rotating diamond tools. Source: Heuer [33]

realized. Hence this system may also upgrade older machine tools with less accuracy. However, in any modern machine tool the in-built x-axis provides the infeed. A further technique for contact detection was introduced in [34]. The measurement of the rotational speed change of the high-frequency dressing spindle, which gives a maximum number of revolutions of $60\,000$ min^{-1}, was used to determine not only the first contact, but also the whole dressing process. After contact detection of any of the mentioned systems, the conditioning program is continued until the desired number of strokes and infeed are reached. Depending on the type of system it is possible to monitor the course of the signal on the whole width of the grinding wheel. Thus uneven macro-geometric wear of the wheel can be registered, if the measured signal does not exceed the defined threshold reference level over the whole width. This strategy also assures a perfect macro-geometric shape after conditioning. After finishing the conditioning process the final position of the machine x-axis is stored to initialize the next operation.

The use of AE sensors for contact detection of the conditioning and dressing operation can be regarded as state of the art. Many different systems are available. New grinding machine tools with self-rotating conditioning tools are usually equipped with an AE system already in the delivery state. Also the system with dressing spindle rotational speed monitoring has found acceptance in industry, because this system is regarded as very robust and is not influenced by coolant supply or bearing noise, which is still regarded as the major limitation for all AE systems. However, the importance of this last method is declining, because in modern machine tools the control loops for the main drives are extremely fast and thus a deviation in rotational speed is no longer a suitable signal source. Hence the monitoring of the electrical power consumption of the dressing spindle is becoming more attractive, because it has also reached sufficient sensitivity and is installable with the least effort.

4.4.6.2 Sensors for Coolant Supply Monitoring

Relatively large contact areas characterize abrasive processes and especially grinding operations. The large number of cutting edges generates a considerable amount of heat in the zone of contact. Hence the reduction of friction and cooling of the interacting parts is often necessary to avoid thermal damage. Therefore, in almost all cases coolants are used to reduce heat and to provide sufficient lubrication. These are the main functions of any coolant supply. Furthermore, the removal of chips and process residues from the workspace of the machine tool, the protection of surfaces, and human compatibility should be provided. Modern coolant compositions also try to fulfil the contradictory demands of long-term stability and biological recyclability.

At least with the wider use of superabrasives such as CBN, the possibility of high-speed grinding, and highly efficient deep grinding, a closer view on the coolant supply began. Coolant pressure and flow rate measured with a simple flowmeter in the coolant supply tube before the nozzle are now often part of the parameter descriptions. Different authors have also worked on the influence of different nozzle de-

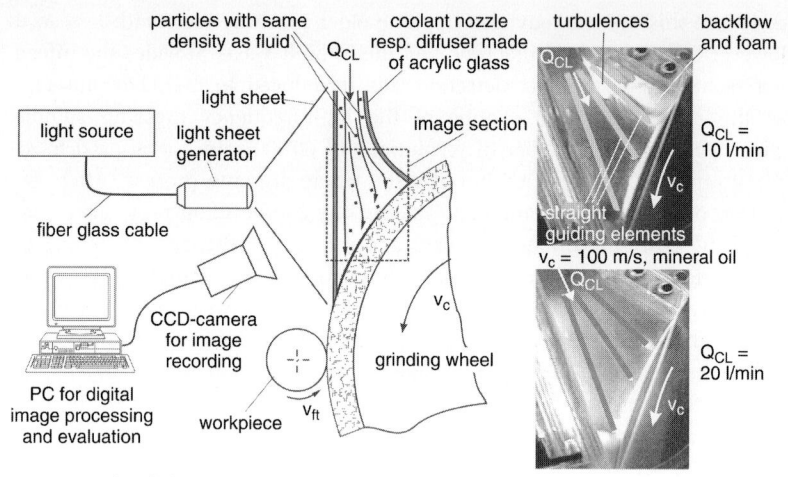

Fig. 4.4-17 Flow behavior monitoring by means of particle image velocimetry. Source: Brinksmeier et al. [37]

signs (eg, [33]). In most cases the influence of different supply options such as conventional flooding nozzles, shoe, spot jet, or spray nozzles, or even internal supply through the grinding wheel, is described by using the already-mentioned process quantities such as forces or temperature [35]. However, in addition to the technological demands, environmental aspects of manufacturing have also attracted significantly more attention. Especially the last mentioned point has led to a detailed investigation of coolant supply and the possibility of reducing or avoiding coolants in grinding completely [36]. Heinzel and co-workers made a very systematic approach to investigate the coolant-related influences and to optimize the relevant parameters and designs. For the development of a suitable shoe nozzle design a special flow visualization technique was used (Figure 4.4-17) [35, 37].

Tracer particles with almost the same density are added to the transparent fluid. All parts of the nozzle of interest are made from acrylic glass and a CCD camera records the flow images perpendicular to the light sheet plane. Although only a qualitative result is available, this technique offers the possibility of systematically studying and improving the whole design of coolant nozzles. As an example, in Figure 4.4-17 (right) the flow behavior of a nozzle with straight guiding elements at two different flow rates is shown. The coolant is mineral oil and the grinding wheel is rotating at a speed of 100 m/s. For the lower flow rate of 10 L/min inhomogeneous flow behavior can be observed. Turbulences, backflow, and foam between top and center guiding elements and at the entry side of the grinding wheel are visible. A doubling of the flow rate leads to steady flow behavior. In [35] it is explained in detail that different flow rates need different adapted guiding elements to achieve the best result.

In addition to this use of an optical monitoring method to optimize the design of coolant nozzles, a special sensor installation for pressure and force investigations was also introduced [35]. The force measurement is done by an already-discussed

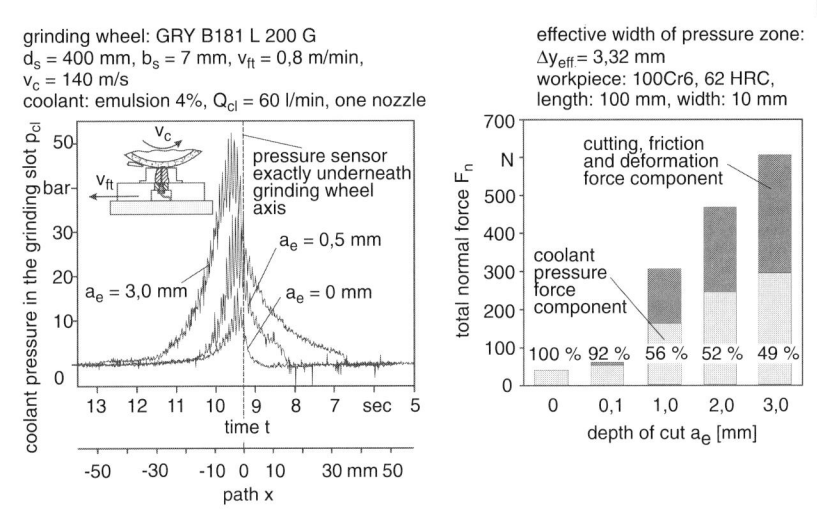

Fig. 4.4-18 Coolant supply monitoring with pressure sensor and dynamometer. Source: Heinzel [35]

piezoelectric dynamometer. During grinding only the total normal force can be registered by this instrument. The idea is to separate the normal force component used for cutting, friction, and deformation from the component that results from the build-up of the hydrostatic pressure between grinding wheel and workpiece and because of the impact of the coolant flow on the surfaces. For this purpose an additional pressure sensor is integrated in the workpiece carrier, allowing the measurement of the pressure course over a grinding path through a bore in the workpiece.

In Figure 4.4-18, results of this sensor configuration are shown [35]. The left part shows a result of the pressure measurement depending on different depths of cut. It is shown that with increasing infeed the maximum of the pressure distribution is shifted in front of the contact zone, which can be explained by the geometry of the generated slot. Higher infeed leads to a geometric boundary in front of the contact zone, resulting in a rise of the dynamic pressure. If the measured pressure distribution is numerically integrated over the corresponding workpiece surface, the coolant pressure force component can be determined, taking some assumptions for the calculation into consideration [35]. Figure 4.4-18 (right) shows results of this combined calculation and measurement. A path with no infeed already leads to a normal force of 34 N, only generated by the coolant pressure. With increasing depth of cut the amount of this force component is, of course, reduced. However, still almost half of the normal force is attributed to the coolant pressure, even under deep grinding conditions of depth of cut $a_e = 3$ mm. This described method is suitable for investigating the influence of different coolant compositions. Especially the efficiency of additives can be evaluated, if the coolant pressure force component is known, and can be subtracted from the total normal force to emphasize the effect on the cutting, friction, and deformation component.

The use of special sensor systems for coolant supply investigations is a relatively new field of activity. First results have shown that these sensors can contribute to a better understanding of the complex thermo-mechanical interaction in the zone of contact. Also direct industrial improvements such as coolant nozzle optimization or additive efficiency evaluations for grinding can be performed. Hence further improvement of the monitoring techniques is desirable.

4.4.7
Sensors for Loose Abrasive Processes

All the processes described so far in this Section 4.4 belong to the group of abrasive processes, but only operations with bond abrasives have been discussed. Some of the introduced sensor systems have also been used to monitor processes with loose abrasives and should be briefly mentioned. Any kind of lapping, abrasive jet cutting, and free abrasive cutting or barrel polishing operations belong to the group of loose abrasive processes.

4.4.7.1 Lapping Processes

The abrasive grains used for lapping are both carried in a paste or within a liquid and applied to the workpiece surface. A tool with usually the opposite of the desired workpiece form moves and the grains follow disordered paths in the gap between the tool and workpiece. The main purpose of any lapping process is the achievement of very high surface finish quality regarding roughness and form accuracy. The process is characterized by small material removal rates and therefore only very sensitive systems are suitable for monitoring. In [38] an investigation of the lapping process using AE signals was reported. Figure 4.4-19 shows the set-up of these investigations and some results achieved.

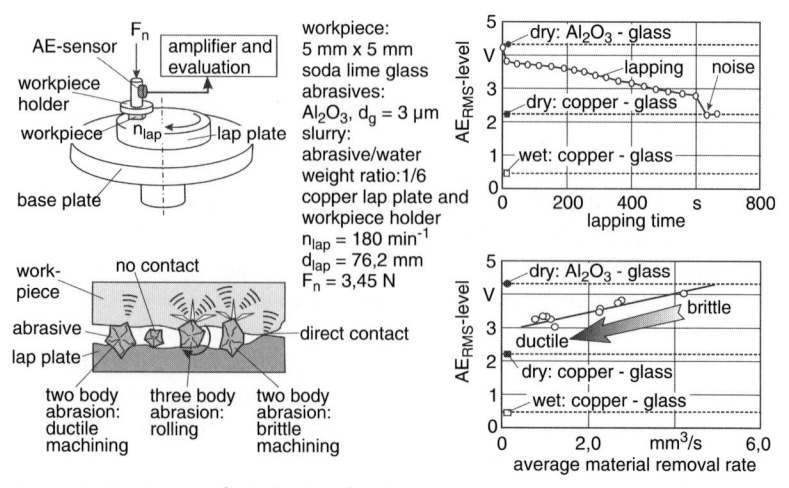

Fig. 4.4-19 Monitoring of single-plate face lapping with acoustic emission. Source: Chang and Dornfeld [38]

The AE sensor was directly mounted on the back side of the work holder. The material removal rate was determined by periodically measuring the workpiece thickness using a dial indicator. In addition to regular conditions, some non-standard situations were also recorded for interpretation. Dry Al_2O_3-glass contact was achieved by rubbing the glass against the same amount of dry abrasives as in the fresh slurry. Dry copper-glass contact was generated by direct contact of the plate and the workpiece without any abrasive or water, while the occurrence of water between these two bodies is called wet copper-glass contact. The monitoring of a regular lapping process shows a constant decrease in the signal level, followed by audible vibration noise. It was reported that in this stage the slurry had dried. Figure 4.4-19 (bottom right) shows the correlation between the AE signal and the material removal rate in this test. High values of the AE signal correspond to high material removal rates at the beginning of the test and are interpreted as brittle machining. The material removal mechanism is then shifted towards ductile machining due to the size reduction of the abrasive grains and their increasing penetration and fixturing in the lap plate, leading to a process with increasing number of fixed abrasives. The authors called this a shift from three-body abrasion with discrete indentations and rolling of the grains to two-body abrasion [38]. The acoustic emission signals reveal the possibility of monitoring this complex process, but further investigations are needed to prove the significance of this technique for different conditions.

Several researchers in Berlin have also worked in the field of lapping processes. Sabotka used an infrared thermography system to measure the temperature distribution on the lapping plate [39]. His aim was to monitor a double plate face lapping process, so the upper lapping plate had to be removed from the workpiece surface before measurements. This delay was taken into account. As the major result, an increase in the lapping plate temperature with lapping time could be stated, but the measured temperatures did not rise significantly over room temperature.

Funck made some special preparations of workpieces for force and temperature measurement. Although his investigation was related to a face grinding process with lapping kinematics, his proposed sensors can also be used for the lapping process [40]. For force measurement he applied four temperature-compensated wire strain gages on a workpiece, which was moving in a guiding carrier system. The carrier system rolls off on inner and outer rings so that the workpieces move in a cycloid motion relative to the rotating plates (similar to a planetary gear motion) [20]. The supply of necessary voltage and the signal transmission was realized by a wireless telemetry system. The temperature measurement was done with thermocouples on a fixed, not moving, workpiece with conventional wire connection. Owing to the great efforts needed for the above-mentioned measurements, the application of these sensor techniques is again restricted to fundamental research.

A special variant of the lapping process is ultrasonic-assisted lapping, where a specially shaped tool, called a sonotrode, oscillates in the infeed direction to the workpiece and is circum-circulated by a suspension consisting of abrasives and carrier liquid. This technique is especially attractive for materials which cannot be machined by electrodischarge machining owing to the lack of electrical conductivity, such as many ceramics and glasses. It was reported that it is possible to monitor an ultrasonic-assisted lapping process with the aid of a suitable AE sensor. The sensor was mounted on the infeed unit near to the sonotrode head. Care has to be taken concerning the influence of the ultrasonic exciting frequency and its harmonics. The results reveal a possibility of controlling the process by AE sensors. An analysis in either the time or frequency domain is possible; the definition of specific thresholds was proposed to identify any process problems. However, a clear correlation with the reason for disturbances is regarded as very difficult [41].

4.4.7.2 Sensors for Non-conventional Loose Abrasive Processes

The application of a liquid (eg, water) or air jet mixed with a specific amount of abrasives is a non-conventional method either to cut materials or to improve the surface quality of workpieces. Especially for brittle, hard, or generally difficult to cut materials the first applications are becoming popular. The major advantages are the low sensitivity to workpiece material properties, the lack of risk of thermal damage combined with minimal stress induction, no tendency for chatter vibration, and the possibility of easy automation. On the other hand, the process is very noisy and unclean, thus generating a significant load on the environment.

Nevertheless it is an effective technique, which is applied, eg, to cut ceramic materials. In [42] the concept of a monitoring system for abrasive water jet machining based on acoustic emission was published. Figure 4.4-20 shows the set-up of the investigations and some first results. The sensor was fixed to the machining table at a relatively large distance from the aluminium oxide ceramic workpiece to avoid damage. It was possible to correlate the energy of the acoustic emission, which is the mean square AE signal, with the water jet pressure, the dis-

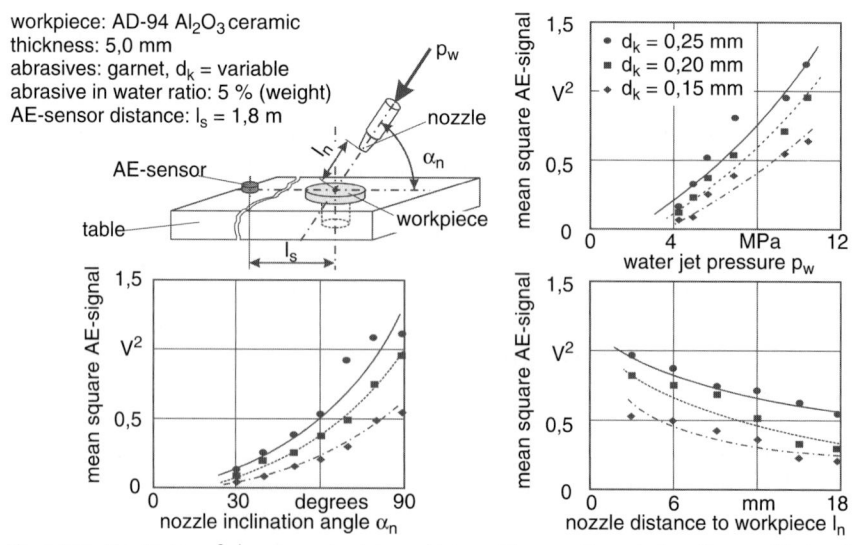

Fig. 4.4-20 Monitoring of abrasive water jet machining with acoustic emission. Source: Choi and Choi [42]

tance between nozzle and workpiece surface, and the attacking angle. In addition, coarser abrasive particles generate higher AE signals. A monitoring strategy for abrasive water jet machining based on AE sensors seems to be possible, and further research to define its potential and limitations will be conducted.

An example of a special non-conventional process related to improving the surface quality of a workpiece is called abrasive flow machining. It is a finishing process that deburrs and polishes workpieces by forcing an abrasive-laden viscoelastic polymer across the surface to be improved. Current applications include cylinder heads and injector nozzles. In [42] the first results were published of the use of an AE sensor system to monitor this complex process.

Obviously it is the first choice of many researchers to try an AE sensor system as a monitoring tool for non-conventional processes. Especially if the material removal rates decrease to very low levels this technique seems to have the greatest potential compared with other methods, which are not sensitive enough. As the demands on surface quality will increase further and on the other hand the size of parts to be machined will decrease further, AE research will be of even more importance in this field.

4.4.8
Adaptive Control Systems

In this section, the use of the already-introduced sensor systems for automated feedback of their signals to the control unit of the abrasive machine tool will be discussed. The schematic set-up of a control loop and the difference between ACC and ACO systems was explained in Section 4.3.7. The main problem with any grinding operation is the unstable process behavior. For example, in ID grinding the limited stiffness of the small grinding spindle leads to significant deflections, accompanied by a fast change of the grinding wheel topography and thus cutting performance. These changing process conditions require the fast measurement of relevant quantities for on-line control of the operation through feedback of necessary information. To explain this idea, the situation in ID and OD grinding is considered [44, 45]. In a so-called conventional grinding process the infeed motion of the grinding wheel is a setting parameter in the control unit of the machine tool. No measurement of effective process quantities and no feedback of signals are installed. This is a pure path-related speed control (Figure 4.4-21, left). Depending on the oversize variations, the first cut contact changes. Owing to the elasticity of the system grinding wheel, workpiece, and machine tool, the normal force in the zone of contact rises very slow. Also the force level at the end of the roughing process is different, leading to different levels of deflection and resulting in different size deviations even after finishing and spark-out. In comparison, it is the purpose of an adaptive control system to keep one or more process quantities at a defined level throughout the entire grinding process. In Figure 4.4-21 (right), an example of a force-controlled adaptive grinding process is shown. The air grinding time is reduced to a minimum, because the grinding wheel approaches the workpiece surface with a higher infeed speed.

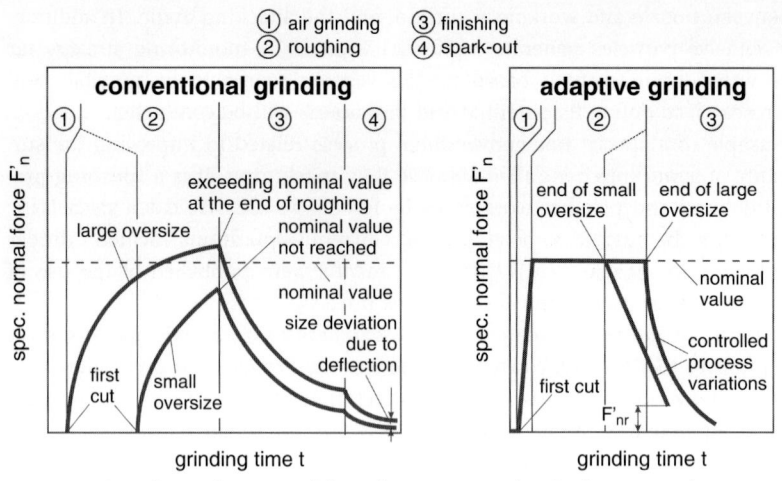

Fig. 4.4-21 Specific grinding normal force for conventional and adaptive grinding. Source: Popp [44], Walter [45]

After the contact is detected, the infeed speed is switched to reach the given nominal value of the force in a very short time. The force is then kept constant until the end of the roughing position of the grinding wheel is reached. Independent of the different grinding times, according to the varying oversizes, a constant defined process condition is assured. Different variations of the force profile during finishing are possible to reach the end of cycle, eg, a linear decrease to a defined force level without spark-out [45].

The advantages of adapted process control were discussed in the 1960s, and major research efforts were made in the 1970s and 1980s. Two main reasons were responsible for the limited industrial acceptance of these ideas at that time. On the one hand, the available computer hardware was not yet sophisticated enough to meet the requirements of the application. After the fast improvement of hardware components there was still a restriction because of the closed structure of the CNC control units. Only additional systems could work as adaptive control modules, and their connection to the machine tool control unit was always related to additional costs and technical problems. The complexity of the supplemented system was another limitation for industrial breakthrough. The other significant drawback was the lack of suitable and reliable sensor systems to measure the necessary process quantities during operation to close the machine internal control loop.

In general, the AC systems are either related to a geometric or a technological desired condition, but also mixtures are known. Geometric systems are designed to achieve a desired workpiece geometry by measuring the actual diameter of the part and by controlling the infeed. Sensor systems were introduced in Section 4.4.5 and belong to the group of tactile macro-geometric sensors. Technologically oriented systems usually use the feed speed of the grinding wheel as the regulated quantity and the grinding forces or spindle power as the regulating quantity. With this concept the quality of the workpiece and economic needs can be met.

Such sensors were introduced in Section 4.4.3. A detailed description of all the specific achievements of the mentioned early work of AC researchers is not possible here. Instead, the focus will be put on the latest developments, taking state-of-the-art computers and sensor systems into account. In addition, the increasing use of open system controllers will drastically reduce the former problems to integrate adaptive control systems in machine tools.

With the development of digital controllers it was possible to close automatically the control loop of the machine tool. Zinngrebe was able to develop an ACC system for ID grinding by applying a self-adaptable controller [46]. He proved the efficiency either by using a piezoelectric dynamometer as a force measurement device or by monitoring the grinding spindle power. In this work only the improvement of the roughing cycle and the reduction of the air grinding time was investigated.

In [44, 47] the OD grinding process was investigated. Again, a self-adaptable controller was used to reduce the grinding time significantly and keep the achieved workpiece quality within close tolerances. Popp [44] compared the grinding normal force measured with ring-type piezoelectric dynamometers with AE signals measured with a wireless set-up on the grinding spindle nose as regulating quantity. Although the normal force was considered to be the superior measuring quantity, the AE system offers advantages concerning investment and mounting efforts and was considered suitable. Additionally, the layout of an OD axial grinding AC system was introduced. Fuchs [47] also used different systems such as workpiece size gages and force sensors for his approach. He developed a complex model of the whole grinding process including the machine tool itself. He recommended the spindle power monitoring as suitable force sensor system.

The use of grinding spindles with active magnetic bearings has been investigated [48]. These spindles were used in ID grinding because of their capability to allow high cutting speeds for small grinding wheel diameters. No additional sensors are needed to measure grinding forces and for contact detection, because this information can be derived from the power demand from the different magnetic bearings to balance the spindle and to adjust the air gap. Although these spindles offer many advantages, their industrial use is limited because of the high investment. Rowe et al. developed an adaptive control system for OD grinding based on diameter gaging and power monitoring [49]. The workpiece accuracy could be maintained or even improved while the productivity was increased.

Walter was one of the first researchers to combine the development of an adaptive control system with the use of fuzzy logic [45]. He developed an ACO system for ID grinding with a subordinate ACC system. Normal force measurement with a piezoelectric dynamometer and spindle power monitoring were used as process quantity sensor systems, and the workpiece macro- and micro-geometric quantities were measured after grinding to train the optimization control. The regulated quantity was the radial feed speed of the grinding wheel. Fuzzy logic theory was applied to describe the complex interaction of the setting quantities and the resulting output quantities. Although the effort to develop the optimization control manually was high, the results showed that a desired constant workpiece quality was maintained while the grinding time was reduced to a minimum without spark-out.

The consequent next step in applying AC systems was made by Memis [50]. She concentrated on the exclusive use of AE signals as regulating quantity for OD grinding. The sensor was mounted at the center of the grinding spindle and wireless data transmission was used. The developed system was tested on different machine tools. The determination of the limit of the AE signal as the essential quantity to control different processes was supported by the use of artificial neural networks. With this technique, automated adaptation of the controller could be achieved. A well-trained and optimized neural network offers the possibility to users of starting new grinding operations with best-adapted control parameters based on the measured process quantities.

The latest development in adaptive control application for grinding is related to out-of-roundness grinding. With the development of modern and fast control units and drives it is now possible to grind crankshafts or camshafts by path-controlled movement of the grinding spindle head (see Section 4.4.5). However, a time- and money-consuming procedure for parameter adjustment is still necessary before workpieces of the desired quality can be ground. A feedback of process quantities is not yet established. To overcome this problem a hybrid system for AC out-of-roundness grinding was developed. In [51] the layout and first test results of this work were published.

To summarize this section, the development of AC systems already reached a high academic level a long time ago. The earlier problems of poor computer and control features are no longer limiting factors. Also, suitable sensors systems are now available. Geometric AC systems using in-process diameter gages are state of the art. The advantages of systems based on force measurement are well known, only the high investment for dynamometers still prevents more general use. Power monitoring has the disadvantage of limited dynamic response and a significant time delay, but is nevertheless used in industrial applications because of the easy sensing and low investment. AE sensors also offer high potential in this field. With robust data processing, industrial use as a measurement system to provide a sufficient regulating quantity for AC systems will be possible.

4.4.9
Intelligent Systems for Abrasive Processes

Any kind of sensor technology for monitoring purposes is applied to ensure the availability of manufacturing systems and to guarantee a desired output from the process. The examples of sensor applications in the preceding sections have revealed wide variations in these applications. The emphasis was put on different technical solutions for obtaining the desired information on the process, tool, or workpiece. All the introduced sensors can be attributed to two different kinds of control loops (Figure 4.4-22) [52].

Sensors for measuring process quantities and sensors for any kind of in-process measurement are related to the machine internal control loop. Sensors which are not installed in the working space of the machine tool or which only measure in auxiliary times belong to the post-process control loop. In the superior control

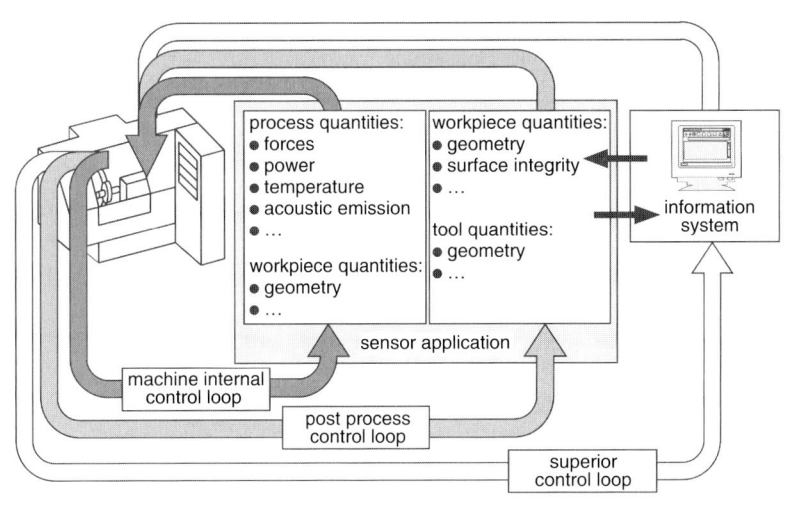

Fig. 4.4-22 Hierarchical order of control loops in manufacturing

loop, the direct use of sensors is not scheduled. In this highest level, any kind of intelligent information system is used either to download control tasks to the sensor systems together with threshold values or to collect measuring data for further processing. In [53] the layout of such an information system was given, consisting of modules for process set-up, process optimization, quality assurance, and information and advice. The use of large databases in connection with a suitable software platform made it possible to assist the planner in parameter selection and the operator in failure analysis. By applying different types of models to describe the complex grinding process, the selection of process parameters to achieve specific output quantities was also supported. This grinding information system was further developed, also including techniques of fuzzy logic and artificial neural networks. In [54] a survey of different approaches to the use of these artificial intelligence methods in grinding was given.

Although the sensor application is effected at a lower level according to the hierarchy of the control loops, it can still be regarded as an essential part of a whole so-called intelligent system. The major task is to obtain as much information as possible from the current process. With single-phenomenon monitoring this aim often cannot be met, so the application of multiple sensors in one process is part of many activities to achieve an intelligent system [55]. In general, the sensor application itself is the same as for a single-purpose monitoring task. A major difference can be seen in the sensor choice. Some systems use several sensors to detect the same type of phenomenon, eg, power and force monitoring to avoid overload of the main spindle. Although the sensors are not necessarily similar, their field of application is redundant. This strategy is chosen to increase the reliability of sensor systems, which is still one of the highest demands from industrial users. The other approach is to combine the data from different types of sensors to increase the flexibility of the monitoring system. In this case the term of sensor fusion is also used, but usually it is not a fusion of sensors but a fusion of data from different sensor sources.

In Figure 4.4-23, an attempt is made to show the field of sensor application in the surroundings of an intelligent grinding system. The term 'intelligent' is not clearly defined. It is most often related to the application of artificial intelligence techniques such as knowledge-based systems, neural networks, and fuzzy logic. The processing of the sensor data is usually still done in a conventional way, but of course using the latest available equipment for analog-to-digital conversion, filtering, sampling, and further calculations. It is one of the major tasks to provide parallel processing of different sensor signals as fast as possible for extremely short response times in case of problems. Different hardware solutions are available but will not be discussed in detail. In principle, there is a trend from stand-alone computers to PC-based solutions, because this is also the kernel of modern CNC machine tool control units. In addition, a lot of work is directed towards a modular set-up of monitoring systems, which should help in implementing these systems in a machine tool surrounding. The sensor system is connected to other modules of the set-up. First, monitoring limits and threshold values have to be fixed depending on the experience of an expert. Databases as part of a knowledge-based system can support this essential step. The measured and processed sensor data are transmitted to succeeding modules such as grinding databases and control modules. Depending on the type of application, direct feedback to the process according to adaptive control strategies is possible (see Section 4.4.8). There might be a direct connection to the CNC machine tool control unit to initiate an interrupt in case a specific limit such as spindle maximum power is exceeded. Also integrating the sensor data in databases can support tasks in the superior control loop such as planning and quality control. Special attention has to be paid to the use of process models. These models are used either to gain knowledge

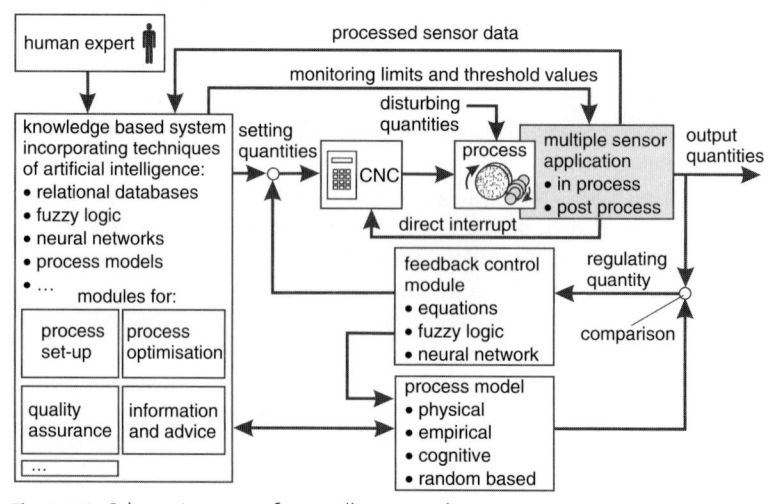

Fig. 4.4-23 Schematic set-up of an intelligent grinding system

about the process or to simulate or optimize the process input parameters [1]. Different types of models are known [53]. All sensor data from in-process or post-process applications can be compared with the output of the different mentioned models to optimize and train them. These models directly support the process set-up and optimization modules of the knowledge-based system. Also, the output of the feedback control unit may be used to improve the accuracy of the process model. Above all these parts of a schematic intelligent grinding system, the human expert is responsible for the configuration of the system including all hardware and software development.

4.4.10
References

1 BRINKSMEIER, E., *Habilitationsschrift;* Universität Hannover, 1991.
2 BYRNE, G., DORNFELD, D.A., INASAKI, I., KETTELER, G., KÖNIG, W., TETI R., *Ann. CIRP* **44** (1995) 541–567.
3 BRUNNER, G., *Dissertation;* Universität Hannover, 1998.
4 FRIEMUTH, T.,., *Dissertation;* Universität Hannover, 1999.
5 MEYEN, H.P., *Dissertation;* RWTH Aachen, 1991.
6 WEBSTER, J., DONG, W., LINDSAY, R., *Ann. CIRP* **45** (1996) 335–340.
7 WAKUDA, M., INASAKI, I., OGAWA, K., *J. Adv. Autom. Technol.* **5** (1993), 179–184.
8 CHOI, H.Z., *Dissertation;* Universität Hannover, 1986.
9 LIERSE, T., *Dissertation;* Universität Hannover, 1998.
10 UEDA, T., HOSOKAWA, A., YAMAMOTO, A., *J. Eng. Ind. Trans. ASME* **107** (1985) 127–133.
11 BUCHHOLZ, H., *Maschinenbautechnik* **25** (1976) 364–365.
12 WESTKÄMPER, E., KLYK, M., *Prod. Eng.* **1** (1993) 31–36.
13 GOMES DE OLIVEIRA, J., DORNFELD, D.A., WINTER, B., *Ann. CIRP* **43** (1994) 291–294.
14 KARPUSCHEWSKI, B., WEHMEIER, M., INASAKI, I., *Ann. CIRP* **49** (2000) 235–240.
15 WESTKÄMPER, E., HOFFMEISTER, H.-W., in: *Arbeits- und Ergebnisbericht 1995–1997 des Sonderforschungsbereiches 326*, Universität Hannover and TU Braunschweig, 1997, pp. 345–401.
16 PIEGERT, R., RONNEBERGER, E., *Maschinenbautechnik* **31** (1982) 105–108.
17 GOTOU, E., TOUGE, M., *J. Mater. Process. Technol.* **62** (1996) 408–414.
18 WERNER, F., *Dissertation;* Universität Hannover, 1994.
19 FOTH, M., *Dissertation;* Universität Hannover, 1989.
20 TÖNSHOFF, H.K., KARPUSCHEWSKI, B., MANDRYSCH T., INASAKI, I., *Ann. CIRP* **47** (1998) 651–668.
21 RIO, R., *Am. Machinist* July (1998) 66–70.
22 SALJÉ, E., *Ann. CIRP* **28** (1979) 189–191.
23 VON SEE, M., *Dissertation;* TU Braunschweig, 1989.
24 TÖNSHOFF, H.K., BRINKSMEIER, E., KARPUSCHEWSKI, B., presented at the 4th International Grinding Conference, 9–11 October 1990. Dearborn, MI, paper MR90–503.
25 BRODTMANN, R., GAST, T., THURN, G., *Ann. CIRP* **33** (1984) 403–406.
26 PAULMANN, R., MÖHLEN, H., *Ind.-Anzeiger* **108** (98) (1986) 32–33.
27 KÖNIG, W., KLUMPEN, T., in: *Internationales Braunschweiger Feinbearbeitungskolloquium*, 2–4 March 1993, pp. 11.01–11.23.
28 WESTKÄMPER, E., KAPPMEYER, G., in: *Seminar des Sonderforschungsbereiches 326*, Universität Hannover and TU Braunschweig, 1992, pp. 41–58.
29 INASAKI, I., *Precision Eng.* **7** (2) (1985) 73–76.
30 MASKUS, P., *Dissertation;* TU Braunschweig, 1992.
31 AVERKAMP, T., *Dissertation;* TH Aachen, 1982.
32 INASAKI, I., *Ann. CIRP* **34** (1985) 277–280.

33 Heuer, W., *Dissertation*; Universität Hannover, 1992.
34 Tönshoff, H.K., Falkenberg, Y., Mohlfeld, A., *IDR Ind. Diamanten Rundsch.* **1** (1995) 43–48.
35 Heinzel, C., *Dissertation*; Universität Bremen, 1999.
36 Karpuschewski, B., Brunner, G., Falkenberg, Y., *Jahrbuch Schleifen, Honen, Läppen und Polieren*; Essen: Vulkan-Verlag, Ausgabe 58, 1997, pp. 146–158.
37 Brinksmeier, E., Heinzel, C., Wittmann, M., *Ann. CIRP* **48** (1999) 581–598.
38 Chang, Y.P., Dornfeld, D.A., *Ann. CIRP* **45** (1996) 331–334.
39 Sabotka, I., *Dissertation*; TU Berlin, 1991.
40 Funck, A., *Dissertation*; TU Berlin, 1994.
41 Bönsch, G., *Dissertation*; RWTH Aachen, 1992.
42 Choi, G.S., Choi, G.H., *Int. J. Machine Tools Manuf.* **37** (1997) 295–307.
43 Williams, R.E., *J. Manuf. Sci. Eng. Trans ASME* **120** (1998) 264–271.
44 Popp, C., *Dissertation*; Universität Hannover, 1992.
45 Walter, A., *Dissertation*; Universität Hannover, 1995.
46 Zinngrebe, M., *Dissertation*; Universität Hannover, 1990.
47 Fuchs, A., *Dissertation*; TH Darmstadt, 1992.
48 Ota, M., Ando, S., Oshima, J., presented at the 2nd International Symposium on Magnetic Bearings, Tokyo, July 12–14, 1990.
49 Rowe, W.B., Allanson, D., Thomas, A., Moruzzi, J.L., presented at the CIRP Workshop of STC 'G'. Paris, January 1993.
50 Memis, F., *Dissertation*; RWTH Aachen, 1996.
51 Tönshoff, H.K., Karpuschewski, B., Hinkenhuis, H., Regent, C., in: *VDS-Fachtagung 'Schleiftechnik im Wettbewerb'*, 9–10 October 1997, Aachen, pp. 8/1–13.
52 Tönshoff, H.K., Karpuschewski, B., Paul, T., *VDI-Z.* **133** (7) (1991) 58–63.
53 Paul, T., *Dissertation*; Universität Hannover, 1994.
54 Rowe, W.B., Yan, L., Inasaki, I., Malkin, S., *Ann. CIRP* **43** (1994) 1–11.
55 Inasaki, I., *VDI-Ber.* **1179** (1995) 31–45.

4.5
Laser Processing
V. Kral, O. Hillers, *Laser Zentrum Hannover, Hannover, Germany*

4.5.1
Introduction

The field of laser manufacturing has been expanding rapidly in the last 20 years. New materials and new laser sources have considerably increased the potential of laser applications. This expansion has led to a necessity for higher quality and reproducibility when using lasers. To satisfy this demand, a considerable amount of research has been expended into the use of sensing for laser applications.

The goals of implementing sensing technology in laser manufacturing can be categorized into two groups.

1. *Sensors that monitor the process parameters.* These measure the external variables that affect the laser process. This may mean monitoring of the beam characteristics, the workpiece-head distance, the geometrical accuracy of the workpiece, the process gas, filler material feed, the material quality of the workpiece, and the laser mode, for example. All these parameters have an undeniable effect on the process quality and can be monitored using the appropriate sensors. If

these parameters are supervised and controlled, then theoretically no decrease in quality should occur. Currently, such sensors are being used for monitoring purposes (ie, preventative measures). A trend towards closed-control systems will only be possible when all the process parameters can be individually controlled.
2. *Sensors that monitor the process quality.* These sensors are used to monitor the process by measuring its effects. A common approach is to measure the process radiation using optical sensors. Other methods such as the measuring of process-induced vibrations, thermal effects, or meltpool dynamics are also being extensively researched. By correlating signal variation with quality faults, these systems can monitor the quality of the process. The use of such systems can eliminate the need for destructive off-line quality control methods.

Unfortunately, laser processes are so complex that no full proof system has yet been discovered that guarantees 100% quality. This problem can only be solved by mixing parameter control sensors and process control sensors.

4.5.2
Parameter Monitoring Sensors

The goal of implementing these types of sensors is to measure all the relevant process parameters so that any changes in the quality due to parameter variations can be predicted. To achieve this, the source of error in the process has to be identified. Common sources of error are due to changes in workpiece geometry, workpiece material quality, laser beam characteristics, the focus position, and process gas pressure.

4.5.2.1 Sensors for Identifying Workpiece Geometry
Imaging systems, such as the SCOUT [1], are used as seam tracking devices. A camera is mounted on the robot arm and allows the system to adjust the robot motion relative to the seam position. Camera-based systems often have problems when dealing with highly reflective surfaces, since scratches and reflections can be misinterpreted. Another approach is the use of a mechanical guide to follow the seam [2], as shown in Figure 4.5-1. The use of this method is limited by the feed rate, since the mechanical guide is usually spring loaded. The advantage is that a certain robustness is guaranteed.

4.5.2.2 Sensors for Identifying the Workpiece Quality
Both the material quality and the surface cleanliness can considerably affect laser welding or cutting processes. Errors due to unacceptable material quality are difficult to detect on-line. One approach is to use the process radiation from a high-power laser pulse to classify the type and quality of the steel. The process radiation is analyzed using a spectral analyzer, and the individual material-specific

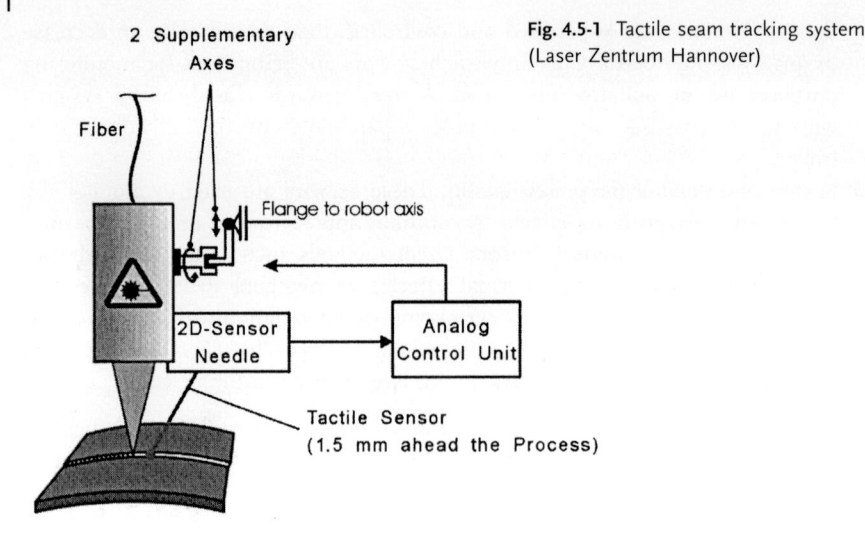

Fig. 4.5-1 Tactile seam tracking system (Laser Zentrum Hannover)

emission peaks are correlated with that of a reference steel. Using this method, undesirable changes in alloy concentration can be detected.

4.5.2.3 Sensors for Beam Characterization

In laser processing, the quality of the beam plays a considerable role in achieving high process quality. For this purpose, sensors are being developed for monitoring the beam power distribution. One common approach is to use a shaft with a pinhole that scans across the beam [3]. Coupled to a mirror, the radiation is reflected to a sensor. With such a system, the intensity distribution across the beam is measured. Another system requires the placing of a thin wire grid in the beam path [4]. From the temperature change of the individual wires, a complete beam power distribution can be reconstructed.

4.5.2.4 Focal Position and Gas Pressure

A drift between the workpiece and the focusing head is often the cause of process faults. Particularly in cutting applications, keeping the focus within the material is important. To solve this problem, capacitive sensors measure the distance to the workpiece and using a control loop, a compensating motion is applied when surface irregularities appear [5]. In addition, more modern systems control the process gas pressure so that no fluctuations can affect the final cut [6]. These sensors are common for two-dimensional applications but have not yet been used in three-dimensional applications.

4.5.3
Quality Monitoring Sensors

Research has been most active in the field of process monitoring. All measurable effects of the laser process are being studied. The sensors used concentrate on optical, acoustic, and visual-based sensing.

4.5.3.1 Optical Sensors

These sensors are used in the most common systems found on the market today. The principle is based on analyzing the process radiation [7]. Depending on the manufacturer, the process radiation is acquired either in- or off-axially from the process point.

Some research has shown that a combination of off- and in-axis radiation can be used for higher reliability [8]. The wavelengths that are being monitored can vary from source to source, but most sensors used are photodiode-based systems monitoring radiation within the visual and near-infrared regions of the spectrum (ie, 400–1000 nm). Other sensors measure the reflected laser radiation. A more costly, but more thorough approach is to use on-line spectral analysis, since individual wavelengths can be taken into account. Such a solution increases the complexity of the failure identification algorithms, and has so far not been industrially implemented.

The optical sensor approach is the most promising approach to identifying faulty laser processes. Unfortunately, even though variations in the process radiation can be correlated with faults, some faults are not always detectable by observing the process radiation signals.

4.5.3.2 Acoustic Sensors

Another approach is to use microphones or ultrasonic sensors to measure process faults. This is particularly useful in the piercing phase of a cutting process [9]. Using these sensors, the piercing time can be optimized. Other approaches in laser welding have shown that it is possible to correlate the acoustic signals with weld-pool vibrations when a pulsed laser process is being used. In micro-structuring using excimer lasers, ultrasonic measurements on the workpiece have shown that the removal rate for each pulse can be monitored [10].

4.5.3.3 Visual-based Sensing

This approach deals with the complete visualization of the effects of the process. In welding, for example, key-hole diameter and weld-pool dynamics [11] can be monitored using charge-coupled device (CCD) cameras. The only difficulty lies in choosing the proper filter so that the bright process radiation does not overexpose the image. This is particularly important when dealing with processes that pro-

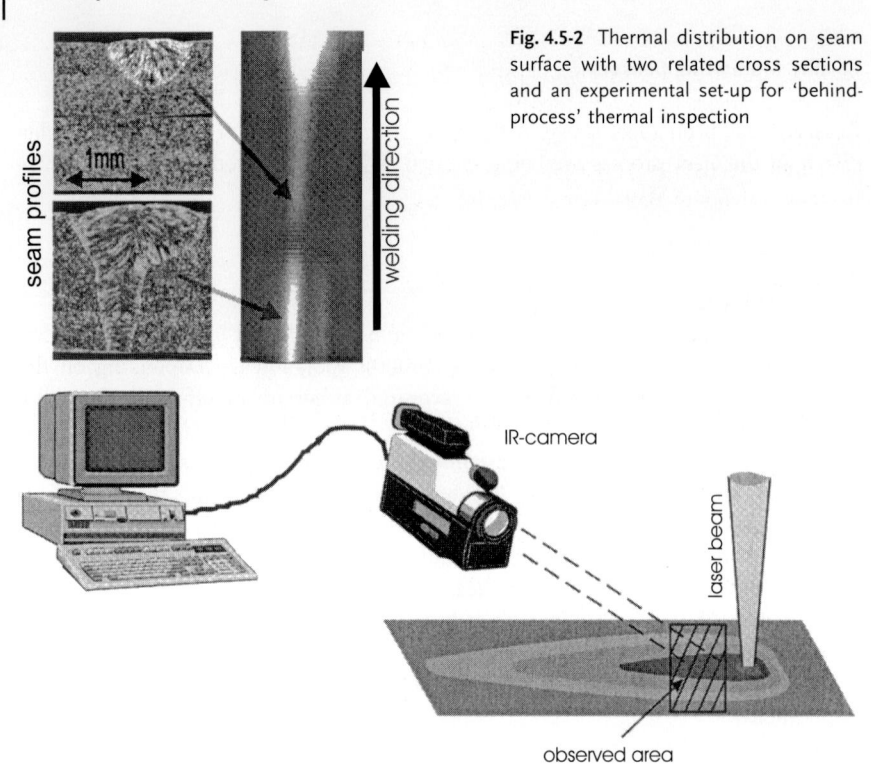

Fig. 4.5-2 Thermal distribution on seam surface with two related cross sections and an experimental set-up for 'behind-process' thermal inspection

duce considerable plasma radiation. Another approach is to use thermal imaging after the process. From the thermal distribution, irregularities in welding processes can be identified (Figure 4.5-2). To increase the signal processing speed, infrared line-arrays can be implemented [12]. Another purely scientific approach is to use X-ray imaging to identify the vertical melt-pool dynamics [13]. Such a research tool has been used to identify the formation of pores in welds.

Unfortunately, visual systems are complex and expensive. Fast feature extraction software is necessary. For this reason they are rarely used in industrial applications.

4.5.4
Conclusion

As can be seen, the sensors that can be used in laser processing applications are varied. Ultimately, what needs to be known is the goal of the sensing system. A well-identified quality problem could be monitored by some of the sensors described above if the fault symptoms are known (eg, fluctuation of the process radiation due to gap variations in overlap welding). To identify all the symptoms, one would ultimately require a 'complete sensing' of the process. This is complex and expensive; currently such a solution cannot be realistically envisioned for use in industrial applications.

The main problem is that the cause and effect of certain faults are still unclear. More research is required to find out which sensors are more adequate for monitoring specific faults so that redundant information can be minimized. Furthermore, more work has to be done to reduce the size and cost of the sensors so that a 'complete sensing' approach can be seriously taken into account.

4.5.5
References

1 BARTHEL, K., TRUNZER, W., *Laser-Praxis* (1994) 10.
2 ANON., *Opto Laser Eur.* (1997) 3.
3 Schwede, H., Kramer, R., in: *Proceedings of ICALEO 1998*, Orlando, FL; www.primes.de.
4 Eg, Prometec UFC60; www.prometec.com.
5 Www.precitec.com.
6 FRIEDEL, R:, *EuroLaser* (2000) 1.
7 TÖNSHOFF, H.K., SCHUMACHER, J., in: *Proceedings of ICALEO 1996*, Detroit, MI.
8 IKEDA, T., KOJIMA, T., TU, J.F., OHMURA, E., MIYAMOTO, I., NAGASHIMA, T., TSUBOTA, S., ISHIDE, T., in: *Proceedings of ICALEO 1999*, San Diego, CA.
9 Www.precitec.som/PPS130E.htm.
10 RINKE, M., *Dissertation;* Düsseldorf: VDI-Verlag, 2000.
11 GRAUMANN, C., *Dissertation;* Düsseldorf: VDI-Verlag, 1998.
12 TÖNSHOFF, H.K., ALVENSLEBEN. V. F., OSTENDORF, A., HILLERS, O., STALLMACH, M., in: *Proceedings of Photonics East, 1999*, Boston, MA.
13 SETO, N., KATAYAMA, AS., MATSUNAWA, A., in: *Proceedings of ICALEO 1999*, San Diego, CA.

4.6
Electrical Discharge Machining

T. MASUZAWA, *University of Tokyo, Tokyo, Japan*

4.6.1
Introduction

Electrical discharge machining (EDM) has been known as one of the non-traditional machining processes since its invention in early the 1940s. However, currently, EDM is one of the most commonly used processes in various kinds of factories. Most EDM machines are used in the field of die and mold making.

The material removal mechanism in the case of EDM is completely different from that in the case of highly conventional machining processes such as cutting and grinding. It is based on melting and vaporization of workpiece materials by heating introduced by a spark or, more precisely, by a transient arc discharge. Thus the machining force is a minor or, rather, a negligible parameter in the control of this process. On the other hand, precise, dynamic control of the spark gap is essential for stable repetition of the discharge. These features make the control system unique and, consequently, the sensing strategy must be different from that in the case of conventional machining processes.

4.6.2
Principle of EDM

From the viewpoint of shape specification of the workpiece, EDM is included in the same group as cutting where the shape of the tool is copied on to the workpiece or further modified according to the trajectory of the tool movement (relative to the workpiece). The tool is referred to as the electrode.

The main difference between EDM and conventional machining processes such as cutting is in the mechanism of material removal. The material removal process in EDM is illustrated in Figure 4.6-1.

A voltage pulse from a pulse generator initiates and maintains a short arc discharge through the dielectric fluid in the gap between the electrode and the workpiece. The discharge heats the surface layer of the workpiece locally and melts (and also partially vaporizes) a small part of the workpiece. The simultaneous vaporization of the fluid produces a high pressure and the molten material disintegrates into microsize particles, forming a shallow crater on the workpiece surface.

The process leading to this crater formation is repeated over the entire workpiece surface facing the electrode, producing a thin cavity layer on the workpiece. By continuing this process with the feed of the electrode facing downward, the entire electrode shape is replicated on to the workpiece.

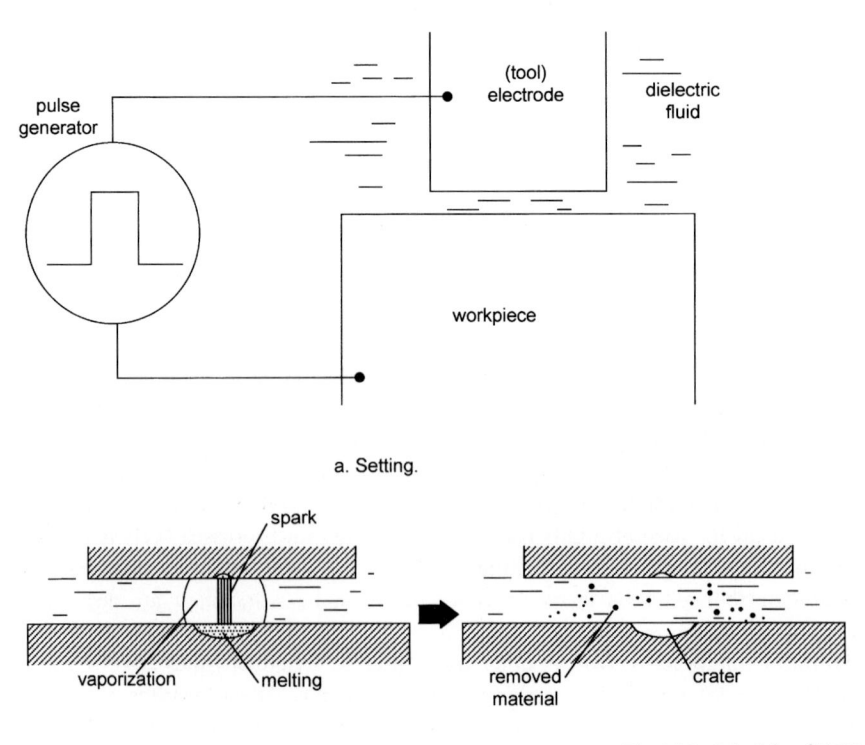

Fig. 4.6-1 Principle of EDM

The discharge also forms a crater on the electrode. However, specifically chosen discharge conditions can minimize the size of the crater to, for example, less than 0.1% of the crater size on the workpiece.

4.6.3
Process Control

Discharge occurs only when the gap distance is within a certain range. If the distance is too large, the discharge cannot be initiated. If the distance is too small, a short-circuit occurs after the discharge, because a bridge is formed by the molten material. This short-circuit terminates any further discharge.

Because of the above characteristics, gap control is the main issue in the case of EDM control

A constant electrode feed rate is not usually adopted, because the occurrence of the discharge is a more or less random process and the speed of removal in the electrode feed direction is not constant.

Therefore, the control strategy involves maintaining a constant gap distance throughout the operation. However, direct in-process measurement of the gap distance is almost impossible. The main reasons for this are that (a) the gap is very narrow, 0.5–50 µm., and (b) the gap distance varies throughout the machining area and the nearest point determines the occurrence of the discharge and short-circuit. For these reasons, indirect detection of the gap distance is adopted in all EDM machines.

The basic control algorithm is shown in Figure 4.6-2.

4.6.4
Sensing Technology

In EDM, sensing technology is mainly required for obtaining information on the gap condition. As described previously, the information finally required is on the gap distance, which cannot be obtained directly. In order to overcome this difficulty, relationships between the gap distance and other detectable parameters have been analyzed. The parameters that represent the state of the gap condition are used as control parameters.

Fortunately, as the removal mechanism of EDM is based on an electrical phenomenon, useful information is obtainable from the electrical parameters of the circuit for discharge. The most frequently used parameters are the gap voltage and the current through the gap, as shown in Figure 4.6-3.

Some other parameters can also provide information on the machining status. The more promising ones are the electromagnetic radiation from the arc and the acoustic radiation caused by discharges.

In the detection of these parameter values, sensors may play a useful role. To date, sensors have played a minor role in EDM control, because the more useful parameters, such as the voltage and the current, can be detected without special sensors. However, in some cases, sensors offer merits from the viewpoints of improved technology and economics.

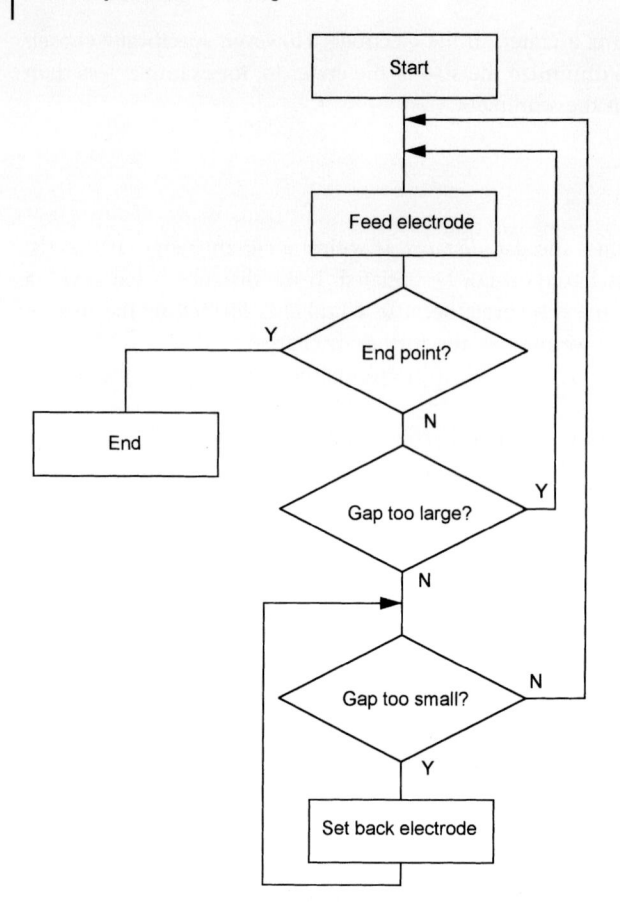

Fig. 4.6-2 Basic control algorithm for EDM

The following is an overview of the major technologies used in obtaining gap information.

4.6.4.1 Gap Voltage

In EDM, the gap voltage contains numerous information. The status of the gap is indicated by the gap voltage as shown in Figure 4.6-4.

The figure shows only typical examples but there also occur various transient and combined types of voltage waveforms. These voltage pulses are input into the control system and statistical calculation is performed which provides the necessary values that can be compared with the set values for proceeding according to the flow chart shown in Figure 4.6-3.

In this detection of gap voltage, no special sensor is usually required. A conventional high-impedance probe commonly used for measuring equipment such as an oscilloscope can be used without any problems. The voltage range is usually from 50 to 200 V.

Fig. 4.6-3 Parameters used for controlling EDM

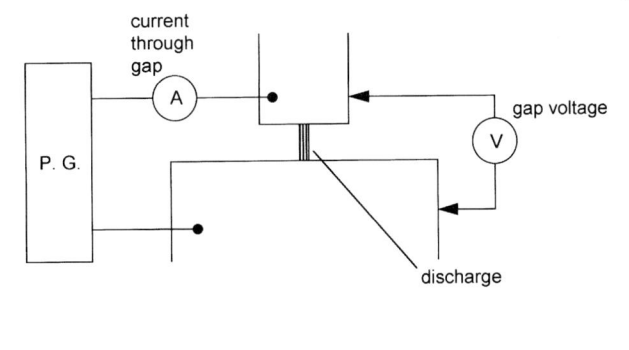

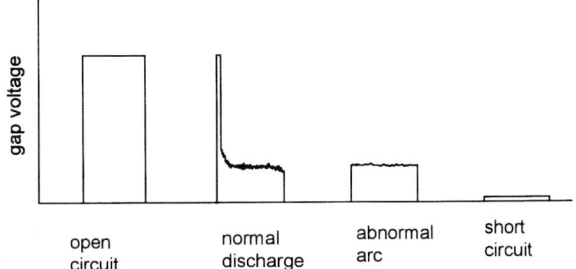

Fig. 4.6-4 Voltage change according to the status of gap

open circuit   normal discharge   abnormal arc   short circuit

When very short pulses are applied for micromachining or very fine finishing, a special pulse circuit called a 'relaxation-type generator' is often used. In such a case, obtaining the input from a probe is difficult because the impedance of the gap is too high when it is open-circuited. Moreover, the input capacitance of the probe may change the energy of the discharge. Special voltage sensors may provide a solution to this problem, but in most cases the detection of the gap voltage is abandoned and replaced by the detection of the current through the gap. This method is discussed in the next section.

4.6.4.2 Current Through Gap

Since discharges for EDM are the current flow from the electrode to the workpiece (or in the reverse direction), the dynamic value of the current through the gap contains numerous information. The typical relationship between the current waveform and the status of the gap is shown in Figure 4.6-5. As is apparent when Figures 4.6-4 and 4.6-5 are compared, the current waveform can be as useful as the gap voltage described earlier. Since the difference between the cases of normal discharge and abnormal arc is only the high-frequency component in the waveform, control using current information requires more careful data processing than that using gap voltage.

For detecting the current waveform, two types of devices are currently used.

(A) Shunt resistor. The most conventional and reliable method involves using a low-resistance resistor inserted in the loop of the gap current flow, as shown in

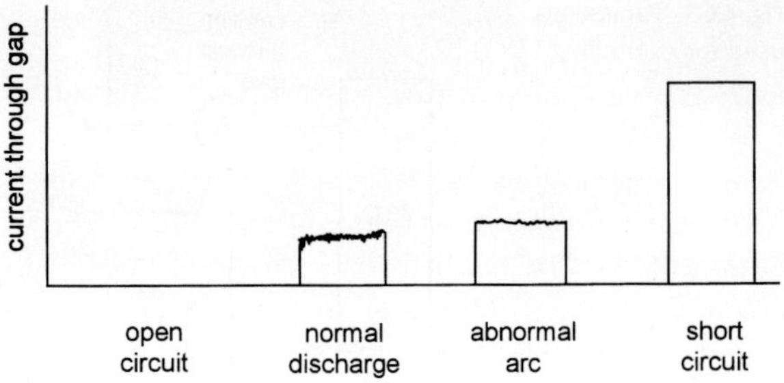
current through gap
open circuit
normal discharge
abnormal arc
short circuit

Fig. 4.6-5 Current change according to the status in gap

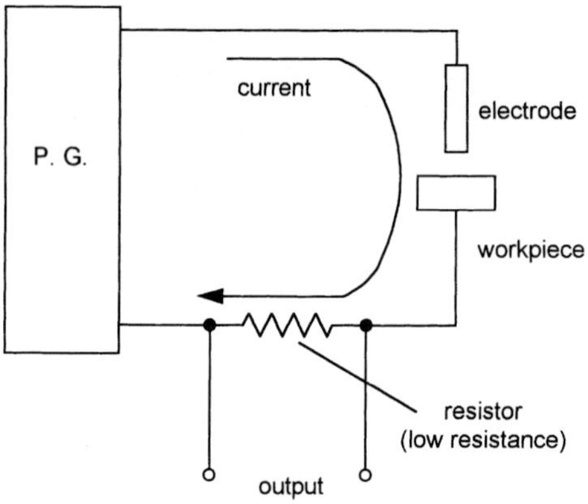
P. G.
current
electrode
workpiece
resistor (low resistance)
output

Fig. 4.6-6 Circuit for current detection

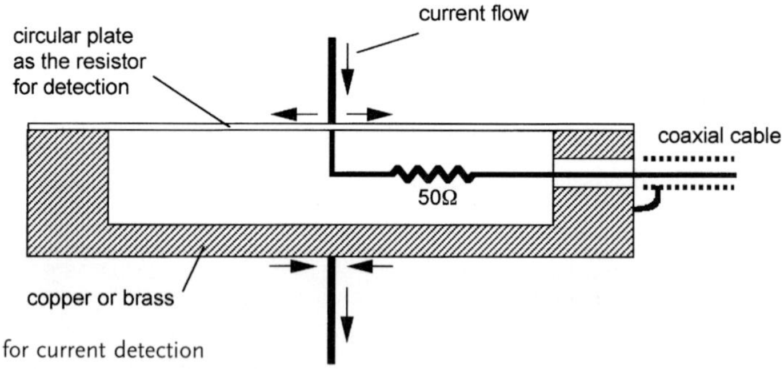
circular plate as the resistor for detection
current flow
coaxial cable
50Ω
copper or brass

Fig. 4.6-7 A shunt for current detection

Figure 4.6-6. In a simple control system, a general-purpose solid resistor can be used as a detector. However, when more detailed information is required, a specially designed shunt is necessary because the current signal contains a very high-frequency component.

A suitable design for a shunt is illustrated in Figure 4.6-7, where the residual inductance can be sufficiently suppressed.

(B) Current probe with Hall sensor. Another method involves using a commercial current probe which outputs current waveforms only by clipping one of the connecting wires to the electrode and the workpiece. In this application, careful atten-

tion is required because the direct current (DC) level is important in EDM control. The probes customized alternative current (AC) cannot be used, or at the very least requires complicated data processing.

4.6.4.3 Electromagnetic Radiation

Since transient arc discharges radiate a wide spectrum of electromagnetic waves, medium-frequency (MF), high-frequency (HF), or very-high-frequency (VHF) receivers can offer useful information concerning discharges. A small dipole (centimeters long) fixed near the machine head can detect the radiation. The signal is processed through bandpass filters and analyzed. This method is particularly effective in distinguishing normal discharges from other types of current such as abnormal arcs and short-circuits, because the radiation mainly occurs during normal discharge pulses except during the rise and fall of pulses.

Arc discharge also radiates high intensity light. The spectrum and intensity of this light contains information on the discharge status. However, this information has not yet been practically applied.

4.6.4.4 Acoustic Radiation

The rapid rise and fall of temperature at the point of discharge produces shock waves and pressure pulses. This causes acoustic radiation. The spectrum contains the audio-frequency range, and one can audibly determine whether discharges are taking place in the gap. This signal can be detected using conventional microphones. However, the dynamic range of a microphone is usually not sufficient to cover the rough machining to very fine finishing range, because the pulse energy ranges from 1 J to 0.1 µJ.

Detecting acoustic emission by attaching sensors on the workpiece or on the electrode (or its holder) is another approach for obtaining acoustic information. Since the generated acoustic spectrum in fine finishing tends to be more intense in the higher frequency range, a bandpass filter designed to transfer HF or VHF may be effective in obtaining information covering a wider range of discharge energy distribution. In this type of application, commercially available acoustic emission (AE) sensors can be used. This type of detection technique is, however, also still under development.

4.6.5
Evaluation of Machining Accuracy

In micromachining, evaluation of machining performance is sometimes difficult, particularly in the case of machining precision. Drilling microholes is the simplest typical example of micromachining. However, suitable means for dimensional measurement are not provided by the manufacturers of measurement equipment, if the object is a hole with a diameter of less than 0.5 mm and a depth of more than 1 mm.

Since recent applications often require holes of diameter 100 or 50 µm, the development of technology for measuring the internal dimensions of such holes is an urgent requirement in production engineering. However, in the measurement of such small-sized objects, errors in fixing the object often strongly influence the result of the measurement.

This leads to the requirement for on-the-machine measurement (OMM). OMM means the measurement is performed without removing the object from its clamped position on the machine tool which machines the microhole on the object. In OMM, no extra error is introduced after machining until measurement is performed.

On the other hand, the tools for measurement must be installed on the machine, because the common use of coordinates between the machining system and measuring system is essential in OMM. This type of set-up is usually difficult because of the wide gap in the design concept of the equipment between machine tools and measuring machines.

A recently developed measuring technique called the vibroscanning (VS) method can provide a solution to this problem in some cases of application.

4.6.5.1 VS Method

The VS method utilizes a probing system to obtain the surface profiles, using the principle of electrical contact between the probe and the object. It is well known that the detection of electrical contact is not precisely reproducible. Therefore, this type of detection has not been applied for dimensional measurement. Several examples of application have been found only in the positioning of workpieces in machine tools with medium accuracy.

The difference between the VS method and the conventional method using electrical contact is that the VS method applies an intermittent contact derived from the vibration of the probe for contact.

Figure 4.6-8 illustrates the basic principle of the VS method. The cantilever for surface detection is vibrated at a low frequency so as to avoid mechanical resonance. The amplitude is adjusted to approximately 2 µm. When the cantilever is

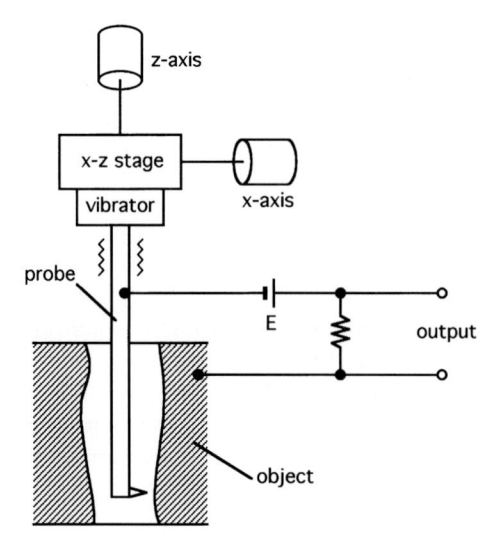

Fig. 4.6-8 Principle of VS method

Fig. 4.6-9 Relationship between probe movement and output signal

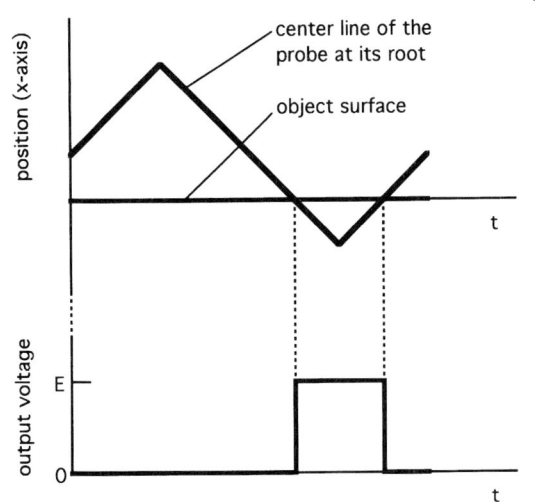

close to the surface, its tip contacts the surface intermittently. This contact produces a series of rectangular pulses as output. The relationship between the probe movement and the output signal is illustrated in Figure 4.6-9. The pulse duration changes according to the distance between the center of vibration and the surface, or the x-coordinate of the probe position. If the probe position is controlled so as to maintain a constant duty cycle of the output, eg, 50%, the vibration center follows the surface curve when the probe is driven in the longitudinal direction, or y-direction. Storing the x and y data of the probe position is equivalent to storing the data for the surface curve.

Owing to the dynamic contact, the output reproducibility of this method can be in the submicron range. Since the deformation of the cantilever at its tip is only around 1 μm, it is possible to measure the internal dimensions of a microhole provided that the cantilever can be inserted in the hole.

4.6.5.2 Application of Micro-EDM

The simple mechanism of the VS method realizes OMM. A practical example of the application of the VS method for measuring EDMed microholes is outlined below.

The set-up of the measuring attachment (VS head) on a micro-EDM machine is shown schematically in Figure 4.6-10. The center lines of the machining head (z'-axis) and VS head (z-axis) are fixed in parallel at known positions on the machine. After a microhole has been machined in the workpiece, the workpiece is moved horizontally to the VS head position and measurement can be performed immediately. This simple and rapid transition from machining to measurement on the same machine introduces two advantages. First, the internal surface of the microhole is fresh after machining and provides a good electrical contact for carrying out VS method. Second, the result of measurement can be fed back to the

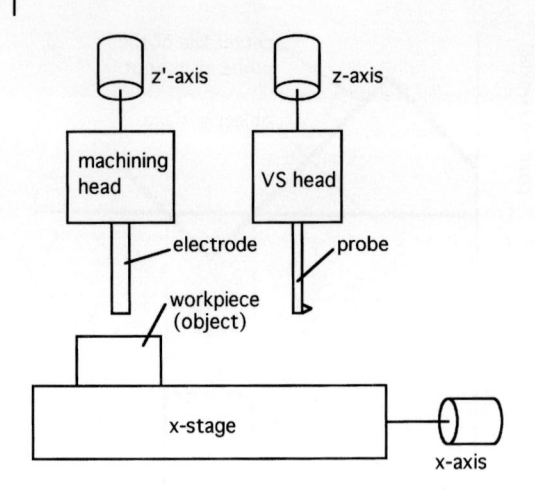

Fig. 4.6-10 Set-up for on-the-machine measurement

machining system by moving the workpiece back to the machining position and performing additional machining. In other words, the dimensional error of machining can be corrected before demounting the workpiece from the machine.

Concerning the VS method, applications for the measurement of practical parts and improvements in reproducibility and accuracy are currently being investigated.

4.7
Welding

H. D. HAFERKAMP, *Universität Hannover, Hannover, Germany*, F. v. ALVENSLEBEN, *Laser Zentrum Hannover*, M. NIEMEYER, *Universität Hannover*, W. SPECKER, *Laser Zentrum Hannover*, M. ZELT, *Universität Hannover*

4.7.1
Introduction

Quality in welding depends not only on the various process parameters such as welding speed, current, and voltage, eg, for arc welding. In most cases, the welding process is affected by:

- tolerances and mismatch of the workpiece geometry;
- tolerances in machines and clamping devices;
- variations in groove shape;
- tack beads; and
- welding deformations.

In manual welding, trained welders are able to compensate for all these influences, because their senses, especially the eyes and ears, give them the information they need to produce a high-quality weld. For mechanical and automatic welding, all this information must be detected by sensors. Such a sensor for welding can be defined as follows [1]:

'A detector, if it is capable of monitoring and controlling welding operation based on its own capacity to detect external and internal situations affecting welding results and transmit a detected value as a detection signal, is called as a sensor. Moreover its whole control device is defined as a sensor system (control system)'.

In this definition, the external situation refers to all workpiece-related geometric values such as changes in dimensions of the welding groove, position of the welding line, and presence of component obstacles or tack welds. The internal situation covers factors such as the shape of the welding arc and molten pool, the penetration depth, and all kinds of effects related to the welding process itself [2].

In general, every physical principle which is able to deliver information about an object's shape and position may be the basis of a sensor. For welding sensors, the special ambient conditions and the industrial constraint for economic efficiency, however, cause many additional restrictions, such as:

- process-induced disturbances, such as light, heat, fume, spatter, and electromagnetic fields, must not influence the sensor;
- the sensor must be satisfactorily durable for welding ambience;
- it must be compact in size and light weight, so that there are no restrictions in handling and accessibility;
- the sensor system must only generate low costs;
- it should have easy maintenance.

Owing to this and because of the very complex process, so far no universal sensor for welding is available which meets all these requirements and is able to detect all the various kinds of information by which the welding process is influenced. For the user, it is necessary to select the most satisfactory sensor type for every special welding task [2–5].

In general, a classification of welding sensors can be made by their functional principle. In Figure 4.7-1 such a classification of sensors for welding is shown in accordance with [3]. Further classification is made by the physical principle on which the sensor is based.

4.7.2
Geometry-oriented Sensors

Geometry-oriented sensors gather their information from the geometry of the welding groove itself or from an edge or a surface which has a defined relation to the seam. Geometry-oriented sensors can be divided into contact and non-contact types.

4.7.2.1 Contact Geometry-oriented Sensors
Contact sensors permit the detection of the welding start/end point and seam tracking with comparatively low technical expenditure. The first seam tracking

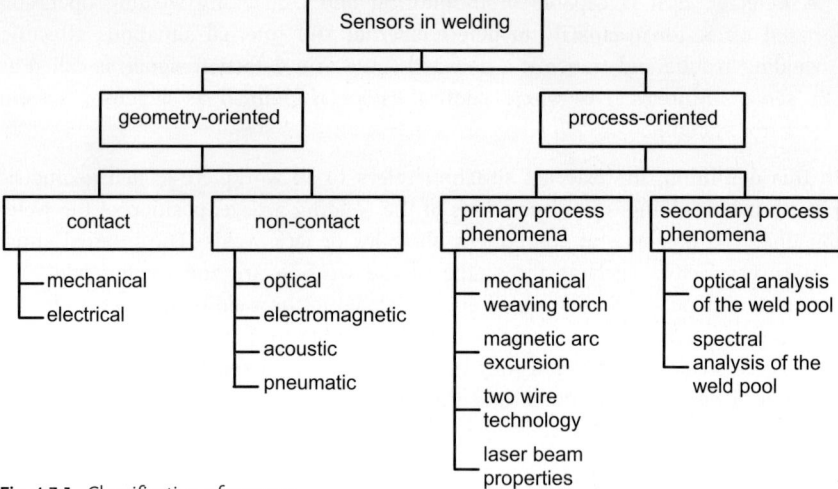

Fig. 4.7-1 Classification of sensors

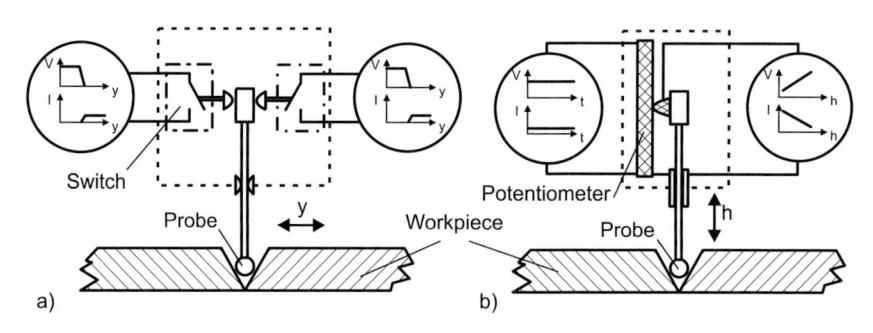

Fig. 4.7-2 Contact probe sensors: (a) limited switch type; (b) potentiometer type

systems in welding were based on the mechanical tracing of a gap or an edge by a probe and the direct transformation of movements to the torch by way of fulcrums. The form of the probe, eg, as a ball, roll, or ring, must be adjusted to the groove geometry. Further technical development leads to electromechanical sensor systems, which convert the probe movement into electrical signals. Generally, there are two different kinds of these sensors. One group operates with limited switches, which deliver an on/off output signal, to track the seam stepwise. The other group uses potentiometers or differential transformers to generate a distance-proportional output signal, which allows continuous seam tracking. In Figure 4.7-2, the principle of these two sensor groups is shown [1, 6, 7].

The probe may have one or two degrees of freedom, so it is able to compensate for most two-dimensional deviations of the weld seam. Usually, the contact probe sensor is in a fixed position ahead of the welding torch, which causes some limitations in the shape of the welding seam. Because of the distance between sensor and welding torch, which leads to deviations on curved seams, this method is commonly used for welding straight-lined seams.

In order to apply contact sensors to welding curved seams, several techniques have been developed. Bollinger [8] described a method that is based on the turning of the complete fixed torch-sensor unit around defined axes. From the measured turn angles, the path feed rate of the different axes is calculated. Nevertheless, this system leads to some deviation on seams with a small radius of curvature. To avoid this, it is necessary to monitor the welding seam using the sensor, prior to welding, and store the deviation values of the seam. With the stored coordinates, the system allows the welding of small curve radii with constant torch orientation. However, the system is not suitable for welding closed contours. The maximum deviation angle to the mean welding direction is given by the author as $\pm 60°$.

Another method for contact sensors for welding curved seams is based on the mechanical decoupling of the sensor and torch motion [9]. In these systems, called memory delay playback, the sensor is mounted on a separate x-y drive block, which allows tracing of the shape of the groove independent of the torch movement. The groove deviation values are stored, and based on the welding speed and the distance between sensor and torch, the correct position of the welding torch is adjusted when it is moving towards the former sensor position. This sensor system leads to satisfactory results, even in welding small bending radii. Nevertheless, this system also is not capable of welding closed contours. It is just able to compensate deviation angles of $\pm 30°$ to the mean welding direction.

Another way of using a contact sensor for welding any curved seam is detection of the seam deviation prior to welding, and compensation of the programmed seam line by the measured deviation values, as described by Schmidt [10]. Prior to the first weld, a contact probe sensor is mounted in lieu of the torch gas cup, and the welding robot senses the deviation of the workpiece weld line to the programmed one. After this sensing cycle, the normal gas cup is mounted, and the robot starts to weld the rectified seam line. This method calls for a separate measure cycle prior to every weld, and is not able to compensate for deviations that occur during welding, eg, due to thermal distortion.

A seam tracking system has been described [11, 12] which was developed for laser beam welding of three-dimensional fillet welds using an industrial robot. The mechanical sensor consists of a metal tracer pin, which is dragged along the fillet joint at a fixed distance ahead of the laser spot. Positional changes induce potentiometric variations at the head of the pin. The sensor feeds these variations to an electronic controller as a stream of analog data. The controller then guides the laser spot accordingly, by means of two servo motors which give the vertical and transverse motion of the laser head, using commercial systems, within a range of ± 7.5 mm and with an accuracy of ± 0.1 mm. Welding speeds up to 6 m/min are possible with current systems.

Generally, the use of contact probe sensors is limited by the wear of the probe itself. Because of permanent contact to the workpiece, there is marked friction wear, which decreases the probe lifetime. Further limits are caused by the accessi-

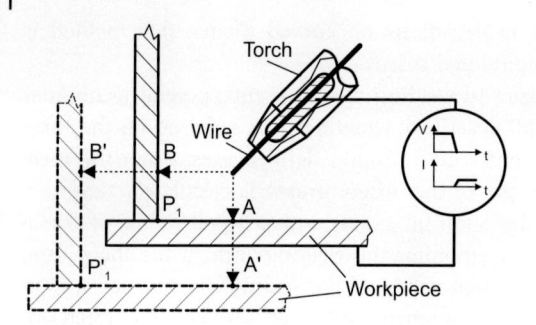

Fig. 4.7-3 Principle of an electrode contact sensor

bility of the joint. The additional sensor mounted in front of the torch means a significant enlargement of the tool, and so welding in confined areas is restricted.

Nevertheless, contact probe sensors have been widely used in many industrial applications for a long time, especially owing to their simple design and easy handling and maintenance.

Another special contact sensor is the electrode or wire contact sensor. This is a sensing method which was developed for arc welding robots. It is able to detect deviations between a taught point of the robot path and the present position of the welding torch. In Figure 4.7-3, the principle of this sensing method is shown.

The basic idea of this sensor is to use the torch as a switch in an electric circuit. In this circuit, the workpiece surface and the welding wire have different polarity. When the wire comes into contact with the workpiece, a change in electric current or voltage can be detected. The difference of the taught point and the actual position can be calculated, and the real position of the workpiece is defined. For using the welding wire as a probe, the stick-out length of the wire must be defined. Therefore, wire extension may be determined by automatically cutting it to length prior to sensing, or it can be calculated by sensing a machine reference point, which has no initial deviations prior to workpiece sensing. Another method is the use of the welding torch gas cup as the contact dip [13–16].

The electrode contact sensor is industrially used in robotic welding to detect variations of the starting and end points of welds and of the length of welds and in sensing the form of the welding gap prior to welding. They are simply designed, easy to use, and not subject to wear. This sensor type is able to achieve an accuracy of ±0.2–0.3 mm in position detection. Beyond that, they cause no restrictions in accessibility of the joint, because there are no additional extensions to the torch [14–17].

In general, the use of these kinds of sensor can be limited by all kinds of insulating coatings on the workpiece, such as primer or oxide layers. Furthermore, the electrode contact sensor is not able to allow for deviations which occur during welding, eg, due to thermal distortion. Hence, it is usually used for short welds or in combination with an additional seam tracking sensor, eg, a through-the-arc sensor [13–17].

4.7.2.2 Non-contact Geometry-oriented Sensors

A further development in sensor systems is the non-contact geometry-oriented sensors. These sensors are based on various physical principles of measurement (see Figure 4.7-1). Generally, they deliver information about the workpiece shape and its position in space. Depending on the sensitivity and accuracy of the sensing system, non-contact geometry-oriented sensors are able to detect the start and end points of welds and track weld seams. The most commonly used types in this category of sensors are based on optical, electromagnetic, and acoustic measurement. The fourth category of pneumatic sensors from the list in Figure 4.7-1 use the impact pressure of a gas nozzle to detect the distance between the workpiece surface and the sensor. This sensing method is not commonly used in welding processes at present.

In laser welding long seams, the problems resulting from the geometric accuracy of the workpiece become a decisive factor. Industrial robots are often used to guide the laser head or the welding torch along the workpiece. In laser welding the robot-guided beam must follow the (three-dimensional) seam geometry accurately, because focus diameters are typically in the range 0.15–0.5 mm. Additionally, any movement out of the focal plane (eg, the distance workpiece lens changes) can cause a defective weld. The robot is usually programmed manually, using a time-consuming point-by-point basis, so that curves are often estimated. In addition to that in arc welding, the process caused thermal distortion of the workpiece, often leading to geometric deviations of the joint line.

Optical sensors, which use the topography of the workpiece surface in order to detect the weld seam, belong to the non-contact geometry-oriented type. The basic principle of the optical measurement used in this sensor group is a light-section procedure. Using a laser diode, a line-shaped laser beam is projected on to the

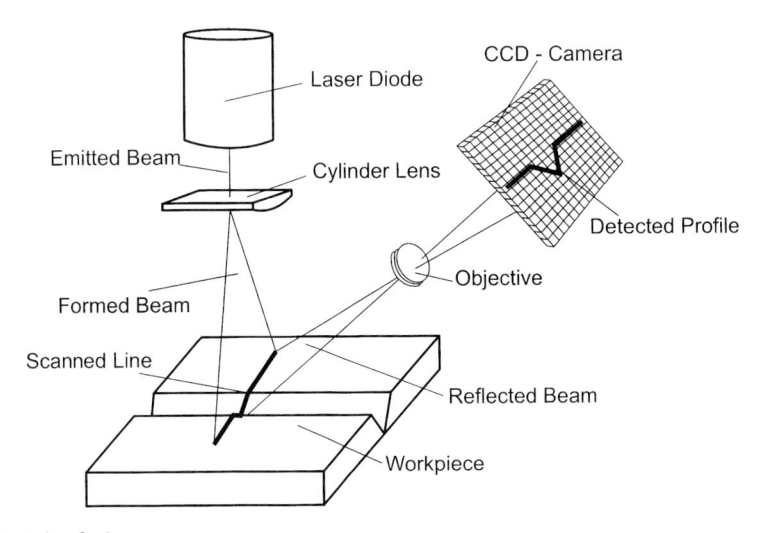

Fig. 4.7-4 Principle of a laser-stripe sensor

workpiece (see Figure 4.7-4). A variation in the distance between sensor and workpiece leads to a change of the reflected beam position. This reflected beam is measured by a charge-coupled device (CCD) camera whose data are processed by a PC, in order to calculate the workpiece surface contour. These data can be used for seam tracking, groove shape detection, and detecting weld start/end points [18–23].

The data of the sensing system are transmitted to the handling system, in order to correct the beam or torch position on the workpiece. Usually, it controls, eg, the robot directly via CNC commands. The measurement accuracy of commercial systems is 0.025 mm, and these systems are suitable up to maximum welding speeds of 15 m/min. The positioning accuracy also depends on the handling system. Both optical components, laser diode and CCD camera, are adapted to the laser head and to the welding torch, which makes it sensitive to alignment, dust, fumes, and spatter. The optical method has the drawback that reflections and scratches on the workpiece surface may cause the system to go astray.

Electromagnetic sensors are non-contact geometry-oriented sensors, which gain their information by the effect of metallic materials on electromagnetic fields. These sensors, used to detect position or displacement, are classified into capacitance and eddy current types. Capacitance sensors measure the capacity between the workpiece and a small electrically conductive plate. They offer the possibility of distance sensing. Matthes et al. [24] described a capacitance sensor for seam tracking in V-grooves. The sensor signal of capacitance sensors is heavily vitiated by deviations in flatness or parallelism of the workpiece surface. Hence this kind of sensor ordinarily is not used in welding, but sometimes is in thermal cutting [2, 3].

The eddy current type is based on the interaction of metallic materials and alternating magnetic fields. The sensor induces eddy currents in the near-surface range of the workpiece. These eddy currents influence the inductance of the sensor coil, depending on the distance between sensor and surface. From this influence, a distance-dependent electrical signal is obtained [2, 3, 25–28]. The principle is shown in Figure 4.7-5.

Depending on the frequency of the eddy current, the sensor reacts in different ways to the various magnetic characteristics of the workpiece materials. Sensors

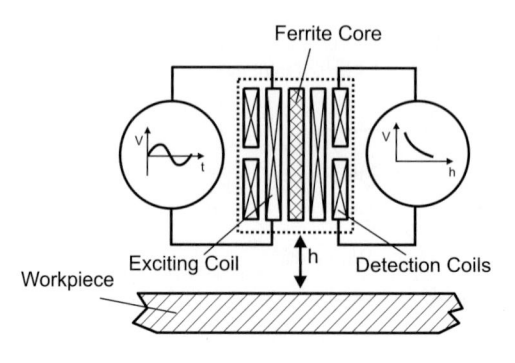

Fig. 4.7-5 Electromagnetic sensor, eddy current type

with low-frequency eddy currents are only suitable for ferromagnetic materials, whereas high-frequency sensors are applicable to both ferromagnetic and non-magnetic materials [27, 29].

Electromagnetic sensors with one coil system are limited to detecting the distance to the workpiece in one direction. Hence they are only able to adjust the torch's height or lateral deviation. Sensors with a combination of several coil systems, however, allow sensing a welding groove in every direction. In addition to height and lateral deviation, these systems can detect changes in the direction of the welding groove, the beginning and the end of a groove and some changes in the setting angle of the welding torch [29–32].

Because of the geometric distance between the torch and the electromagnetic sensor, these systems are affected by some deviations on curved welds (compared with contact probe sensors). To avoid these deviations, several methods have been developed. In one system [29], the sensor rotates around the torch, and the sensor signal is connected to the direction by a turn angle transmitter. Considering the welding speed and direction, the deviation between the sensor and torch can be compensated by the control system. This allows one to track curved seams in every direction with satisfactory precision.

For tracking fillet welds, another possibility is to sense the weld flanks by a collateral arrangement of two sensors to the torch [2, 33]. Every sensor is arranged perpendicular to one flank. In that way, by sensing and adjusting the distance to them, the torch follows the seam. An eddy current sensor has been described [34] which is concentric to the torch and integrated in the gas cup. This leads to a very compact design, so accessibility problems are minimized.

In general, eddy current sensors are able to compensate deviations with an accuracy of ±0.15–0.5 mm. They are suitable for detecting almost every kind of groove shape. In butt joint welding they are able to track gaps up to a width of 0.05 mm. Nevertheless, the use of these sensors is limited in several ways. In general, some additional extension of the torch is necessary, so the accessibility of seams is limited. When welding butt joints, filler and cover passes are difficult to track using eddy current sensors. Edge misalignment on butt joints causes deviations to the center of the weld, so very exact preparation of the workpiece and reliable fixture is essential for accurate seam tracking. Moreover, eddy current sensors are affected by any foreign magnetic field in the sensing area. Even geometric changes in the region of the seam, such as clamping fixtures, tack welds, workpiece thickness, and material non-homogeneities, can influence the sensor signal [3, 26, 29, 34–36].

In spite of these disadvantages, eddy current sensors are widely used in many industrial applications. Because of their robust design, universal application capability, and comparatively low cost, their application is economical for a great variety of sensor tasks.

Another type of non-contact geometry-oriented sensor utilize ultrasonic signals to gather information. The principle of this kind of sensor (see Figure 4.7-6) is based on the fact that ultrasound waves are reflected from material surfaces, and that the propagation of these waves in air is related to the distance between the

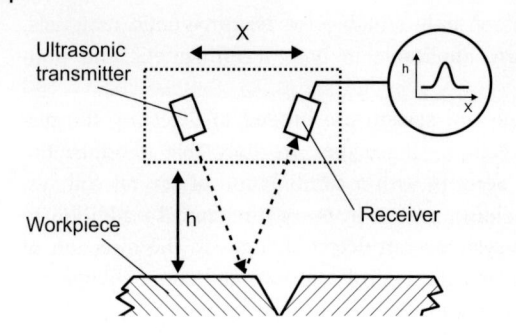

Fig. 4.7-6 Principle of non-contact ultrasonic sensor

ultrasonic transmitter and the receiver. For tracking weld seams, it is possible to use either the reflected energy amplitude or the range information, or both [37, 38]. In the first case, the sensor scans the area in front of the welding torch and finds the seam by detecting a modification of the reflected energy in the course of a scan cycle. This energy modification is caused by any change in the wave's angle of incidence to the reflecting surface, eg, on groove edges. In the second case, the distance between the sensor and the workpiece surface is monitored by timing the interval between wave transmission and echo return. Based on these distance measurements in the scanned area, the workpiece profile along the scanning path can be determined.

For tracking curved seams using ultrasonic sensors, commonly the same methods are used as for electromagnetic sensors. In most applications, the ultrasonic sensor moves on a circuit path around the welding torch, and the seam direction is determined considering the sensor's angular position, and the measured distance values [37–42].

In order to increase the sensitivity and lateral resolution of non-contact ultrasonic sensors, different methods are used. Zhang et al. [39] described a sensor system that uses high-frequency ultrasonics of 1.15 MHz to improve tracking accuracy. In general, the width and wavelength of an ultrasonic beam increase as the frequency decreases. Lower frequencies correspond to poorer resolutions, but longer travel distances. Thus, for traditional applications of non-contact ultrasonic sensors, such as long distance measurement of large objects, low-frequency sensors are suitable. However, in weld seam tracking, small gaps, eg, in butt welding, are difficult to detect. On the other hand, unlike low-frequency ultrasonics, high-frequency ultrasonic signals deteriorate significantly in air. So the range of these signals is limited. To overcome these limits, the transmission efficiency of the high-frequency ultrasonic transmitter was improved and an adjusted transducer was developed to focus the beam [39]. The system shows a tracking accuracy of 0.5 mm.

Another method for improving the tracking accuracy of non-contact ultrasonic sensor systems was presented [40]. The system uses the level of the reflected energy from the workpiece surface to detect the welding seam. The ultrasonic transmitter operates with an ultrasonic frequency of 150 kHz, because at this frequency the sensor is least sensitive to the arc noise in welding. For improving the resolution of the acoustic signal detected, and to eliminate the influence of noise from other than the scanned direction, a waveguide is used. This waveguide improves the sensor system's tracking accuracy using several effects. On the one hand, it concentrates the receiver's sensitivity to a small area, being positioned near it. On the other hand, a matched dimensioned waveguide acts as a resonator, and thereby increases the receiver's sensitivity. In addition, it can be used as a filter to attenuate interfering signals which arise from, eg, process noise. This system also shows a seam tracking accuracy of 0.5 mm.

The major limitations in using non-contact ultrasonic sensors in welding are caused by the significant enlargement of the torch. The application of the sensor itself and its guiding mechanism leads to accessibility problems in welding small and complex structures. Further, for seam tracking on butt welds, there must be at least a groove of 0.5 mm depth and width that the sensor is able to detect.

In general, ultrasonic sensors are distinguished by their simple configuration and low cost. The sensors are robust in harsh welding environments with arclight, fumes, dust, and sputter, without any degeneration in sensing sensitivity. In addition to common industrial welding conditions, ultrasonic sensors are also suitable for underwater wet welding [41, 42].

4.7.3
Welding Process-oriented Sensors

Process-oriented sensors gain their information from the primary and secondary process phenomena. In this classification, primary process phenomena data related to laser welding are the beam quality and the laser power. In arc welding, primary process phenomena are arc related and they can be acquired directly in the welding circuit, such as welding current and voltage. Arc welding sensors, for example, use this information. Secondary process phenomena data, on the other hand, are gathered by observation of the joining area while welding. From the radiation of the arc and welding pool and from the geometry of the joining area, information for torch positioning is generated.

4.7.3.1 Primary Process Phenomena-oriented Sensors
Referring to the classification of welding sensors in Figure 4.7-1, the through-the-arc sensor is a primary process phenomena-oriented sensor. This sensor type uses the electrical characteristics of the welding arc in order to detect the distance between the torch and workpiece surface. Generally, in arc welding the ohmic resistance in the welding circuit is closely related to the arc length. Depending on the welding process and the characteristic curve of the power source used, the welding voltage or current is more influenced by this [43].

The general principle of a through-the-arc sensor in gas-metal-arc (GMA) welding with consumable wire electrodes is shown in Figure 4.7-7. While the torch is

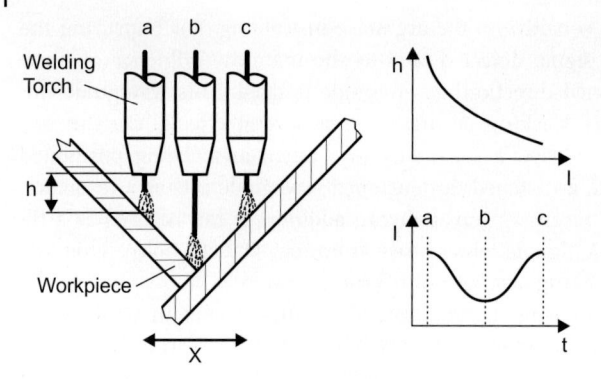

Fig. 4.7-7 Current waveform in an oscillating consumable electrode arc

moving laterally to the welding path, as shown from position (a) through (b) to (c), the distance between the contact tube and the arc root changes. Because of the arc length self-adjustment effect, the stick-out length of the wire electrode, and hence the ohmic resistance in the welding circuit, also change. This causes a variation in the welding current and voltage, according to the source characteristics. In GMA welding, usually constant-voltage power sources with a slight falling characteristic are used. This leads to a in rise welding current when the distance between contact tube and arc root decreases (see Figure 4.7-7). Thus, by monitoring the welding parameters while the torch is oscillating laterally to the welding path, the joint geometry can be detected. For seam tracking, for example, a comparative measurement of the welding current in the stationary points of the oscillating path leads to the determination of the torch position relative to the joint [13, 44, 45].

In contrast to GMA welding, in tungsten inert-gas (TIG) welding usually power sources with a constant current characteristic are used. Here, any change in arc length leads to a variation of the welding voltage. This is utilized to control the torch's distance to the workpiece surface by keeping the arc voltage constant. Hence the workpiece surface can be scanned by the arc itself. For seam tracking, the torch oscillates laterally to the welding path along the groove. The stationary points of the oscillating path are determined by comparing the instantaneous torch height with a reference value. Thus, the torch follows the seam by weaving from one groove face to the other. By measurement of the torch oscillation amplitude, a bead height control in addition to seam tracking is possible [46, 47].

For the different arc welding processes (submerged arc (SA), GMA, and TIG) and the distinct metal transfer modes in GMA welding such as short, spray, or impulse arc, several analysis methods for the determination of the torch height, seam line, groove, and weld pool geometry by the measured welding parameters have been developed. These methods are based on, eg, comparing the instantaneous welding data in the turning points of torch oscillation, comparing integrated welding data, comparing frequency components of the welding data, or using multivariate analysis of the hysteresis loop, formed by the torch position and the welding data [44–54]. This illustrates the importance of the signal processing techniques for the performance of a through-the-arc sensing system. Further, the sensitivity and the accuracy of such a system depend to a high degree on a steady, stable, and trouble-free welding process, because any discontinuity in arc burning affects the sensor immediately.

For seam tracking using an arc sensor system in arc welding processes which use only one arc, oscillating motion of the arc laterally to the welding path is necessary. This oscillation can be achieved by mechanical movement of the torch, or by excursion of the arc by virtue of electromagnetic forces. Most industrial applications use mechanical movement of the torch. This can be achieved by an additional axis or by using, eg, all six axes of a welding robot. Because of the torch's inertia or even the robot's mass, the oscillation frequencies of such systems are limited to a maximum of 10 Hz. Therefore, the maximum welding speed, at which these systems can track a seam with satisfactory accuracy is 60 cm/min [4].

In order to minimize the problem of inertia and to increase the possible oscillation frequency of the arc, systems that use a high-speed rotating arc have been developed. In these systems, the arc is forced to make a circular motion by specially designed torches. Several torches with motor-driven rotating contact tubes or torch necks have been described. They can operate with oscillation frequencies up to 50–150 Hz. A high oscillation frequency improves the accuracy and sensitivity of through-the-arc sensor systems in two ways. First, the signal rate increases, so it is possible for the sensor to correct the torch position faster, and the maximum welding speed for these sensor systems increases to more than 100 cm/min. Second, the detectable welding parameter variations increase, because the effect of arc length self-adjustment is suppressed. This self-adjustment effect usually needs 50–200 ms to compensate for arc length variations by adjustment of the wire stick-out length. Hence changes of the torch height for shorter periods lead only to a change in the arc length, without influencing the wire stick-out. As the ohmic resistance of the arc is much higher than that of the welding wire, the effect on the welding parameters is larger. According to this, it is even possible to use through-the-arc sensor systems with high-speed rotating arcs for seam tracking with filler metals of very low ohmic resistance, eg, in the welding of aluminium [17, 46, 55–58].

Another method to reach high oscillation frequencies uses the effect of magnetic forces on the arc. Arc sensor systems with an electromagnetic excursion of the arc without torch movement have been described [59, 60]. In these systems, the arc is deviated by magnetizing coils that are mounted on the torch. Because of the nearly zero-mass properties of the arc, very high oscillation frequencies can be reached. In [61], four magnetizing coils are arranged around the torch in such a manner that, in addition to deviation of the arc, they can also be used as an eddy current sensor system for detecting the starting point of the seam. In general, all the systems that use electromagnetic forces for arc deviation need some additional equipment near the torch. Thus, one of the major benefits of the through-the-arc sensing system, the use of torches without any additional add-on pieces, is compensated for.

In contrast to the one-arc welding processes, torch oscillation is not necessary for seam tracking in processes that use more than one arc. In multi-arc processes the torch can be adjusted in such a way that the arc roots are placed on the different groove faces. By comparing the process data of the arcs, seam tracking is possible. Torch height control can be achieved by comparing process data with a reference value. For tracking curved seams, these systems need an additional axis to adjust the torch orientation. This axis can be used to control the bead height with respect to the groove width, and also the welding speed [62–64].

In general, all through-the-arc sensing systems are limited to detecting the welding groove while welding, because the information for seam tracking is generated by the welding arc itself. Hence another sensor system must be used to detect the starting and end points of the weld. Most popular in robotic welding is a combination with an electrode contact system. Further, they are only suitable for tracking seams which show a detectable groove. Therefore, a minimum edge of 1 mm height to the groove face is necessary [65]. Another limitation concerns especially the use of through-the-arc sensing systems in GMA welding. Here, depending on the analysis method used, the free choice of the welding parameters is restricted. For example, an analysis method for a spray arc process may not be suitable for short arc welding.

Despite these limitations, through-the-arc sensing systems are the most commonly used sensors in robotic steel welding. This is due to several unique benefits of these systems. For example, they are able to sense the welding data in real time and exactly in the arc area without any deviation. There are no additional extensions to the torch and therefore the accessibility of welding seams is not restricted by this sensing system. Further, through-the-arc sensors are the most robust sensor systems because they need no special equipment in the harsh environment near the welding torch. Finally, in comparison with other sensor systems, they are inexpensive.

In laser welding, the primary process phenomena essentially refer to the beam power and quality. These are very important factors in laser machining. It is difficult to identify the source of any process fault if the laser source itself does not behave in a repetitive and reliable manner. Unfortunately, laser sources and beam guidance systems are complex. Any change in the behavior of an optical, electrical, or mechanical component could have grave repercussions on the output beam. Such a change could cause a fault in the welding process whose source can only be detected if the laser is off-line and checked. For this reason, the implementation of condition monitoring sensors is useful for providing complementary information about the condition of the beam.

A beam monitoring system is usually used off-line for checking the status of the laser. A beam monitoring system can measure the beam caustics (eg, the beam diameter in and around the focal plane), alignment, and power distribution. With such a system, misalignments, mode, and overall beam quality factor can be measured. Unfortunately, such systems work off-line. A common measuring principle is based on using a rotating needle that reflects a small fraction of the beam on to a very accurate power measuring device. As this needle scans the beam a power profile of a specific plane can be acquired. By varying the position of this plane, the complete caustic can be acquired. Such systems are slow and have to be applied in the raw beam and therefore cannot be used on-line [66].

However, other simpler systems can be used, but the information is not as exhaustive as for the off-line systems. They supply the position of the beam, its diameter, and its rough power distribution for both Nd:YAG and CO_2 lasers.

One of these systems is the grid-based sensor. This sensor is based on the use of a grid structure made up of thin wires that are positioned within the raw beam. The wires in the grid are at an angle of 30° to each other. By measuring the temperature of the wires at the specific angles, a complete image of the power distribution can be calculated. The loss of power to the raw beam is of the order of 1%. As the system is thermally based, high-pulsed processes (maximum 5 kHz) cannot be observed. Another problem is based on the fact that the grid can cause diffraction effects and thereby reduce the beam quality. The advantage is that its positioning in the beam path is both practical and simple [67].

Another system is a thermopile-based sensor. This sensor is based on the use of thermopiles that are typically used for off-line checking of laser powers. The sensor is usually made up of a series of thermopiles so that geometric information can be attained. Currently, an eight-quadrant thermopile-based sensor is available with which the beam position, beam power, and beam diameter can be calculated. By using a system for extracting a portion (1%) of the raw beam, the laser source can be monitored at a sampling rate of 1 Hz. A larger number of thermopiles would mean a more accurate measurement, but its thermal-based nature would still mean a low sampling rate. The mechanisms for uncoupling a portion of the beam involve the use of a partially transitive mirror or a diffractive mirror. The set-up of this sensor is not as practical as that of the above-mentioned grid-based sensor, but the measuring mechanism is robust and simple.

Especially in Nd:YAG laser welding a so-called cover slide monitor is used. This cover slide protects the focusing lens from splashes that could damage or contaminate the lens. With CO_2 lasers, such a cover slide is just as expensive as a lens and is only rarely used. In Nd:YAG laser welding, this cover slide collects all the splashes and gradually darkens. This contamination has effects on the beam profile and power distribution. These effects could cause process faults not related to the process parameters, but related to the quality of the incoming beam. To solve this problem, a cover slide monitor has been developed. It consists of two sensors located on the side of the cover slide that collect the stray reflections and the temperatures of the cover slide. From the signals, the level of contamination on the cover slide can be monitored. Since the sensors react only to the raw beam and are not affected by the process radiation, the geometry of each laser pulse can be monitored. Obviously the signal level depends on the contamination level [68].

4.7.3.2 Secondary Process Phenomena-oriented Sensors

Referring to the classification of sensors (Figure 4.7-1), the optical analysis of the weld pool belongs to the secondary process phenomena-oriented sensors. In the development of welding sensors, many efforts have been made to gain information from the appearance of the weld pool, especially referring to weld bead size and penetration control. The principle of a weld pool viewing and control system is shown in Figure 4.7-8. A CCD camera observes the molten pool and from the captured images in an image-processing unit relevant data, eg, concerning the pool size or shape, are extracted. According to these data, a CPU controls the welding power source and the guiding system in order to adjust current, voltage, and welding speed. Such a system is able to keep the weld pool size and shape constant, so that a constant penetration of the weld can be assumed. In addition to weld penetration control, vision-based molten pool viewing systems can also be used for seam tracking. However appropriate control software must be developed [2, 69].

In arc welding, a main problem in viewing the molten pool by a sensing system, eg, a photodiode array or a CCD camera, is the extremely intense arc light. This leads to irradiation of the image sensor, so that the molten pool itself cannot be detected without a special sensing technique. In general, two different principles of reducing the influence of arc light on the sensor system are used. The first is based on the use of optical filters, which allow only selected wavelengths to reach the image sensor. Hence the image sensor is shielded from the intense arc light and an image of the molten pool can be detected. This principle is suitable for all kind of welding arcs. The second principle is based on image capturing only during periods of low arc light intensity. For that purpose a shutter in front of the image sensor is synchronized with the variation of the arc current in short-circuiting and pulsed-arc welding. This shutter clears the way to the image sensor only during short-circuiting or the base current period of the arc, in which the arc light intensity is low enough to get a clear image of the molten pool. This

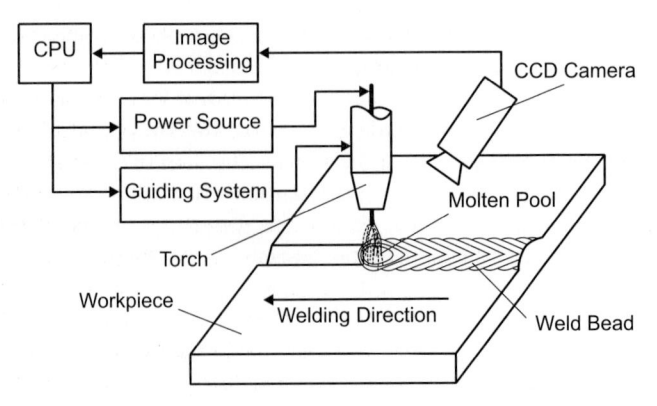

Fig. 4.7-8 Principle of a weld pool viewing and control system

principle is restricted to welding arcs that shows a recurrence of low arc light intensity periods, as in short-circuiting and pulsed-arc welding [16, 70–74].

In order to increase the accuracy of the weld pool viewing and control systems, several methods have been developed. A sensing system has been described which uses specular reflection to measure the three-dimensional shape of the free weld pool surface [74]. In order to do that, a pattern of structured-light strips is generated on the workpiece surface by projecting a pulsed laser beam through a special grid. In the molten pool area, the reflected pattern is deformed by the profile of the weld pool surface. This reflected pattern is used to compute a three-dimensional image of the weld pool surface. From these data, conclusions concerning the weld penetration are possible. In order to increase the accuracy of the system and to suppress the influence of arc light, a combination of a shutter and a narrow-band optical filter in front of the CCD camera is used. The shutter is synchronized with the laser pulse and the optical filter matches the laser wavelength. The major restriction of this system is the relatively long calculation time for the computation of the weld pool surface data. An average time of 1 s is given, so that at present real-time sensing and control of the weld are not possible. Further, the system is limited to GTA welding. In GMA welding the weld pool surface is very unsteady because of the periodic impacts of droplets. Therefore, in order to trace this surface, a significantly higher imaging and processing speed is necessary.

Another method of controlling the weld bead has been described [70]. This control system for pulsed GTA welding is based on fuzzy logic and artificial neural network theory. In order to increase the accuracy of the sensing system, a shutter in front of the CCD camera is synchronized with a special step pulse form of the welding current. It was found that the pool image is clearest if it is taken when the welding current is maintained at an intermediate stage between pulse and base values. The captured image of the weld pool area is edited by special digital filter algorithms, so that the weld bead width can be detected. From these data a fuzzy neural network controls the bead face width by regulating the welding parameters. The whole image processing time of this system is about 0.7 s, which restricts the field of application to relatively slow welding speeds. Nevertheless, the suitability of fuzzy logic and artificial neural networks for controlling the welding process is demonstrated.

Another image-based monitoring and control system for GTA welding of thin mild steel plates has been described [73]. In this system, the brightness difference between the molten pool and base metal is used to detect the molten pool edges on a scanning line. The accuracy of the system is increased by using the so-called self-arc-lighting method. It is found that the brightness difference between the molten pool and base metal increases according to the increase in the lighting intensity in the weld pool area. As the arc itself has an extremely bright and intense light, it is applied to illuminate the weld pool area, using reflecting mirrors in front of and lateral to the torch.

Further, a shutter in front of the CCD camera is synchronized with the peaks of the pulse current used in order to use the highest possible arc light intensity during image taking. To prevent the direct influence of the arc on the image sensor, a special filter is applied. The system is able to control the penetration in welding mild steel plates. Its application is limited to pulsed GTA welding. Further, the accessibility of the joint is restricted by the significant enlargement of the torch and the influence of dust and fumes on the mirrors may cause problems in lighting the pool area.

In general, all sensing and control systems which are based on viewing the weld pool from the top face in order to control the penetration depth have the advantage that they can be used in applications, in which the back side of the weld is inaccessible. The main restrictions at present relate to the image processing time of these systems. For very unsteady processes or high welding speeds, the image processing speed and hence the sampling frequency of the sensor system must be significantly increased.

In laser welding, the quality of a weld is influenced by many parameters, such as the laser source, the beam guidance system, the handling system, the clamps, and the workpiece. Changes in one of the numerous parameters, caused by unnoticed wear, thermal effects, or the material properties, influence the process and may accidentally lead to a decreased quality of the weld. One practical method to determine the weld quality, on-line during the welding, is to detect and evaluate the emission of secondary process radiation [75–77].

During deep penetration laser welding, a keyhole begins to be generated by the reaction force of the vapor escaping from the surface with high velocity. Different sources (molten material, vaporized metal, and plasma) cause the emission of secondary process radiation at different wavelengths.

A photodiode with suitable spectral sensitivity can be used as a detector for the light emitted due to the beam-material interaction. The measured signal depends on the radiation flux density of the process. This means that changed conditions for the laser source, the beam guidance system, the cover slide, and the focal position can be recognized by the signal amplitude. One can look into the keyhole if the photodiode is integrated into the laser source and into the beam delivery system so that the process is observed coaxially to the laser beam. In this case, the signal amplitude is decreased when there is no sufficient connection during overlap welding, due to a gap between the two sheets [78].

Investigations in production have shown that the influence of the laser beam intensity on the welding results and on the measured signal could be verified under the conditions of 3-D welding processes. Even a decrease in the laser power at the workpiece, caused by spatters on the cover slide, could be recognized with the decreased signal amplitude. However, it is not possible to predict the reason for the failure. A distinction can only be made between a good weld and a weld failure, depending on the signal amplitude.

Depending on the type of laser, the detector can be positioned at different locations. For practical applications, special attention must be paid to inadvertent misalignment and contamination of the sensor, in order to ensure a reliable measurement.

The installation of the sensor at the welding head is simple but reduces the accessibility of the workpiece. The reliability of the sensor is also reduced by dust

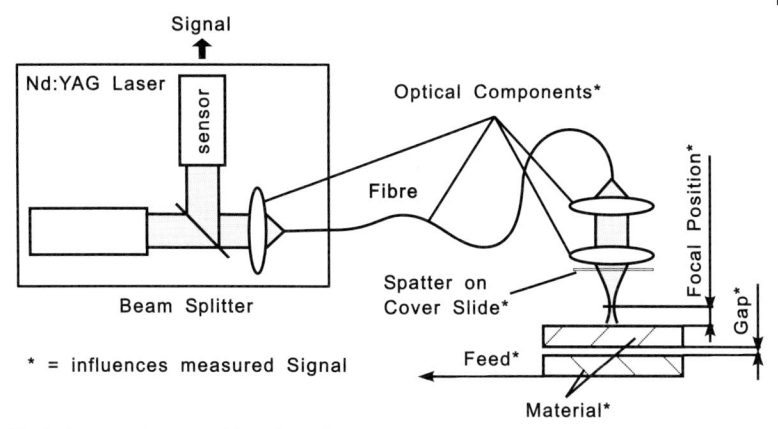

Fig. 4.7-9 Optical sensor integrated into laser beam source

and spatter. The sensor can be used for 3-D laser welding to observe the process through the same optical path which is used for the laser radiation. This provides a signal independent of the direction of the feed, and permits a 'look' into the keyhole [79].

Scraper mirrors fulfil this essential condition, and so they are usually used for CO_2 lasers without a limitation on the output power. For a low output power of < 3 kW, reflective and transitive beam splitters made of zinc selenide can also be used for this type of laser. The sensor cannot be integrated into the laser source, because the other optical components of CO_2 lasers heavily reduce the transmission of the process radiation on the laser beam path.

When Nd:YAG lasers are used, both reflective and transitive beam splitters can be used to outcouple the process radiation without a limitation on the laser power, due to suitable surface coatings. The process radiation is sufficiently guided from the workpiece to the laser source by all commonly used optical components, including quartz glass fibers. Therefore, the favorable position of the sensor for Nd:YAG welding processes is inside the laser source (Figure 4.7-9). The reflective and the transitive beam splitter set-ups have been successfully long-term tested for different power Nd:YAG laser sources. The costs for the subsequent integration are comparatively low.

Owing to the correlation between the process radiation and the welding result, PC-based systems have been developed to recognize weld faults automatically. For the recognition of the faults, different methods can be used. Good results for the overlap joint geometry have been achieved using a dynamic comparison between the amplitude of the current signal measured and the amplitude of recorded reference welds [80]. Thus, systematic fluctuations are not evaluated as faults. The reason for the occurrence of an error cannot be detected using this method. A software low-pass filter can be used to damp high-frequency changes of the measured signal which are not correlated with the seam quality. The sensitivity of fault recognition is adapted to the requirements of the specific application using the set-up screen. There, it is possible to determine the allowed deviations between the reference recording for different sectors of the weld separately. The reference welds for the calculation of the dynamic reference can be chosen at will from a file list. The signal of each weld can be stored automatically, and a protocol of the evaluation results is generated, which can be stored in a file or printed.

Another sensing technique in welding is based on the non-contact thermal sensing of the workpiece. Relating to the classification made in Figure 4.7-1, this is also a secondary process phenomena-oriented sensing method. The principle of such a sensing system corresponds to a high degree to the viewing system, which is shown in Figure 4.7-8. The main difference is that in lieu of the CCD camera an infrared (IR) camera is used.

The principle of the non-contact thermal sensing method is based on the fact that all objects in the temperature area above 0 K emit radiation. The intensity of the emitted radiation is connected with the radiation's wavelength and the object's temperature. Especially arc welding processes are connected with a significant rise in workpiece temperature in the area around the welding arc. Thus, from non-contact measurement of the IR emissions in the welding area, information about the condition of the weld can be extracted. Basing on the assumption that the surface temperature distribution is an indicator of the internal conditions of the weld, this technique is used for weld form and penetration control. Moreover, IR sensing systems are suitable for seam tracking and weld bead width control. They can be used for all kinds of position, process, and joint type [81–83].

Restrictions for non-contact thermal sensing techniques in welding are given by the application of the IR camera near the torch. This means a significant enlargement of the tool, so accessibility problems may occur. Further, the accuracy of this sensing method is influenced by changes in the surface emission characteristics of the workpiece, eg, by any coating, oxides, or even a change of the surface finish. The sampling rate of industrially used IR cameras is limited to 25–50 frames/s. This, in addition to the required image processing time, restricts the welding speed.

Another sensing technique in welding uses the spectral analysis of the weld pool. Referring to the classification in Figure 4.7-1, this is a secondary process phenomena-oriented sensing method. It is suitable, eg, for dilution determination in weld coating. Especially in surfacing mild steel components with special materials, the properties of the coating are closely related to the degree of parent metal dilution. By means of a spectrometer, the metallic elements of the weld pool are detected on-line during welding. From the measured element composition, the quality of the coating can be inferred. Further, the welding parameters can be adjusted in order to achieve the desired element composition by influencing the amount of dilution from the base metal [84, 85].

4.7.4
Summary

A review of the currently available and studied sensor systems for welding applications has been given. A possible classification of the sensor types was given according to their functional principle. In addition, factors that affect or restrict the different sensor systems have been discussed.

The review shows that there is currently no sensor system for welding applications commercially available or just studied which satisfies all requirements of an ideal system. Each of the systems reviewed has limited capabilities in some way or causes some restrictions to the welding process. Hence it is necessary to select the most satisfactory sensor system for each application. To overcome individual limitations of sensor systems, a combination of two or more different systems working together in one welding application could be used.

4.7.5
References

1 Masumoto, I., Araya, T., Iochi, A., Nomura, H., *J. Jpn. Weld. Soc.* **52** (1983) 339–347.
2 Nomura, H. (ed.), *Sensors and Control Systems in Arc Welding*; London: Chapman & Hall, 1994.
3 *Sensoren für das Vollmechanisierte Lichtbogenschweißen*; DVS-Merkblatt 0917-1, August 1996.
4 Haug, K., Schilf, M., *Praktiker* (12) (1999) 490–493.
5 Nakayama, S., *Key Technology for Automation of Arc Welding and its Applications*; USA: American Welding Society, 1998, pp. 327–338.
6 Faber, W., Lindenau, D., *ZIS-Mitt.* **27** (12) (1985) 1290–1296.
7 Dilthey, U., *DVS-Ber.* **73** (1982) 72–74.
8 Bollinger, J.G., *Sens. Rev.* **7** (1981) 136–141.
9 Araya, T., Fujiwara, O., Udagawa, T., Yoshida, T., *Development of Automatic Welding Line Trace Control System*; IIW-Document XII-K-89-78, 1978.
10 Schmidt, D., *Ind.-Anzeiger* **109** (29) (1987) 41–43.
11 *Opto Laser Eur.* (38) (1997) 37.
12 Tönshoff, H.K., von Alvensleben, F., Ostendorf, A., Schumacher, J., *Laser Mag.* (3) 1996.
13 Ranke, P., *DVS-Ber.* **134** (1991) 51–56.
14 Dilthey, U., Reisgen, U., *DVS-Ber.* **160** (1994) 123–137.
15 Schnell, G., *Technika* (19) (1992) 85–91.
16 Drews, P., Willms, K., *Maschinenmarkt* **98** (11) (1992) 50–55.
17 Oster, M., *Verbesserung der Nachführgenauigkeit von Lichtbogensensoren für das MSG-Schweißen durch Schnelle Kreisförmige Auslenkung der Drahtelektrode*; Aachen: Verlag Shaker, 1995.
18 *LaserVision, Nahtverfolgung und Inspektion*; Rodgau: Jurca Optoelektronik.
19 *Scout:3D Nahtführung*; Munich: Dr. Barthel Sensorsysteme.
20 Wu, J., Smith, J.S., Lucas, J., *IEE Proc. Sci. Meas. Technol.* **143** (2) (1996) 85–90.
21 Haug, K., Pritschow, G., in: *IECON Proceedings (Industrial Electronics Conference)*, Vol. 2; Los Alamitos, CA: IEEE Computer Society, 1998, pp. 1236–1241.
22 Braggins, D., *Sens. Rev.* **18** (1998) 237–241.
23 *Lasers Optron.* 18 February (1999) 27–28.
24 Matthes, K.-J., Schurich, K., Herrich, J., *DVS-Ber.* **143** (1992) 57–61.
25 Faber, W., Nitzsche, R., Schauder, V., Lindenau, D., *ZIS-Mitt.* **30** (1988) 1183–1191.
26 Drese, J., Hinrici, R., *Ind.-Anzeiger* **107** (33) (1985) 29–31.

27 Matthes, K.-J., Schurich, K., ZIS-Mitt. 26 (1984) 1169–1175.
28 Spur, G., Duelen, G., Pörschke, H., Tech. Mess. 51 (1984) 250–254.
29 Schmal, K.H., in: Proceedings of the Aachen Conference: Schweißen mit Robotern, Aachen, 30–31 March 1987.
30 Treuenfels, A., Ind.-Anzeiger 106 (8) (1984) 35–38.
31 Florian, W., Ohlsen, K., Krause, H.J., DVS-Ber. 118 (1989) 147–151.
32 Florian, W., Ohlsen, K., DVS-Ber. 109 (1987) 118–126.
33 Blumke, F., Schaffrath, W., Tech. Mess. 51 (1984) 427–431.
34 Goldberg, F., in: Automation and Robotisation in Welding and Allied Processes, International Conference, Strasbourg, 2–3 September 1985, pp. 393–400.
35 Goldberg, F., Karlen, R., Met. Construct. 12 (1980) 668–671.
36 Schmal, K.H., Automobilindustrie 27 (1982) 435–442.
37 Estochen, E.L., Neumann, C.P., IEEE Trans. Ind. Electron. IE-31 (1984) 219–224.
38 Umeagkwu, C., Maqueira, B., IEEE Trans. Ind. Electron. 36 (1984) 338–348.
39 Zhang, S.B., Zhang, Y.M., Kovacevic, R., J. Manuf. Sci. Eng. 120 (1998) 600–608.
40 Mahajan, A., Figueroa, F., Robotica 15 (1997) 275–281.
41 Suga, Y., Machida, A., in: Proceedings of the 1996 6th International Offshore and Polar Engineering Conference, Part 4, Los Angeles, CA, 26–31 May 1996, pp. 128–132.
42 Tanaka, M., Morita, T., Kitamaru, N., Tohno, K., Irie, T., Matsushita, H., Ogawa, Y., Sumitomo, T., Proceedings of the 1997 7th International Offshore and Polar Engineering Conference, Part 4, Honolulu, HI, 25–30 May 1997, pp. 508–514.
43 Puscher, P., DVS-Ber. 83 (1983) 124–128.
44 Dilthey, U., Stein, L., Oster, M., DVS-Ber. 170 (1995) 233–239.
45 Dilthey, U., Stein, L., Oster, M., Int. J. Joining Mater. 8 (1) (1996) 6–12.
46 Sugitani, Y., Mao, W., Weld. Int. 9(5) (1995) 366–374.
47 Nomura, H., Sugitani, Y., Suzuki, Y., Trans. Jpn. Weld. Soc. 18 (2) (1987) 35–42.
48 Panarin, V.M., Weld. Int. 13 (1999) 155–158.
49 Orszagh, P., Kim, Y.C., Horikawa, K., Sci. Technology Weld. Joining 3 (1998) 139–143.
50 De Boer, F.G., Jonkman, M.E., Australas. Weld. J. 42 (First Quarter) (1997) 36–38.
51 Mao, W., Ushio, M., Sci. Technol. Weld. Joining 2 (1997) 191–198.
52 Eichhorn, F., Borowka, J., Habedank, G., DVS-Ber. 118 (1989) 40–42.
53 Drews, P., Liebenow, D., Niessen, K., Betriebsleiter (7–8) (1991) 56–62.
54 Xiao, Y.H., den Ouden, G., Weld. Met. Fabr. January (1996) 18–22.
55 Dilthey, U., Gollnick, J., in: Proceedings of the 24th Annual Conference of the IEEE Industrial Electronics Society, Aachen, 31 August–4 September 1998, pp. 2374–2377.
56 Nomura, H., Sugitani, Y., Murayama, M., Development of Automatic Fillet Welding Process with High Speed Rotating Arc; IIW-Document XII-939-86, 1986.
57 Nomura, H., Sugitani, Y., Eur. Pat. 0 066 626, B23K 9/12, 1985.
58 Sugitani, Y., Kobayashi, Y., Murayama, M., Weld. Int. 5 (1991) 577–583.
59 Iceland, W.F., US Pat. 32 04 081, 1965.
60 Puscher, P., Ger. Pat. DE-PS 25 33 448, B23K 9/12, 1983.
61 Hangmann, N., Trösken, F., Ger. Pat. DE-OS 37 23 844, B23K 9/12, 1989.
62 Hirsch, P., Ger. Pat. DE 25 46 221, B23K 9/12, 1979.
63 Hirsch, P., Ger. Pat. DE 26 11 377, B23K 9/12, 1982.
64 Lüttmann, U., Schweisstech. Forschungsber. 38 (1991).
65 Eichhorn, F., Pfalz, J., Ind.-Anzeiger 105 (58/59) (1983) 29–31.
66 Coutouly, J.-F., Deprez, P., Vantomme, P., Deffontaine, A., Opto Laser Eur. (58) (1999) 34.
67 Cremer, V., Laser Mag. (3/99) (1999) 30–32.
68 Cover Slide Monitoring Device; Gothenburg: Permanova.
69 Anderson, P.B.C., TWI J. 6 (1997) 654–697.

70 CHEN, S.B., WU, L., WANG, L., LIU, Y. C., *Weld. J. (Miami)* **76** (1997) 201–209.

71 EGUCHI, K., YAMANE, S., SUGI, H., KUBOTA, T., OSHIMA, K., in: *Proceedings of the 24th Annual Conference of the IEEE Industrial Electronics Society, Aachen, 31 August–4 September 1998*, pp. 1182–1185.

72 DREWS, P., WAGNER, R., WILLMS, K., *VDI-Z.* (2) (1991) 88–92.

73 SUGA, Y., MUKAI, M., USUI, S., OGAWA, K., in: *Proceedings of the Seventh International Offshore and Polar Engineering Conference, Honolulu, HI, 25–30 May 1997*, pp. 502–507.

74 KOVACEVIC, R., ZHANG, Y.M., *Proc. Inst. Mech. Eng., Part B: J. Eng. Manuf.* **210** (1996) 553–564.

75 SUN; A., KANNATEY-ASIBU, JR., E., GARTNER, M., *J. Laser Appl.* **11** (1999) 153–168.

76 KOGEL-HOLIACHER, M., JURCA, M., DIETZ, C., in: *Proceedings of ICALEO 98, Orlando, FL,* Vol. 85, 1998, pp. 168–177.

77 MÜLLER, M., DAUSINGER, F., HÜGEL, H., in: *Proceedings of ICALEO 98, Orlando, FL,* Vol. 85, 1998, pp. 122–132.

78 TÖNSHOFF, H.K., SCHUMACHER, J., in: *Proceedings of ICALEO 96, Detroit, MI,* 1996, pp. 45–54.

79 TÖNSHOFF, H.K., VON ALVENSLEBEN, F., OSTENDORF, A., GÜTTLER, R., in: *Proceedings of ICALEO 98, Orlando, FL,* Vol. 85, 1998, pp. 78–85.

80 RICCIARDI, G., CANTELLE, M., MARIOTTI, F., CASTELLO, P., PANESA, M., *Ann. CIRP* **48** (1999) 159–162.

81 CHIN, B.A., ZEE, R.H., WIKLE, H.C., *J. Mater. Process. Technol.* **89–90** (1999) 254–259.

82 CHIN, B.A., WIKLE, H.C., CHEN, F., NAGARAJAN, S., *J. Chin. Inst. Eng.* **21** (1998) 645–657.

83 MATTHES, K.-J., HERRICH, J., LINDNER, D., DREWS, P., BENZ, S., MATZNER, D., *Schweissen Schneiden* **49** (1997) 170–174.

84 DILTHEY, U., PAVLENKO, A., ELLERMEIER, J., *Schweissen Schneiden* **48** (1996) 227–230.

85 BOUAIFI, B., GÜNSTER, J., HILLEBRECHT, M., *DVS-Ber.* **176** (1996) 117–120.

4.8
Coating Processes

K.-D. BOUZAKIS, N. VIDAKIS, *Aristoteles University of Thessaloniki, Thessaloniki, Greece*
G. ERKENS, *CemeCon GmbH, Würselen, Germany*

4.8.1
Coating Process Monitoring

4.8.1.1 Introduction

One of the most promising, generally applicable and well-adopted methods to enhance the surface performance of materials is the use of coatings. The reason is that it is extremely difficult and expensive to meet homogeneous materials having on their surface the ensemble of desired properties, such as high hardness, wear resistance, adequate stiffness, increased ductility, chemical inertness, stainlessness, controllable electrical and thermal conductivity behavior, etc. Some of the aforementioned properties are contradictory to each other and it is impossible to achieve them simultaneously [1–3]. Surface coatings nowadays offer a satisfactory reliable and economical alternative solution to this problem. There are several methods to deposit surface films, based on a variety of physical, chemical, and thermal processes and transformations, taking into account also the substrate material specifications and the duty of the element or tool to be coated.

Coating methods have rapidly gained a significant part of material technology in a variety of application fields. To quantify this fact, the estimated market share of the coated tools was approximately 95% at the end of the 1990s [4]. On the other hand, coatings play an important role in electronics and engineering. Traditional machine elements, such as bearings, gears, etc., are nowadays a major field for research and development for the coatings industry. The doubtless profitable results of the coatings applications led to the development of flexible and sophisticated deposition processes that today offer mono- and multilayer, structural, or compositional films, with excellent mechanical, physical, and chemical properties, but consequently increased the complexity of the corresponding coating deposition equipment.

Coating processes are exceptionally multiparametric functions, so that their successful progress and completion require a strict but flexible monitoring and control system [5, 6]. Even the simplest deposition methods involve more than two parameters that have to be tracked throughout the process, in order to establish the required coating properties. Furthermore, as will be described below, the wide variation of applicable coating strategies sets their control, case dependent. Considering that an optimized deposition process is a consequence of the proper selection of the values for all influencing parameters, deposition monitoring has to be integrated and inclusive.

4.8.1.2 Vacuum Coating Process Classification

The most popular and well-adopted deposition methods are based on vapor transformations and more specifically on physical (PVD) and chemical vapor deposition (CVD) [7]. On the other hand, both of them, especially the former, are the most multi-parametric and sensitive methods, thus demanding strict monitoring and control. Therefore, in the frame of the scope of this section, the sensors required for PVD and CVD processes are presented. Figure 4.8-1 illustrates a rough classification of these deposition strategies, which are even more subdivided, considering the design principles of the corresponding installation manufacturers.

The PVD method consists in removing metallic ions from a target, forming a plasma environment, and making them react with a gas to form a ceramic compound, which is deposited on the surface of the product, called the substrate. Thereby, the deposition species are atoms or molecules or a combination of them. The deposition is effected by condensation, exothermic, and makes extensive use of plasmas. The deposition is performed in a high-vacuum chamber, and the deposition temperatures vary from 150 to 500 °C [1, 8]. The PVD method is very flexible, and offers a variety of single and multilayer coatings (structural or compositional), covering a wide range of physical, chemical, and electrical properties of the film deposited.

In a typical CVD process, reactant gases at room temperature enter a tight reaction chamber. The gas mixture is heated as it approaches the deposition surface, heated radiatively, or placed upon a heated substrate. The CVD method is performed at higher temperatures, >850 °C [1]. Depending on the process and operat-

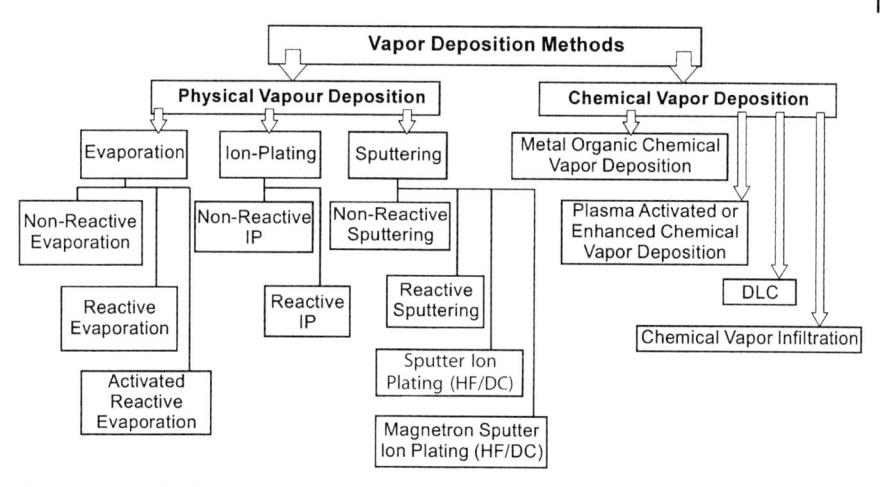

Fig. 4.8-1 Classification of vapor deposition processes

ing conditions, the reactant gases may undergo homogeneous chemical reactions in the vapor phase before striking the surface. Near the surface thermal, momentum, and chemical concentration boundary layers form as the gas stream heats, slows owing to viscous drag, and the chemical composition changes. Heterogeneous reactions of the source gases or reactive intermediate species, formed from homogeneous pyrolysis, occur at the deposition surface forming the deposited material. Gaseous reaction by-products are then transported out of the reaction chamber. Traditional applications of the CVD method, such as the production of coated tools, are being progressively replaced by the PVD method, owing to its flexibility and significantly lower deposition temperatures.

4.8.1.3 Vacuum Coating Process Parameter Monitoring Requirements

As in any other multi-parametric process, there are two main categories of sensing systems that are involved. The first focuses on the run-in phase and includes testing and calibration of the coating device. In this respect, these control systems are used by the research and development and also by the maintenance divisions of coating device producers and research institutes. For these reasons, the equipment utilized is removable, complicated, and expensive, whereas it orients the state of the art in coating process sensing systems. The second category includes fixed sensing systems that are steadily implemented with the coating device and are necessary to monitor the progress of the coating process by the coating producers, ie, the end users of such devices. The first group of sensing systems ensures that the coating device is well constructed and standardized, whereas the second group makes certain of the quality of the final product, ie, of the coated part. In this section, both sensing system types will be presented with respect to each deposition method.

In situ coating process monitoring follows the same fundamental rules as every other automated controlled system. With respect to the importance of the parameter to be monitored and the possibility of adjusting it, in order to optimize the coating process, open- or closed-loop control systems can be designed. Hence there are parameters that have to be continuously checked but not adjusted, whereas others may be integrated in feedback systems and be adjusted throughout the deposition course. The principles of adaptive control find a wide range of applications in this last group of process parameters. In every case, coating processes are very sensitive to any deposition parameter modification, even transient, considering the restricted film thickness and its refined properties. Therefore, every sensing system involved must have increased resolution, accuracy, repeatability and reliability. These requirements concern the sensors, the electronic devices, and the evaluation tools of the measured magnitudes, which compose the sensing systems.

A further important parameter that must be considered by device producers is to ensure that sensors are coated without reducing their sensitivity. For this purpose, process sensors are protected, when necessary, by special filters or ion traps, which are considered as process spare parts. On the other hand, there are sensors, such as quartz crystals for deposition rate monitoring, which modify their indications as they are progressively coated. Therefore, they must be replaced at the termination of each specific coating process, in order to monitor the course of the consequent deposition process.

Coating process parameters are monitored during the deposition with the aid of various types of sensors. A distinctive classification of them is with regard to their output, ie, whether they produce analog or digital signals. Both types are used in coating technology and for the same measurable parameter analog or digital sensors may be applied. Nowadays, advanced data acquisition systems are computer supported and may handle either analog or digital signals. In the first case, analog to digital converters (usually A/D cards) are utilized to let the digital hardware tools evaluate and control the input measurements. These electronic circuits must also satisfy the demands of the whole sensing system.

To simplify the coating process control further, advanced and powerful software tools are exploited today in order to visualize the deposition progress. Coating device producers support their customers with exclusive programs that electronically and remotely control the deposition process. For this purpose they develop software using modern programming languages in Windows® environments, or using existing platforms, such as the LabView® code. In every case, these programs offer visual environments for data acquisition, alerts, parameter adjustment, and reports of the whole process.

4.8.2
Sensors in Vapor Deposition Processes

4.8.2.1 Vapor Process Parameter Map

Physical and chemical deposition methods have various common parameters to be monitored, although the first ones are more complicated and multi-parametric. Therefore, shared parameters and magnitudes are controlled by sensors based on the same principle, whereas their measuring range is suitably adjusted or modified. Figure 4.8-2 illustrates the parameters that influence the deposition processes and therefore must be controlled, for either development or quality control purposes. In addition to the measuring parameter, Figure 4.8-2 summarizes the sensor type per case and their measuring range. It can be seen that it is possible to monitor different parameters using the same sensor type, as will be described more analytically below.

4.8.2.2 Vacuum Control

The required high vacuum, for vapor coating processes, is accomplished by means of medium- or high-vacuum equipment and process instrumentation (chamber pressure $<10^{-4}$ Torr). The pressure must held under control in situ, ie, during the process, and therefore pressure modifications or leaks are completely inadmissible and must be rapidly detectable. Vacuum control covers two stages.

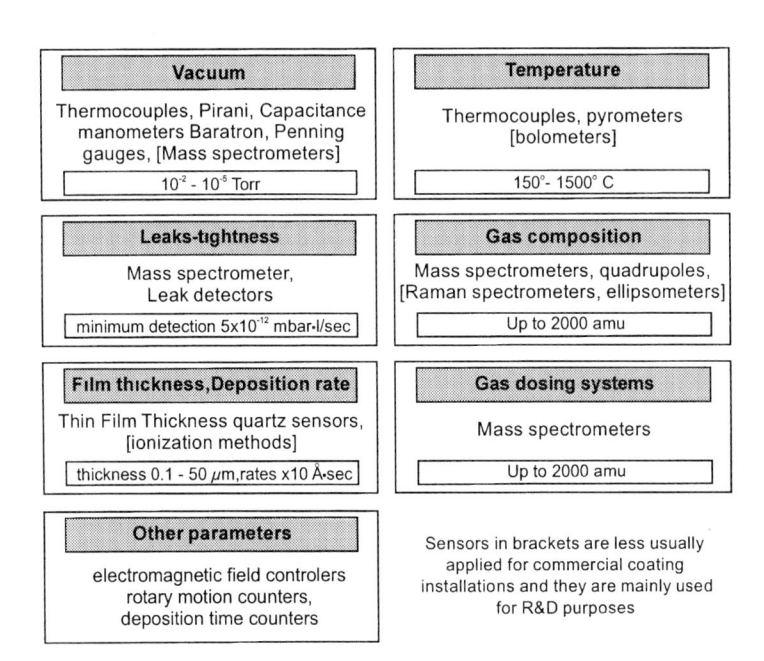

Vacuum	Temperature
Thermocouples, Pirani, Capacitance manometers Baratron, Penning gauges, [Mass spectrometers]	Thermocouples, pyrometers [bolometers]
$10^{-2} - 10^{-5}$ Torr	150°- 1500° C

Leaks-tightness	Gas composition
Mass spectrometer, Leak detectors	Mass spectrometers, quadrupoles, [Raman spectrometers, ellipsometers]
minimum detection 5×10^{-12} mbar·l/sec	Up to 2000 amu

Film thickness, Deposition rate	Gas dosing systems
Thin Film Thickness quartz sensors, [ionization methods]	Mass spectrometers
thickness 0.1 - 50 µm, rates x10 Å·sec	Up to 2000 amu

Other parameters	
electromagnetic field controlers rotary motion counters, deposition time counters	Sensors in brackets are less usually applied for commercial coating installations and they are mainly used for R&D purposes

Fig. 4.8-2 Influencing parameters and sensors for their monitoring in vapor deposition processes

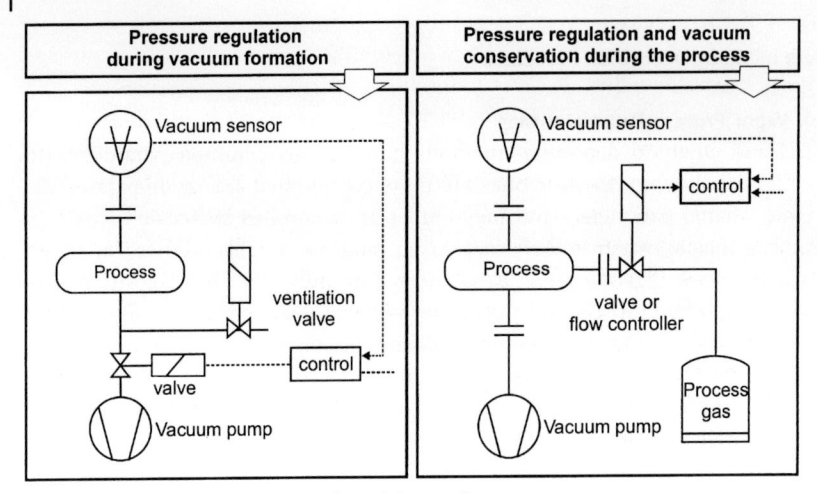

Fig. 4.8-3 Vacuum monitoring initially and during the coating process

The first covers the vacuum formation itself at the beginning of the coating process, whereas the second aims to maintain the pressure stable, also while the process gases are flowing into the deposition chamber [9, 10] (see Figure 4.8-3). These requirements are accomplished by designing special gas flow systems, consisting of vacuum fitments, pipelines, and valves, which are properly controlled by special hardware and software tools.

There are several approaches to vacuum monitoring and control during the PVD process. Measurement of pressure in a vacuum system is done with any of a variety of gages, which for the most part work through somewhat indirect means, eg, thermal conductivity of the gas or the electrical properties of the gas when ionized. The former are typically used at higher pressures ($1-10^{-3}$ Torr) and the latter in lower ranges. Such gages are sensitive to the type of gas in the system, requiring that corrections be made. Accidents have occurred when this was not taken into account. For example, the presence of argon in a system will result in a pressure reading on a thermal conductivity gage (thermocouple or Pirani, for example, as described in the next two sections) that is much lower than the true pressure. It is possible to overpressure a system significantly while the gage is still indicating vacuum. Other indirect vacuum measurements are based on partial pressure determination, ie, through mass spectrometry.

Actually, there is no universal gage that can measure from atmospheric to ultrahigh vacuum pressure. Therefore, the instrument chosen depends on the pressure range and the residual gases in the vacuum. In any case, pressure measurements may be either output sensor signals to be further processed by open- or closed-loop control units or direct indications in special displays. The only solution for a single pressure gage covering a wide measurement range is by combining more than three probes in one control system.

4.8.2.2.1 Thermocouples

Thermocouple vacuum gages offer an economical method of vacuum pressure measurement with accuracy suitable for many monitoring and control applications and general laboratory use. Simplicity of operation, low cost, and the inherent ruggedness of these gages have led to their widespread general use with time-proven dependability and zero maintenance. The pressure range between 10 and 10^{-3} Torr is indicated by measuring the voltage of a thermocouple, spot-welded to a filament exposed to vacuum system gas (see Figure 4.8-4). This filament is fed from a constant current supply, and its temperature depends on the thermal losses to the gas. At higher pressures, more molecules hit the filament and remove more heat energy, causing the thermocouple voltage to change. These sensors are used extensively in foreline monitoring and to provide the trigger signal to switch automatically the main chamber from backing. The pressure indications of thermocouples, as already mentioned, are gas-type dependent.

4.8.2.2.2 Pirani Gages

Pirani gages for medium-vacuum monitoring are usually utilized in PVD installations and their indications are also dependent on gas type. Pirani gages operate on the principle that the thermal conductivity of gas is proportional to the number of residual gas molecules, ie, the pressure. The operating principle of Pirani gages is illustrated in Figure 4.8-5. Two filaments, often made of platinum, are used as resistances in two arms of a Wheatstone bridge. The reference filament is immersed in a gas at fixed pressure, whereas the measurement filament is exposed to the vacuum system gas. Both filaments are heated by the current, passing through the bridge. Thereby, gas molecules conduct heat away from the immersed element and inbalance the bridge. Their measuring range is from 10^{-3} up to 1 Torr. In most cases, these vacuum sensors produce output signals that may

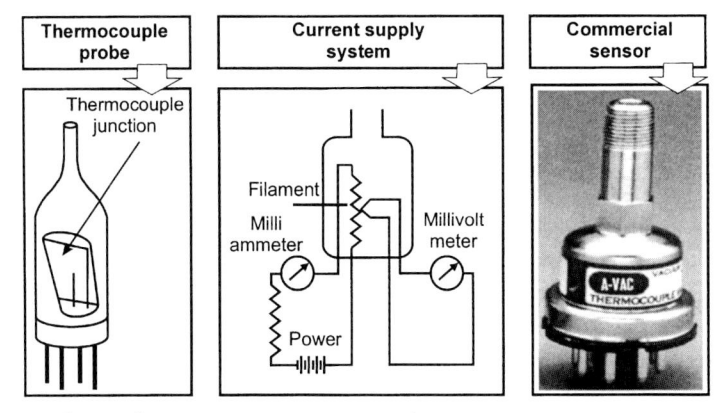

Fig. 4.8-4 Thermocouple gage for vacuum monitoring (sensor by A-VAC)

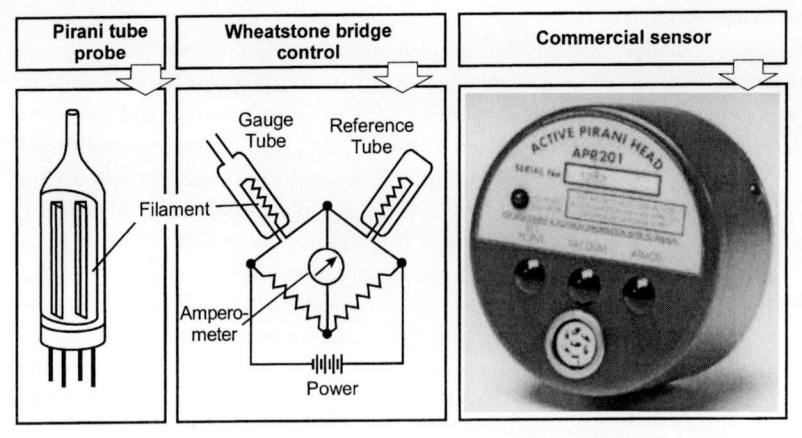

Fig. 4.8-5 Pirani gage for vacuum monitoring (sensor by Creative Group)

be transmitted to data acquisition and control units, and they may be used to track the sub-pressure progress rate.

4.8.2.2.3 Capacitance Manometers

In capacitance manometers, a thin metal diaphragm, called a membrane, separates a known from an unknown pressure and its deflection is a measure of the pressure difference. As the name suggests, the deflection is determined by measuring the electrical capacitance between the diaphragm and the fixed electrodes. Capacitance manometers are the most accurate devices for measuring the differential (known pressure on reference side) or absolute (hard vacuum on reference side) pressure of all gases and vapors at the gage operating temperature. Gage heads are specified by their maximum pressure (from 2.5×10^4 down to 2×10^{-3} Torr). All types of pressure gages are affected by ambient temperature changes, but other error sources are so much larger that temperature is ignored. The capacitance manometer, by contrast, is so accurate that gage-head temperature variation is a critical source of error. The pressure indications of capacitance gages, as already mentioned, are independent of gas type.

The most common capacitance gages used in coating devices are found with the trade name Baratron. Figure 4.8-6 illustrates the operating principle of capacitance manometers and typical commercial products that exhibit improved thermal stability owing to the extremely low temperature coefficients. The Baratron unit, capacitance manometer, is designed for use in the millitorr regime, the pressure normally used during sputtering. After being turned on, the Baratron head takes several hours to warm up, so it is wise to leave it running whenever the chamber is pumped down. Baratron producers recommend that the controller unit be sent back for calibration on a yearly basis. Advanced capacitance diaphragm sensor technology and solid-state electronics are alternatives to make precision pressure

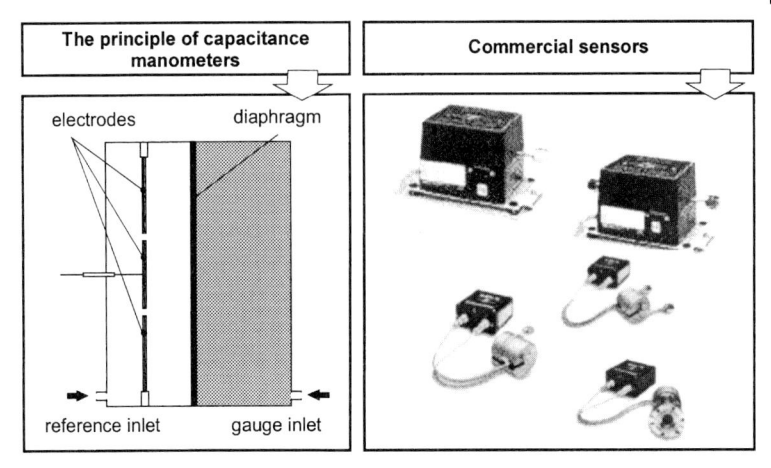

Fig. 4.8-6 Capacitance gage principle and Baratron sensors (by MKS)

measurement easy, whether on the production line or in the research laboratory. Baratron systems are completely modular with a full selection of pressure ranges, levels of accuracy, options, and accessories, and therefore may be met in various applications.

4.8.2.2.4 Penning Gages

Penning gages, as all ionization gages, use roughly the same principle as capacitance manometers. Energetic electrons ionize the residual gases, the ions are collected at an appropriate element, and the current produced is converted to a pressure indication. Hot filament gages, called ion gages, use thermionic emission of electrons from a hot wire, whereas cold cathode gages use electrons induced by glow discharge or plasma. The Penning gage is a cold cathode type (see Figure 4.8-7). Their measuring range varies from 10^{-2} down to 10^{-5} Torr. The lower pressure limit is caused by the difficulty of initiating a discharge at lower pressures with an operating voltage of 2 kV. Higher voltages cannot be used, however, since field emission occurs and electrons leave the anode in a mechanism completely unrelated to pressure.

4.8.2.2.5 Mass Spectrometry

Mass spectrometers use the difference in mass-to-charge ratio (m/z) of ionized atoms or molecules to separate them from each other [9–15]. Mass spectrometry is therefore useful for the quantification of atoms or molecules and also for determining chemical and structural information about molecules. Molecules have distinctive fragmentation patterns that provide structural information to identify structural components. Hence, the general operation of mass spectrometers is to

4 Sensors for Process Monitoring

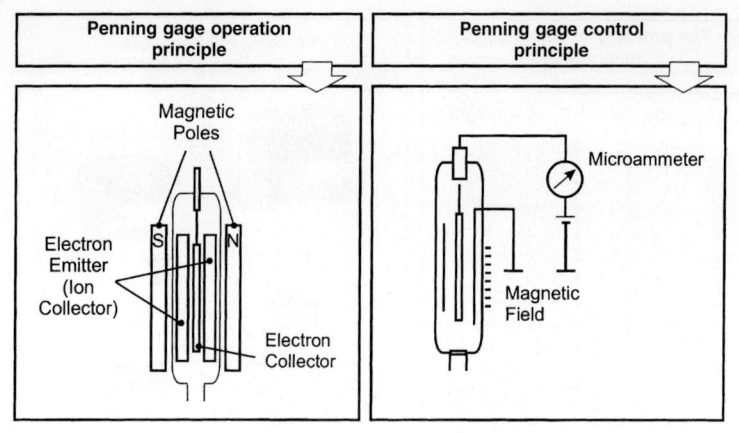

Fig. 4.8-7 Penning gage principle for vacuum monitoring

create gas-phase ions, to separate the ions in space or time based on their mass-to-charge ratio and to measure the quantity of ions of each mass-to-charge ratio. In this instrument, a thin beam of positively charged ions is deflected by an electric field and then deflected in the opposite direction by a magnetic field. The amount of deflection of the particles as registered by the spectrometer is dependent on their mass and velocity: the greater the mass or velocity of the particle, the less it is deflected. The ion masses can easily be converted to partial pressures and therefore to monitor the vacuum status of the coating chamber. The measuring range of such vacuum gages is from 2.5×10^4 down to 2×10^{-2} Torr.

All mass spectrometers have a system for introducing the substance to be analyzed, a system for ionizing the substance, an accelerator that directs the ions into the measuring apparatus, and a system for separating the constituent ions and recording the mass spectrum of the substance. While conventional mass spectrometers can provide a high degree of resolution, their variations, including the tandem mass spectrometer (which consists of multiple mass spectrometers connected end to end) and the tandem electrostatic accelerator (which consists of a series of mass spectrometers connected to an electrostatic particle accelerator), provide far more sensitivity than a single mass spectrometer.

Nevertheless, mass spectrometers are mainly used as gas analyzers, and vacuum monitoring is a minor application. Therefore, further specifications are included in the section on gas composition monitoring.

4.8.2.2.6 Combination Vacuum Gage Systems

In order to create a universal vacuum sensor, a variety of the aforementioned sensors are usually combined into a format, which is able to measure from ultrahigh vacuum to over atmospheric pressure. Figure 4.8-8 illustrates a typical system, which combines the technologies of the cold cathode, Pirani, thermocouple and

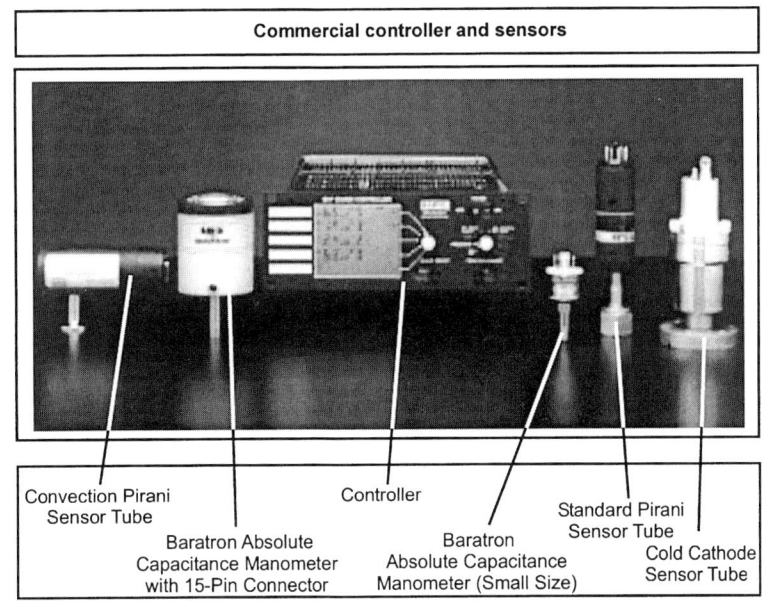

Fig. 4.8-8 A typical five sensors combination gage for vacuum monitoring (sensors by MKS)

capacitance manometer pressure sensors. This gage provides a wide pressure measurement, from 10^{-10} up to 10^4 Torr.

4.8.2.2.7 Leak Detectors

Vacuum destruction by leaks is sufficient to damage the vapor deposition process. Actually, leaks are detected, during the coating deposition, in most cases by the indications of the aforementioned pressure gages, which produce a control system alert. However, coating device producers usually utilize more specialized equipment to detect the tightness of their deposition chambers. Helium mass spectrometer leak detection is the most sensitive method to detect leaks in a product that requires pressure tightness (see Figure 4.8-9). This method works by directing helium to pass through a small leak and then be drawn into a detector, which is basically a mass spectrometer. Very small amounts of helium can be detected with such a system. For coating applications, leak detectors are composed of a mass spectrometer tube and a solid-state electronic control panel, but without a vacuum pumping system. The operation of this instrument relies on the presence of a high-vacuum status of the component to be inspected. The mass spectrometer tube is connected to the vacuum system. More than one mass spectrometer tube, sharing the same control panel, may be attached to different locations of the vacuum equipment system to be inspected for improved detection efficiency.

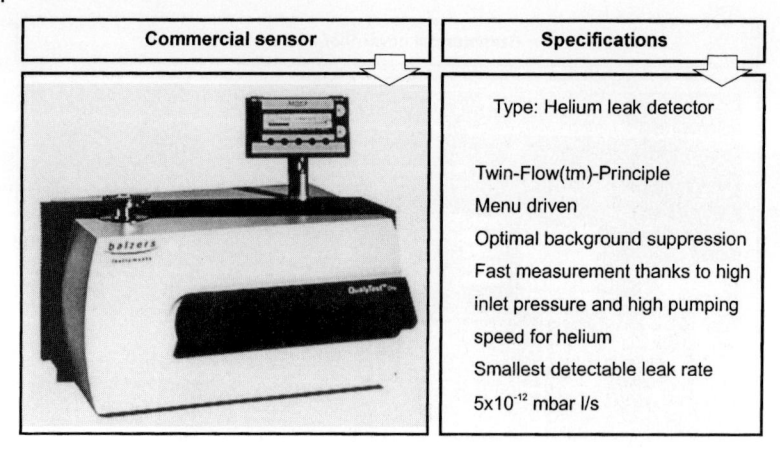

Commercial sensor	Specifications
	Type: Helium leak detector Twin-Flow(tm)-Principle Menu driven Optimal background suppression Fast measurement thanks to high inlet pressure and high pumping speed for helium Smallest detectable leak rate 5×10^{-12} mbar l/s

Fig. 4.8-9 Helium leak detector for vacuum processes (sensor by Balzers)

4.8.2.3 Temperature Control

Vacuum coating process efficiency and quality are strongly temperature dependent. The physical, mechanical, and technological coating properties are very sensitive to the deposition temperature, so this magnitude is usually reported as a coating specification. An increase in the deposition temperature, for the same coating grade, usually enhances hardness and stiffness but on the other hand induces internal stresses and decreases the coating ductility. Furthermore, the deposition temperature may cause crystalline transformations of the substrate, such as uncontrolled annealing. Hence the temperature is a key parameter that has to be closely tracked and adjusted throughout the deposition period. PVD coating processes fall into two groups regarding the deposition temperature, ie, low-temperature deposited (150–250 °C) and high-temperature deposited types (250–550 °C). On the other hand, CVD processes occur at higher temperatures, as already mentioned. To monitor the deposition temperature, two types of thermometers are mainly used, ie, pyrometers and thermocouples.

4.8.2.3.1 Pyrometers

By definition, a pyrometer (see Figure 4.8-10) is a thermometer that measures high temperatures. Such temperature gages are commonly used in vapor deposition chambers to monitor the deposition temperature. Several such sensors may be mounted within the coating device, in order to depict the net temperature at various points of concern. Pyrometers are cost-effective and sensitive solutions either for full production or for pilot plants, since they are reliable tools for in situ, non-contact, and non-destructive monitoring. Pyrometers may be met as analog or digital sensors, with infrastructure for connection to integrated data acquisition systems.

Fig. 4.8-10 Typical pyrometer used for coating process monitoring (sensor by Quantum Logic)

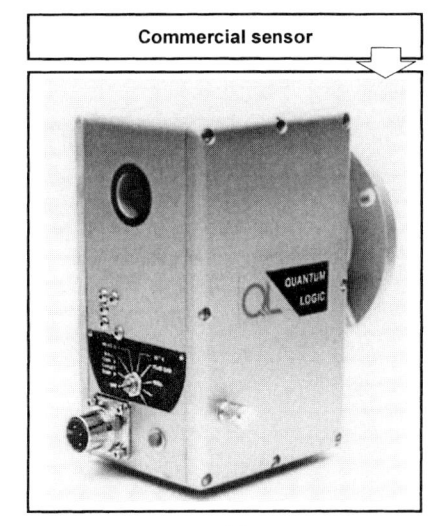

Commercial sensor

Features and specifications

Type: Online laser pyrometer
Temperature range: 200°- 3000°C
Accuracy: -+3°C
Measurement of target emissivity
High speed measurements
(ten readings per second)
Flange or base mounting
Water cooling possibilities

Typical specifications for such pyrometers used in vapor deposition plants are as follows:

Temperature range	200–600 °C
Accuracy probe	Max. of $\{1$ or $0.2\% \, T\}$ °C
Accuracy in situ	Max. of $\{5$ or $1.5\% \, T\}$ °C
Repeatability probe	Max. of $\{0.5$ or $0.1\% \, T\}$ °C
Repeatability in situ	Max. of $\{1.5$ or $0.75\% \, T\}$ °C
Sensing resolution	0.1 °C
Minimum update rate	1 Hz.

4.8.2.3.2 Thermocouples

In addition to pressure measurements, very accurate temperature measurements can be made also with thermocouples, in which a small voltage difference (measured in millivolts) arises when two wires of dissimilar metals are joined to form a loop, and the two junctions have different temperatures. Since the voltage depends on the difference in the junction temperatures, one junction must be maintained at a known temperature. Otherwise, an electronic compensation circuit must be built into the device to measure the actual temperature of the sensor. Special electronics are designed to carry thermocouple instrumentation signals while providing electrical isolation and hermeticity. Lately, in coating installations, non-contact infrared (IR) thermocouples are being used (see Figure 4.8-11), for the following reasons:

- the objects to be measured are usually moving or rotating (PVD chambers with planetary motion capabilities);

Commercial sensor

Specifications

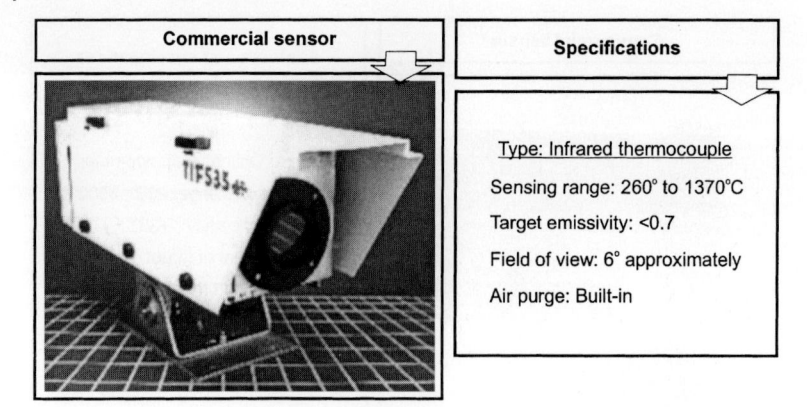

Type: Infrared thermocouple
Sensing range: 260° to 1370°C
Target emissivity: <0.7
Field of view: 6° approximately
Air purge: Built-in

Fig. 4.8-11 Typical infrared thermocouple used in vapor coating process monitoring (sensor by TIF)

- contact alters the object to be measured and usually causes inaccuracies;
- a contact device wears too quickly because of friction or radiation;
- a much wider area must be monitored than can be done with contact devices.

IR thermocouples combine the latest developments in IR technology and use thermoelectric principles to determine the temperature of an object by measuring its emitted IR radiation.

Typical specifications for IR thermocouples used in PVD plants are as follows:

Bakeout temperature	Up to 450 °C with plug disconnected
Vacuum rating	Down to 10^{-10} Torr
Accuracy	Max. of $\{1 \text{ or } 2\% \, T\}$ °C
Resolution	~ 0.0001 °C
Couples	1–18 pairs.

4.8.2.3.3 Bolometers

By utilizing surface machining technology, such as anisotropic etching of silicon, simulation for sensor with optimized responsivity, read-out circuits design, and blackbody coating, a floating membrane responding to IR radiation can be produced. The bolometer can be a standard optical detector in the applications to security, temperature measurement, non-destructive testing, and IR spectroscopy. A bolometer essentially consists of a thin SiN membrane with a superconducting meander on top to read out the temperature. A schematic cross section is shown in Figure 4.8-12. The temperature rise of the membrane is proportional to the power of the IR radiation, which is absorbed by the bolometer. Biased on its transition temperature the superconductor is a very sensitive thermometer for the membrane temperature. In order to enhance the efficiency, an absorbing layer of gold black is deposited over the membrane and the meander.

Fig. 4.8-12 Bolometer principle — Schematic cross section of high-T bolometer

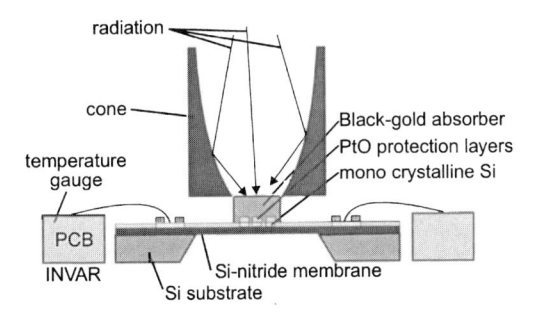

Bolometers are the most sensitive detectors of broadband electromagnetic radiation at wavelengths between 250 µm and 5 mm. Bolometers are extremely sensitive broadband detectors because they are sensitive to electromagnetic radiation at any wavelength. Their high precision and cost make them unsuitable for commercial vapor coating installations, and they are usually used for research and experimental purposes.

4.8.2.4 Gas Analyzers for Coating Process Control

A very important parameter for the coating process monitoring is the in situ analysis and control of the residual and process gases. In this way, using preset experience, operators are alerted by an alarm when selected gas species exceed predetermined limits. Consequently, yield problems, system downtime, and poor product quality can sometimes be avoided. Gas flow data that must be controlled in PVD coating processes are as follows [9–15]:

- base pressure composition;
- gas impurity/composition;
- reactive gas partial pressure.

These measurements are based on mass spectrometry. For gas analysis applications, a so-called quadrupole mass spectrometer is being used. A typical quadrupole mass filter consists of four parallel metal rods arranged as shown in Figure 4.8-13. Two opposite rods have an applied potential of $(U + V\cos(\omega t))$ and the other two rods have a potential of $-(U + V\cos(\omega t))$, where U is a direct current (DC) voltage and $V\cos(\omega t)$ an alternating current (AC) voltage. The applied voltages affect the trajectory of ions traveling down the flight path centered between the four rods. For given DC and AC voltages, only ions of a certain mass-to-charge ratio pass through the quadrupole filter and all other ions are thrown out of their original path. A mass spectrum is obtained by monitoring the ions passing through the quadrupole filter as the voltages on the rods are varied. There are two methods: varying ω and holding U and V constant, or varying U and V (U/V fixed) for a constant ω.

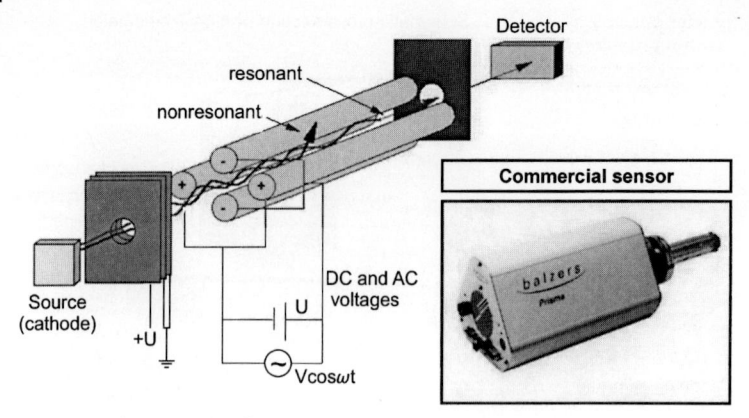

Fig. 4.8-13 The principle of quadrupole mass spectrometer and characteristic systems for coating applications (sensor by Balzers)

This powerful, low-cost tool is useful for massive deposition process monitoring and research and development applications, while offering a general-purpose gas analysis. Such gas analyzers employ appropriate software offering simple operations, providing detailed in situ gas analysis for vacuum systems. Modern commercial quadrupole systems as the shown in the bottom right part of Figure 4.8-13 offer the following:

- economical solutions that may be easily integrated into existing vacuum tools, including coating chambers;
- unrivalled sensitivity and detection limits supported by spectral library aides for identification of gases;
- better product quality, increased productivity, and high dynamic range;
- limited dimensions and weight and advanced peripheral electronics, which allow the real time monitoring of the gas composition within the coating chamber.

A further feature of mass spectrometers is that they may be used to detect quantitatively residual gases, in the coating chamber after the vacuum formation, and before the deposition process initiation. This parameter is very important to PVD processes, since it ensures the coating quality. For coating devices without special leak detectors, mass spectrometers may also serve such a need. In addition to the quadrupole type, other spectrometers are also applied for gas analysis, such as Raman and Auger spectrometers, ellipsometers, etc. [9, 10]. However, these devices are used mainly for research and development purposes.

4.8.2.5 Thin-film Thickness (TFT) Controllers for Deposition Rate Monitoring and Control

Thin-film controllers are usually combined hardware/software systems that automatically control, in real time, vacuum deposition rates [9, 10]. Such systems allow the accurate measurement of the rate and thickness of any deposited

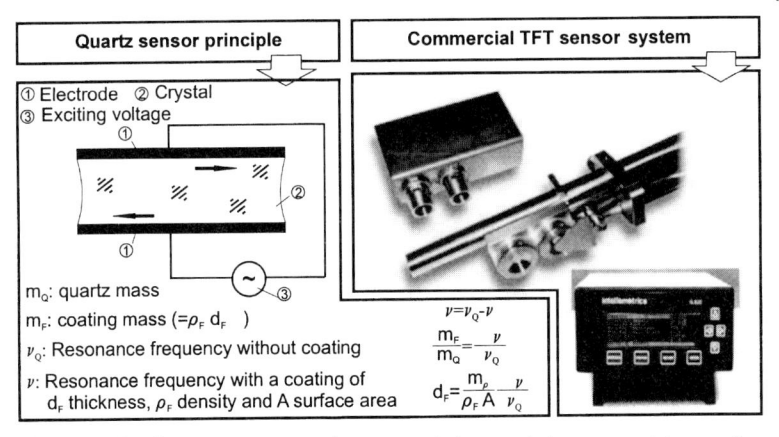

Fig. 4.8-14 The principle of a TFT quartz crystal sensor and characteristic commercial controllers for PVD applications (sensors by Intermetrics)

materials with the aid of quartz crystal sensors, which are usually also coated. Through an AC voltage, a quartz crystal can be excited to mechanical vibrations (piezoelectrical effect inversion). The eigenfrequency of this vibrating system is affected by the coating mass m_F deposited on the crystal (see Figure 4.8-14). The quartz crystal is placed in the vacuum work chamber. One face of the crystal is exposed towards the deposition source such that as material is deposited it will coat the crystal. The crystal is connected to external components and the whole system becomes an oscillator, the output of which is controlled by the frequency of crystal oscillation. As material is deposited on the crystal, its frequency modification is monitored. The output signal processing offers the deposition rate and the coating thickness in any deposition stage. Typical specifications for these sensors are frequency range 4–10 MHz, with a mean resolution of 10 MHz and a sampling time of 100 ms. These closed-loop control systems, with feedback action, keep the deposition rate (Å/s) constant during the steady state and determine the appropriate smooth fading during the transitory stage.

TFT systems include software tools to visualize the coating process. These tools include statistical and analytical functions that evaluate of the sensor data and trigger the adjustment feedback to optimize the deposition rate. Precision fiber optic-based instrumentation facilitates an intrinsically safe, non-contact operation in any environment.

Another possibility for measuring the deposition rate is based on the ionization of the vacuum phase, as presented in Figure 4.8-15. Specific ions from the vacuum phase are directed to a metallic target, with the aid of a specially formed magnetic field. Thereby, a bias voltage is generated, proportional to the ion collisions at the target. In this way, the deposition rate is continuously tracked and the film thickness can be measured in real time. This method is usually applied for research and development purposes, whereas for commercial installations quartz crystals are preferred.

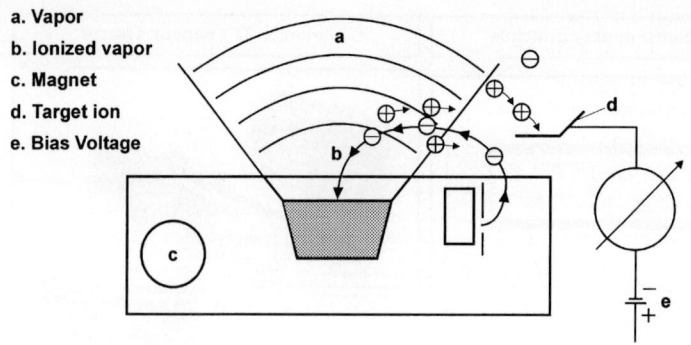

a. Vapor
b. Ionized vapor
c. Magnet
d. Target ion
e. Bias Voltage

Fig. 4.8-15 Monitoring of the deposition rate through ionization

4.8.2.6 Gas Dosing Systems and Valves

The gas supplement to the coating chamber is a major parameter that must be monitored and controlled, usually with the aid of closed-loop control. Thereby, accurate and repeatable automatic upstream pressure and gas flow control is achieved. Modern gas dosing systems include advanced electromechanical parts, whereas for every specific gas a control valve is combined with gas flow sensors, which provides a complete optimized control system. A typical gas dosing system is presented in Figure 4.8-16. It consists of a stepper motor control valve with electronics and position sensors. It can be used as an autonomous component by system builders and can be directly controlled by the system controller. Figure 4.8-16 also includes typical features of such a commercial closed-loop regulation system.

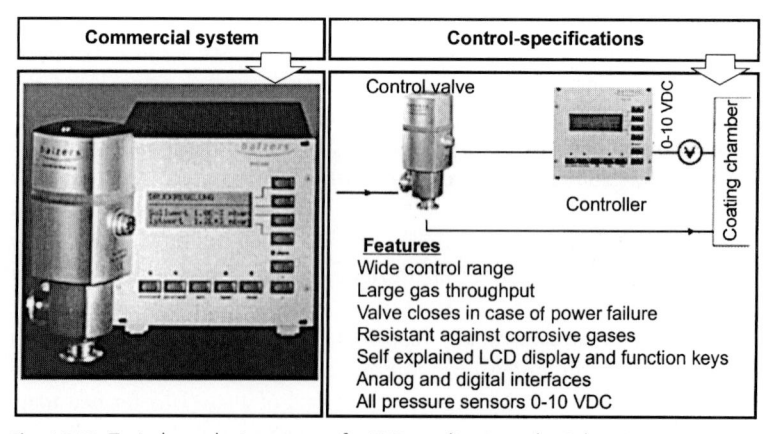

Features
Wide control range
Large gas throughput
Valve closes in case of power failure
Resistant against corrosive gases
Self explained LCD display and function keys
Analog and digital interfaces
All pressure sensors 0-10 VDC

Fig. 4.8-16 Typical gas dosing system for PVD applications (by Balzers)

4.8.2.7 Other Parameters Usually Monitored During the PVD Process

In addition to the aforementioned process parameters, which are monitored during the film deposition, there are further process variables that may also be controlled. These variables depend on the complexity of the coating chamber and might be controlled with the aid of open- or closed-loop control systems. Typical magnitudes that are controlled during the deposition process in coating devices are the plasma status, the deposition time, the accumulated rotations, and the rotation speed of the substrate in chambers with planetary motion capabilities, the electromagnetic parameters of the device, the power supply, etc. Moreover, plasma diagnostics are very a very critical aspect in vapor deposition processes.

The understanding and characterization of plasmas in coating processes is very important, especially for research and development purposes [16]. The ideal tools for monitoring and quantification of plasma processes are special devices, usually called plasma process monitors. This sensing device consists of a differentially pumped mass spectrometer with an integrated energy analyzer. It allows or investigates plasmas, ie, analyzes ions and neutrals (molecules, radicals) with respect to energy and mass. The particles are extracted from the plasma via an extraction orifice. The plasma to be monitored may be for coating, cleaning, and etching processes. In such plasma monitoring devices, special attention is paid to protect either the plasma or sensor from arcing and to prevent any possible vacuum destruction.

4.8.3 References

1 STRAFFORD, K., et al., *Coatings and Surface Treatment for Corrosion and Wear Resistance*; Chichester: Ellis Horwood, 1984.
2 BOUZAKIS, K.D., et al., in: *Proceedings of International Conference on THE Coatings*, Thessaloniki, 14–15 October 1999, pp. 35–66.
3 SPUR, G., STÖFERLE, TH., *Handbuch der Fertigungstechnik, Band 4, Abtragen und Beschichten*; Vienna: Carl Hanser, 1987.
4 TÖNSHOFF, H.K., et al., in: *Proceedings of International Conference on THE Coatings*, Thessaloniki, 14–15 October 1999, pp. 1–20.
5 LEYENDECKER, T., *CemeCon Entwicklungs-Berichte*, Aachen, 1993.
6 ESSER, S., *CemeCon Entwicklungs-Berichte*, Aachen, 1995.
7 HAEFER, R.A., *Oberflächen- und Dünnschichttechnologie. Teil 1, Beschichten von Oberflächen*; Berlin: Springer, 1987.
8 ERKENS, G., *Doctoral Thesis*; RWTH Aachen, 1998.
9 FREY, H., KIENEL, G., *Dünnsichttechnologie*; Düsseldorf: VDI Verlag, 1987.
10 BARTELLA, J., et al., *Vakuumbeschichtung 3 – Anlagenautomatisierung – Mess- und Analysentechnik*; Düsseldorf: VDI Verlag, 1993.
11 EGUCHI, N., et al. *Mater. Sci. Eng.* **A139** (1991) 339–344.
12 JONES, J.G., et al., Presented at the Symposium on Nondestructive Evaluation of Ceramics at the American Ceramic Society 99th Annual Meeting. Cincinatti, OH, 4–7 May, 1997.
13 ERGO, C.A., et al., presented at the 5th European Conference on Diamond, Diamond-like and Related Materials, Tuscany, Italy, 25–30 September 1994.
14 WEBB, A.P., *Appl. Surf. Sci.* **63** (1993) 70–74.
15 REGO, C.A., et al., presented at the 4th International Conference on New Diamond Science and Technology, Kobe, Japan, 18–22 July 1994.
16 HRON, M., et al., *Contrib. Plasma Phys.* (1999) 1–6.

4.9
Heat Treatment
P. Mayr, H. Klümper-Westkamp, *Stiftung Institut für Werkstofftechnik IWT, Bremen, Germany*

4.9.1
Introduction

After mechanical manufacturing, heat treatment is mainly applied in one of the final steps of production, to adjust the workpiece properties to the later mechanical, tribological, and corrosive load. The lifetime of these components is defined through the raw material used, the construction geometries, and the quality of heat treatment.

The great demands for quality control and the documentation of process parameters have led to the increasing importance of sensor applications in heat treatment. Furthermore, increasing automation combined with the idea of integrating the heat treatment into the production line needs many different sensors to make the whole process as reliable as possible.

4.9.2
Temperature Monitoring

The primary process variable to be measured and controlled in heat treatment is the temperature, which directly influences component properties. There are a very large number of different temperature-dependent physical properties that can serve, at least in principle, as the basis for thermal sensors. Only a few of them are used in heat treatment plants to measure the temperature.

Thermocouples are the most commonly used temperature sensors in heat treatment. They consist of two conductors of different metallurgical composition, which are made up into an electrical circuit with two junctions. If the two junctions are brought to a different temperature, a thermoelectric potential is created. The magnitude depends on the composition of the conductors chosen and the temperature difference [1].

Their simple construction allows sensors to be made which are able to provide reliable and accurate data under extreme measurement conditions. With thermocouples a range from −270 to 2000 °C can be covered, depending on the material used. The accuracy depends on the temperature being measured and ranges from 0.1 K at room temperature to ±10 K at very high temperature. The characteristics of various commonly used thermocouples are illustrated in Figure 4.9-1. International, eg, IEC 584, and national standards, eg, DIN 43710 in Germany, have been set up. NiCr-Ni is the most frequently employed thermocouple in industrial plants up to 1100 °C. In most applications they are sheathed. The thermocouple wires are embedded in an isolating powder (eg, MgO, Al_2O_3) and surrounded by a metal sheet (eg, Inconel, stainless steel). At higher temperatures, PtRh-Pt thermocouples are used. They should be protected by ceramic tubes [2].

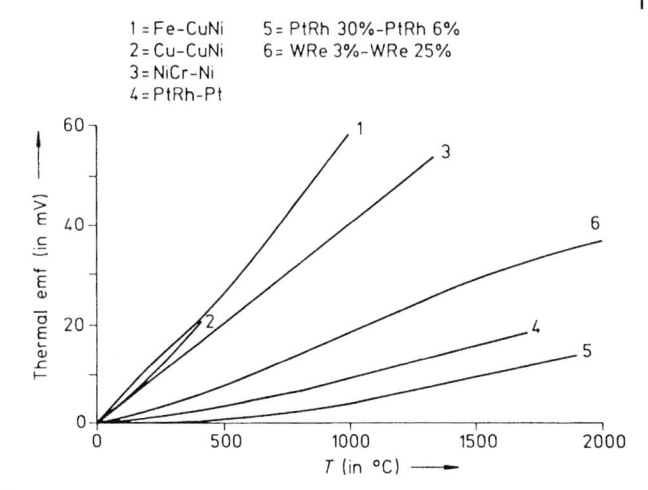

Fig. 4.9-1 Thermal emfs of commonly used thermocouples [2]

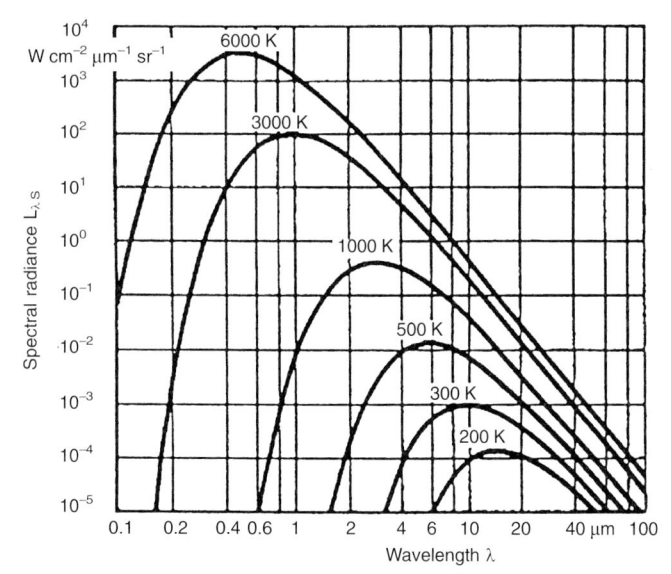

Fig. 4.9-2 Characteristic radiation of a blackbody [3]

According to Planck's law, the spectral radiation of a blackbody at a particular wavelength varies only with the blackbody temperature. The *optical pyrometer* makes use of this principle by measuring ratios of spectral radiances. The radiation characteristic of a blackbody is shown in Figure 4.9-2. The wavelength spectrum utilized for radiation thermometers usually lies between 0.5 and 20 μm. The use of radiation thermometers is simple when dealing with surfaces showing blackbody condition. Unfortunately, in most applications deviations from blackbody condition occur as a result of absorption, emissivity, and reflection effects. The most difficult problems arise with unknown and variable emissivity.

Optical pyrometers can be subdivided into four groups depending on the wavelength and wavebands used:

- total-radiation thermometers;
- single-waveband thermometers;
- ratio thermometers;
- multi-waveband thermometers.

The very common single-waveband thermometers with silicon or germanium detectors use short wavelengths between 0.5 and 2 µm. For very low temperatures longer wavelengths are used because of a higher rate of change of energy with temperature. Ratio thermometers measure the ratio of energy in two narrow wavebands. The advantage is that their readings will be independent of target size, obscurity of optical path, and emissivity, provided that they are the same at the two wavelengths [2].

The electric resistance of some metals increases with increase in temperature in a nearly linear fashion. This property has been applied for temperature measurement. Platinum is the most precious metal used for thermometry. Platinum can be drawn in to very thin wires of high purity. Thick- and thin-film sensors are also available.

Ruggedness and relatively low costs are combined in platinum thick-film thermometers. They are used in industrial applications up to 450 °C when encased in ceramic, glass, or other protective materials. Atmospheres containing carbon or hydrogen are poisonous for platinum and lead to unstable resistance. *Resistance thermometers* have the advantage of high accuracy measurement. Typically, a temperature resolution of $\pm 10^{-3}$ °C can be achieved up to 500 °C whereas thermocouples are limited to ± 0.5 °C. Special geometries are available for surface temperature measurement.

In special applications other metals such as copper, nickel, and iridium are used [2].

A number of physical effects have been checked to realize *fiber-optic temperature sensors* [4]. The main fiber-optic thermometers today use the temperature dependence of the fluorescence decay time. This effect has been utilized in commercial instruments by Luxtron and Sensycon. The main advantage of this temperature measurement method is the capability to allow measurements in severely hostile environments. They can be made in the presence of intense radiofrequency and microwave fields as well as very high voltages and strong magnetic fields. In the range from –200 to 450 °C, an accuracy of ± 0.1 °C can be reached at the calibration point.

In continuously working furnaces, temperature measurements can be made by *furnace tracker* systems. A data logger for thermocouples is positioned in a thermal protection unit made of inert and maintenance-free material. During the furnace travel the data logger stores the measured temperatures, which can be read later. Such a furnace tracker works up to temperatures of 1200 °C. Such systems have been developed and are sold by DATAPRO and Stoppenbrink.

4.9.3
Control of Atmospheres

Heat treatment is done, depending on the aim of the treatment, in different furnaces with different pressures, atmospheres, and plasma application. The scale in pressure ranges from vacuum up to high pressures of several hundred bar. Depending on the aim of heat treatment, different gases and gas mixtures are applied. When there are no requirements on the workpiece surface, heat treatment can be done in air, which means that oxidation takes place during heat treatment. To avoid this oxidation, inert atmospheres with low oxygen partial pressure must be used and controlled.

Another group of processes is the thermochemical heat treatment processes. The purpose is to modify the workpiece by specific reactive atmospheres, so that carbon (carburizing) and/or nitrogen (nitriding) diffuses into the near-surface region and parallel oxidation can take place at the surface. To control the oxygen partial pressure, the oxygen probe was the first developed in situ sensor in thermochemical heat treatment. With the knowledge of the oxygen partial pressure, many other process parameters can be evaluated. Some different constructions of oxygen probes are discussed in [5].

4.9.4
Carburizing

For controlling the carburizing process, the chemical equilibrium between the carburizing gas and steel is used. Under the assumption of unchanged basic components CO and H_2, the simplified value 'CO_2' or 'dew point' is derived, which is the main controlled variable for the carbon potential.

It has been well established for more than 25 years that the partial pressure of oxygen measured by oxygen probes [6] can be used as an exact indicator of the carbon potential. This permits considerable progress with regard to accuracy, responsiveness, reproducibility, and technical handling, so that this sensor is being increasingly used.

The basic principle is an oxygen ion-conducting electrochemical cell of stabilized ZrO_2. The measured potential E depends on the difference in oxygen partial pressure P_{O_2} at the two electrodes. $P_{O_2}^0$ is the known reference oxygen pressure, normally of air. In Figure 4.9-3 a schematic diagram of such a sensor is shown. The temperature T is measured in kelvin [7].

$$E\,(\text{mV}) = 0.0496\,T \log(P_{O_2}/P_{O_2}^0) \tag{4.9-1}$$

In oxygen-containing carburizing atmospheres, the following chemical reaction is the main indicator for the carburizing reaction:

$$CO \rightleftharpoons [C] + \tfrac{1}{2}O_2 \tag{4.9-2}$$

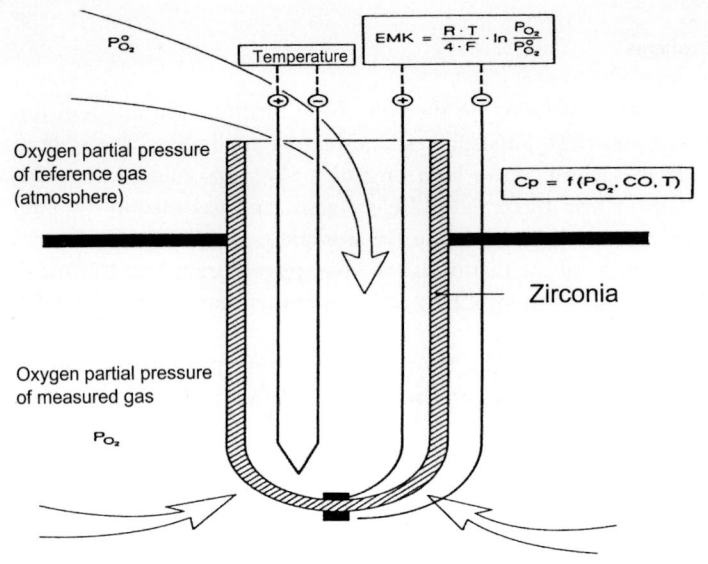

Fig. 4.9-3 Mode of action of an oxygen probe [7]

Under the assumption of thermodynamic equilibrium, the carbon activity a_c is given by

$$\log a_c = \log[P_{CO}/(P_{O_2})^{0.5}] - 5927/T - 4.545 \tag{4.9-3}$$

With known pressure of carbon monoxide P_{CO} and temperature T, the carbon activity can be calculated from the measured oxygen partial pressure. The carbon activity of the carburizing atmosphere defines the dissolved carbon in austenite under the assumption of thermodynamic equilibrium. More information about oxygen-free carburizing atmospheres can be found in [7].

A further innovation is the application of oxygen probes in heat treatment, primarily known from automobiles for measuring the lambda point of combustion. When they are used in heat treatment plants they are normally located outside the furnace and measure the oxygen partial pressure on the basis of an electrochemical unit at temperatures around 500 °C. The measured value has to be recalculated to the furnace temperature and additionally it should be mentioned that owing to the outside mounting, more uncontrolled influences can change the value. As the measured oxygen potential can be regarded in most cases as integral information for the whole furnace, this system can also be useful for controlling the degree of sealing of the furnace. Because of the mass production of this lambda probe it is much cheaper but in many applications not as reliable as the standard in situ oxygen probe.

A direct method for measuring the carbon uptake in almost every atmosphere is the so-called wire sensor [8]. This sensor is based on the resistivity change of a thin, pure iron wire with increasing amount of dissolved carbon, as shown in Fig-

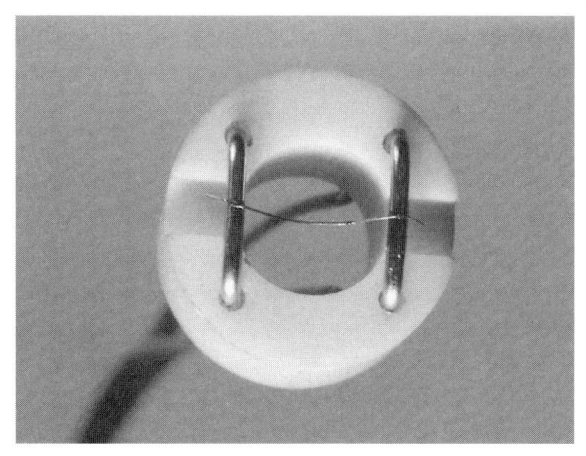

Fig. 4.9-4 Probe tip of the wire sensor (Werkfoto PE-Essen)

ure 4.9-4. Periodically the wire is decarburized with hydrogen so that it can be used again. This in situ sensor is mostly applied for periodic checking of the absolute carbon potential in furnaces. Continuous control of the carbon potential in a furnace is not practicable because of problems with oxidation, formation of carbides, and changing surface of the wire. This in situ sensor is sold by Process-Electronic.

4.9.5
Nitriding

Gas nitriding is usually carried out in ammonia-containing atmospheres. Since Lehrer's paper [9], it is known that the nitriding potential K_N is the driving force of nitriding and defines the phase composition of the white layer. The nitriding potential K_N is defined by the well-known ratio of the partial pressures of ammonia, P_{NH_3}, and hydrogen, P_{H_2}:

$$2\,NH_3 \rightleftharpoons N_2 + 3\,H_2 \tag{4.9-4}$$

$$K_N = P_{NH_3}/(P_{H_2})^{0.5} \tag{4.9-5}$$

In the case when pure decomposition of ammonia dominates the nitriding process, this nitriding potential presents a reliable indicator for the nitriding process. In nitriding atmospheres little is known about the basic level of the partial pressure of oxygen. Established process control have used external gas analyzers based on infrared absorption or heat conduction for many years. Recently, in situ sensors have been developed which are able to measure directly the nitriding potential inside the furnace. They are based on oxygen probes containing a solid electrolyte of stabilized zirconia. The experimental approach to measuring the nitriding potential with oxygen probes is different to that for the control of the carbon potential. For a precise function the concentration of water vapor present in the nitriding atmosphere should never drop below 1 vol.%.

The measuring principle is as follows. The nitriding atmosphere with its residual ammonia passes over the outer zirconia electrode of the sensor. Subsequently it is dissociated totally into hydrogen and nitrogen by a 1000 °C nickel catalyst. This atmosphere with altered composition and lower oxygen potential reaches the inner electrode of the sensor. Between the two electrodes a potential difference according to the ammonia content in the nitriding atmosphere can be measured and the nitriding potential can be calculated [10].

$$\Delta U \,(\text{mV}) = 0.0992 \, T \log(1 + \tfrac{3}{2} P_{NH_3}/P_{H_2}) \qquad (4.9\text{-}6)$$

This method also works in nitrocarburizing atmospheres provided that no additional gas components dissociate in the heated catalyst, changing the thermodynamic equilibrium of the water-gas reaction. Hydrogen, for instance resulting from decomposition of additional methane, shifts the water-gas and therefore influences the oxygen partial pressure.

$$H_2O \rightleftharpoons H_2 + O_2 \qquad (4.9\text{-}7)$$

The utilization of this sensor for nitrocarburizing atmospheres should also work, provided that the inlet gas flow and composition are presicely measured.

Some different types of construction are sketched in Figure 4.9-5. The main difference is that the nickel catalyst is positioned in a second steel tube. Whereas types 2 and 3 pump the reference gas outside the furnace from the catalyst to the second electrode, type 4 works as two oxygen probes measuring two partial pressures (the dissociated and the normal atmosphere) against air.

Some practical results of a complete measuring and control system for nitriding and nitrocarburizing processes were presented in [11], which demonstrate that the system is capable of reliable and reproducible operation. Important values of the nitrocarburizing process are ready for documentation to accomplish the claims of DIN/ISO 9000 ff.

Another method for measuring the hydrogen partial pressure in heat treatment atmospheres is described in [12]. Special membranes of palladium and palladium-silver are permeable only for hydrogen. With a pressure gage the hydrogen partial pressure can be directly measured beneath the membrane.

4.9.6
Oxidizing

Oxidizing is of increasing interest. Since it became recognized that post-oxidation, directly conducted after nitrocarburizing, significantly increases the corrosion resistance of nitrocarburized steel parts, this combination is widely used today [13, 14].

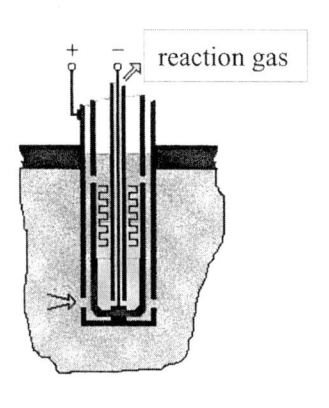

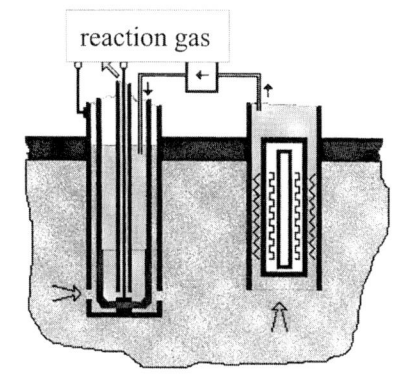

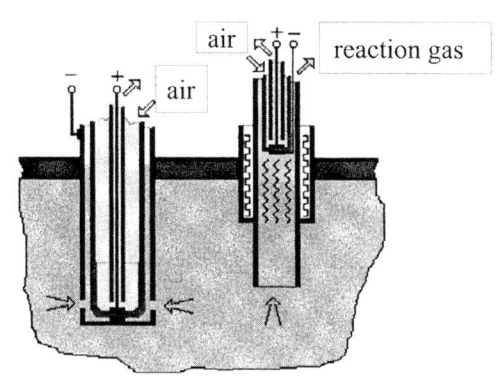

Fig. 4.9-5 Different types of nitriding gas sensors (Werkfoto PE-Essen)

Pre-oxidation is also a process under discussion in order to prepare the surface after mechanical manufacturing for thermochemical treatment such as nitrocarburizing and carburizing. Finally, oxidation as a stand-alone process for increasing corrosion and wear resistance can be a cheap alternative instead of nitrocarburizing or chrome plating.

The oxidizing atmosphere can consist of water, air, CO, CO_2, and additional gas components. If the oxidation is performed in an uncontrolled manner, the type and thickness of the oxidized layer can vary over wide ranges and thus result in different surface properties. In order to provide the workpiece with an optimum magnetite layer on the surface, a certain oxidation potential has to be established.

The main controlled variable for oxidation is the partial pressure of oxygen. This variable is easily measured by an oxygen probe containing a solid electrolyte of stabilized zirconium oxide. New developments in thin-film technology make oxygen sensors based on the change of resistivity much cheaper. The sensing material is TiO_2 [15]. Further materials are still under investigation.

4.9.7
Control of Structural Changes

In order to establish proper thermodynamic conditions for the desired heat treatment result, measurement of temperature and atmospheric constituents are necessary. In addition to the thermodynamic considerations, the kinetics play an important role. This aspect is monitored by a group of sensors which are able to measure the result of mass transfer directly. Such sensors have been developed making use of the eddy current technique [16].

Nitriding treatment not only influences the mechanical and chemical properties of a steel but also modifies the electrical and magnetic characteristics in the near-surface regions. For this reason, measurement of electrical and magnetic changes will yield substantial information about the progress of nitriding and nitrocarburizing. As shown in Table 4.9-1, the electrical resistance, the magnetic permeability, and the Curie temperature are changed. The so-called KiNit sensor is shown in Figure 4.9-6. The main components are the coil and the exchangeable specimen mounted as the coil core. The coil and the mounting unit are specially designed so that temperature and the nitriding atmosphere will not change the electromagnetic properties, even after prolonged exposure. The complex impedance of this configuration is measured and correlated with the nitriding and nitrocarburizing parameters. For each steel group a calibration is necessary to obtain quantitative information about white layer thickness and nitrided case depth. Using temperature-dependent measurements the phase composition of the white layer can also be detected.

Many problems in nitrocarburizing are connected with passivation. A thin layer in the surface prevents mass transfer into the material. Most passivation thin films originate unintentionally during manufacturing and finishing processes or are present due to immediate oxidation on high-chromium containing steels. An example is shown in Figure 4.9-7. At the beginning, after heating up, the sensor signal becomes constant with time, but after chemical activation of the surface the eddy current loss decreases, which means that nitrocarburizing takes place. After 7 h of nitrocarburizing a precipitation layer of about 170 µm is measured.

Tab. 4.9-1 Electrical and magnetic properties of iron and iron compounds [17]

Material	Specific electrical resistance ($10^{-8} \Omega m$)	Curie temperature (°C)	Relative magnetic permeability at 500 °C
Fe	10	770	10^2–10^4
Fe with 1 wt% N	+19	?	Reduced
Fe_4N	Similar to Fe_3N	480–510	1
Fe_3N	4500	294–299	1
Fe_3 (N, C)	Similar to Fe_3N	320–388	1
Fe_3C	90	213–223	1

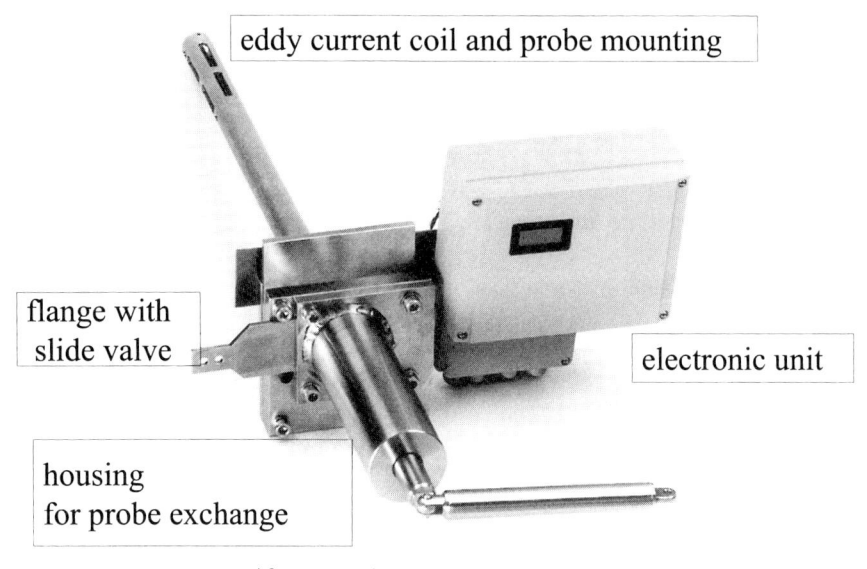

Fig. 4.9-6 KiNit sensor (Werkfoto Ipsen-Kleve)

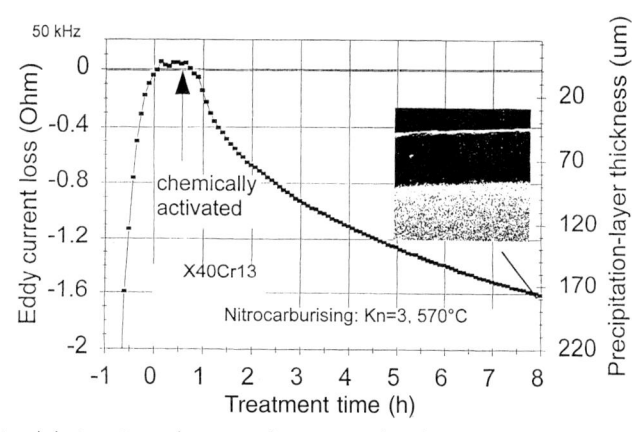

Fig. 4.9-7 KiNit sensor signal during nitrocarburizing of a passivated probe (X40Cr13) [16]

Another application is nitrocarburizing with a well-defined surface layer thickness. Figure 4.9-8 shows an example of how it is possible to nitrocarburize a 6.5 µm thick white layer with any even larger case depth. After the sensor signal shows the defined layer thickness, the nitriding potential is decreased to that level, so that no further white layer grows. In this way, any combination of white layer thickness and case depth can be precisely obtained. Even nitrocarburizing without the occurrence of a white layer is possible and reproducible by using this eddy current sensor.

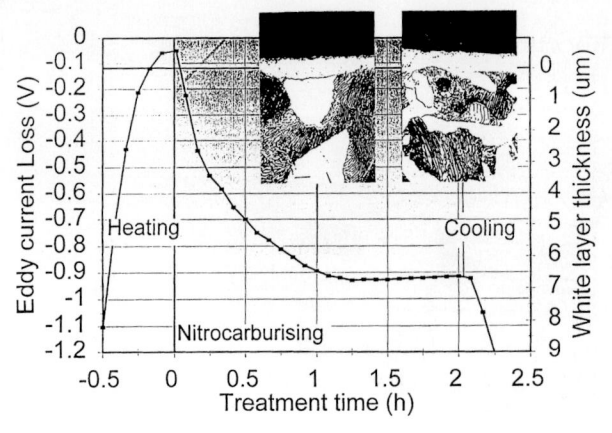

Fig. 4.9-8 Defined nitrocarburizing [16]

A similar eddy current sensor can be applied to *annealing*. Because annealing is often the latest heat treatment step, the final properties of the component are defined by this treatment and document its importance. After hardening and case hardening, annealing is applied to reduce or remove residual stresses and also to adjust the hardness and toughness of the material. Parallel retained austenite is reduced in order to stabilize the dimensions of components. This transformation of retained austenite to martensite or ferrite depends on temperature and time and it is difficult to identify the time when the transformation is finished because several parameters influence the amount of retained austenite, and also stabilization of the austenite can occur.

A typical sensor signal during annealing of steel grade 100Cr6 at 240 °C is shown in Figure 4.9-9. Before annealing 30% retained austenite was measured, which transformed according to the sensor signal. After 16 h of annealing no retained austenite was measured, the sensor signal being unchanged. The reproducibility of this sensor is very good, as can be seen in Figure 4.9-9. Based on x-ray measurement, this sensor can be calibrated according to the transformed retained austenite [18].

Another example of eddy current measurement during annealing of steel grade X210Cr12 is shown in Figure 4.9-10. Because of the higher chromium and carbon contents the structural changes of this material during annealing are more complicated. Many of the structural changes can be measured with eddy current sensors, as shown in Figure 4.9-10. During heating up above T_{ac1} iron carbides form, and their Curie temperature is reached at the maximum. Above T_{ac2} carbides of the type $(Fe,Cr)_3C$ are formed, which is necessary before retained austenite can transform above T_{ac3}. During isothermal treatment at 550 °C, mainly retained austenite transforms. Also during cooling structural changes in form of magnetic transformation and martensitic transformation can be identified. This sensor is therefore a powerful tool, not only for research and development, to obtain information about the microstructure of critical components during heat treatment [19].

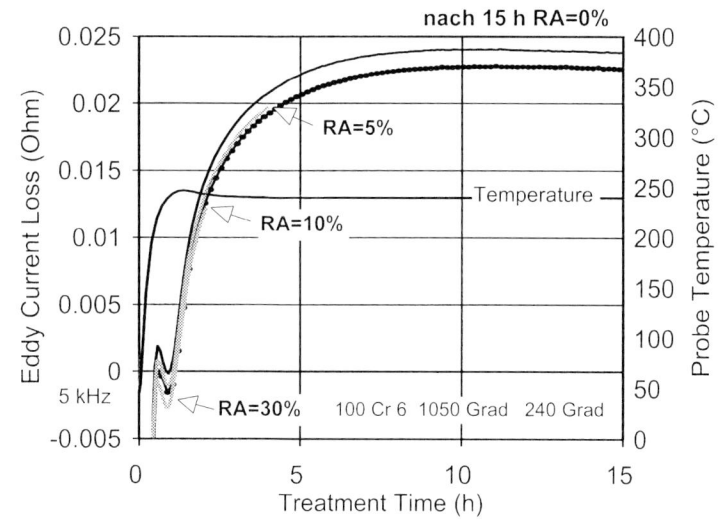

Fig. 4.9-9 Measurement of the transformation of retained austenite during annealing [16]

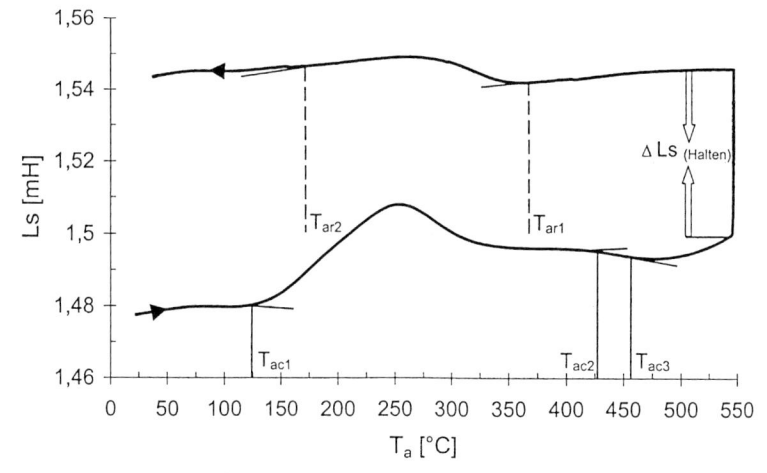

Fig. 4.9-10 In situ measurement of a complete annealing cycle with an eddy current sensor. (X210Cr12, $T_A=1050\,°C$, $t_A=20$ min, $T_a=550\,°C$, $t_a=120$ min) [19]

Similar developments for eddy current measurement during oxidation, recrystallization and bainitic transformation are in progress.

4.9.8
Quenching Monitoring

Martensitic hardening of steel products always includes a defined rapid cooling from the hardening temperature. This step, called quenching, has so far been a neglected stage with respect to process control. Usually only the temperature of the quenching medium is controlled. However, the basic variable related to the quenching severity is the heat transfer coefficient. New methods and sensors have been developed in recent years to determine the heat transfer coefficient in liquid and gaseous quenching media.

4.9.8.1 Fluid Quench Sensor

This sensor is based on the measurement of the difference between the fluid temperature and the higher temperature of the electrically heated wall of the sensor [20]. The wall of the sensor is heated by an internal constant power source. Between the wall and the surrounding fluid a thermal transfer takes place. The thermal transfer is purely convective, because the wall temperature is below the boiling point of the quenching fluid.

For constant quenching conditions, a constant convective thermal transfer and a defined constant temperature difference between the fluid and the sensor wall will be obtained. The design of the sensor and the resulting thermal fluxes are shown schematically in Figure 4.9-11. The sensor is effectively a quality assurance tool which in addition to the usual temperature measurement function enables changes in fluid composition, faults, and incorrect circulation settings to be continuously detected and alarms be generated.

4.9.8.2 Hollow Wire Sensor

This sensor is constructed to measure the heat transfer coefficient of liquids and gaseous quenching media from room temperature up to 1000 °C. The sensor consists of a short piece of thin hollow wire, which is heated by electric conduction to

Supplied Heat:
$Q_{beat} = U^2/R = \text{const.}$

Removed Heat:
$Q_{convection} = \alpha \times A \times (T_1 - T_2)$

Heat Balance:
$Q_{beat} = Q_{convection}$

$\Rightarrow (T_1 - T_2) = C/\alpha = \text{const. if } \alpha = \text{const.}$

$\Rightarrow (T_1 - T_2)$ characterize the convective heat transfer

1. Temperature sensor measuring increased temp. T_1
2. Temperature sensor measuring fluid temp. T_2
3. Fluid
4. Constant heating
5. Protection tube
6. Heat insulation
7. Device for measuring the temperature difference

Fig. 4.9-11 Principle of the fluid quench sensor [20]

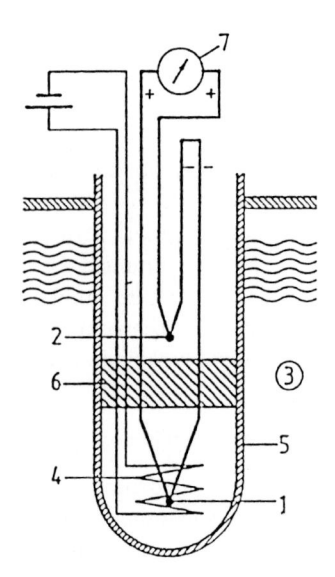

Fig. 4.9-12 Determination of the heat transfer coefficient by the hollow wire method [21]

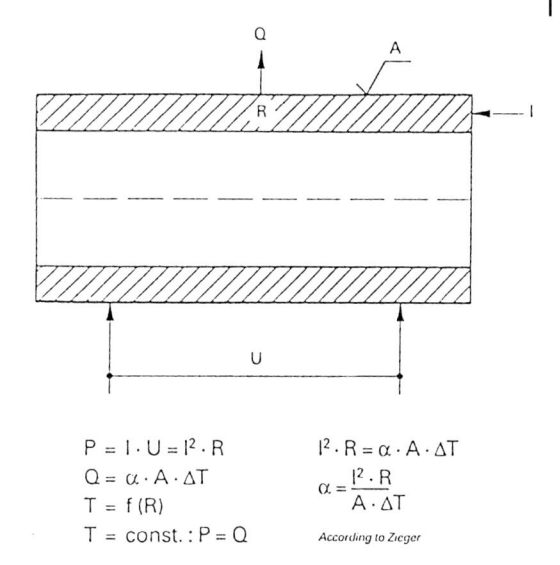

$$P = I \cdot U = I^2 \cdot R$$
$$Q = \alpha \cdot A \cdot \Delta T$$
$$T = f(R)$$
$$T = \text{const.} : P = Q$$

$$I^2 \cdot R = \alpha \cdot A \cdot \Delta T$$
$$\alpha = \frac{I^2 \cdot R}{A \cdot \Delta T}$$

According to Zieger

the hardening temperature of the workpiece [21]. By measuring the resistance of the wire, its temperature is known. To keep the wire in thermal equilibrium at a certain temperature, a certain electric current is necessary. With electric heat going in and convection heat dissipating into the quenching medium, the heat transfer coefficient can be calculated at that temperature (Figure 4.9-12). This method simulates the heat transfer of a common workpiece. However, the sensor signals are not stable over long periods so this method needs further improvement before industrial utilization.

4.9.8.3 Flux Sensor

This sensor was developed for qualifying gaseous quenching media, as they are used in high-pressure gas quenching systems in combination with vacuum furnaces. The sensor consists of a cylindrical probe of stainless steel with two thermocouples near the surface and one at the center as shown in Figure 4.9-13. Additional measurement of the temperature of the cooling gas allows the determination of the heat transfer coefficient. With the knowledge of this heat transfer coefficient, temperature profiles in real workpieces can be calculated with sufficient accuracy [22]. The comparison of the temperature profile with TTC diagrams of the specific steel grade makes it possible to calculate hardness profiles.

4.9.9
Control of Induction Heating

Electromagnetic induction heating is done by placing the electrically conducting material in an alternating electromagnetic field. This process is used in the metal-

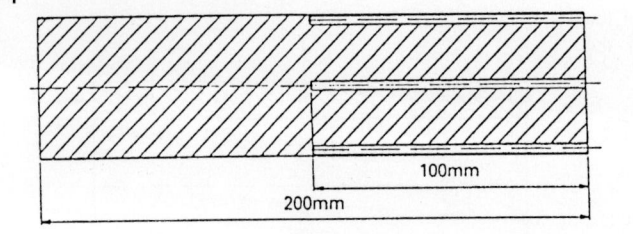

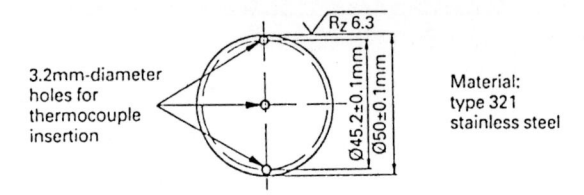

Fig. 4.9-13 Flux sensor measuring the convective heat flux [22]

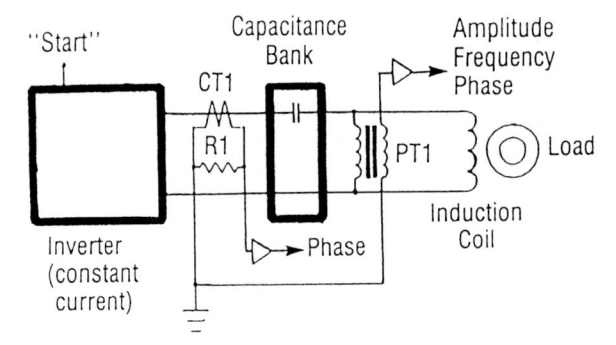

Fig. 4.9-14 Interfacing of the induction heating to the load analyzer [24]

working industries throughout the industrialized world. Eddy currents are defined as those currents which are induced into a body of conducting material by variation of magnetic flux. Both heating and testing are based on eddy current processes. Electric resistivity and magnetic permeability are the major influences, and both are affected by temperature.

The load analyzer [23] is a computer-based instrument which measures the electrical characteristics in the load circuit of an induction heater and correlates them with the temperature in the load or workpiece. This information is obtained instantaneously from the same electromagnetic field as used for the heating process. A potential transformer (PT) and a current transformer (CT) are installed in the heater output circuit (Figure 4.9-14). Their secondary windings are connected to the load analyzer, which measures simultaneously the amplitude, the phase, and the frequency. The amplitude and the frequency reveal information about the power supply and the phase about the load or workpiece to be heat treated [24, 25].

The mentioned process is primarily viewed as a comparative evaluation. However, it has the capacity to detect further conditions such as part size, microstructural differences, cracks, and heating and cooling conditions of the metal load itself, so further advantages in quality control and process monitoring are possible.

4.9.10
Sensors for Plasma Processes

Many heat treatment processes, especially thermochemical heat treatment processes, run easier by applying an additional electrical plasma. Also etching, sputtering, and cleaning is done by plasma processes. The main macroscopic parameter influencing the result is the plasma current density. This current density can be measured by a PCD sensor. It is a separate cathode of known area situated inside the plasma reactor. Measurement of its current and dividing by the known area of the cathode yields the plasma current density. This value can directly be used for regulating the electrical parameters for different workloads in the plasma reactor [26].

More detailed information about the plasma can be measured by a Langmuir sensor. Local plasma parameters such as electron and ion density, electron and ion temperature, and plasma potential of an inert plasma can be evaluated. In practical applications, Langmuir sensors showed significant signals when the sensor surface was modified by coatings or contaminants. This effect is utilized to measure important plasma parameters in coating processes. Subjecting an electrode of small diameter to the same coating conditions as the material to be coated will provide information about progress. This measurement technique is still under development, especially for applications to plasma nitriding and plasma carburizing processes to make them more reliable and reproducible [27].

4.9.11
Conclusions

In the last 25 years, many sensors have been developed and applied to heat treatment. Some of them are well known, some of them are ready for application, and some of them are still under development. There is a great demand for new sensors in the different steps of heat treatment to realize better automation and quality control.

4.9.12
References

1 ASTM Committee E20 on Temperature Measurement, *Manual on the Use of Thermocouples in Temperature Measurement;* Philadelphia: American Society for Testing and Materials, 1993.

2 GÖPEL, W., HESSE, J., ZEMEL, J. N. (eds.), *Sensors – A Comprehensive Survey. Volume 4: Thermal Sensors,* Ricolfi, T., Scholz J. (eds.); Weinheim: VCH, 1990.

3 GRUNAR, *Grundlagen der Berührungslosen Temperaturmessung ,Strahlungsthermometrie';* Raytec Firmenschrift.
4 GRATTAN, K.T.V., *Meas. Control* **20** (1978) 32.
5 BEURET, P., DENZER, A., GÖHRING, W., HANZLIK, K., HOFFMANN, R., KÖRTVELYESSY, L., KUNTZMANN, B., *Härterei Tech. Mittl.* **44** (1989) 278–284.
6 MÖBIUS, H.-H., *J. Solid State Electrochem.* (1997) 2–16.
7 *AWT-Fachausschuß 5, Arbeitskreis 4, Die Prozeßregelung beim Gasaufkohlen und Einsatzhärten;* Renningen-Malmsheim: Expert-Verlag, 1997.
8 WÜNNING, J., *Härterei Tech. Mitt.* **43** (1988) 266–270.
9 LEHRER, E., *Z. Elektrochem.* **36** (1930) 383.
10 BERG, H.-J., SPIES, H.-J., BÖHMER, S., *Härterei Tech. Mitt.* **43** (1991) 375.
11 WEISSOHN, K.-H., *Härterei Tech. Mitt.* **53** (1998) 164.
12 LOHRMANN, M., *Härterei Tech. Mitt.* **54** (1999) 271.
13 TAYLOR, E., *Met. Prog.* **124** (2) (1983) 21–25.
14 DAWES, C., TRANTER, D.F., *Heat Treat. Metal* **9** (4) (1982) 85–90.
15 BELING, S., MEHNER, A., KLÜMPER-WESTKAMP, H., HOFFMANN, F., MAYR, P., in: *9th CIMTEC – World Forum on New Materials, Symposium IX: Solid State Chemical and Biochemical Sensors,* Vincenzini, P., Dori, L. (eds.); 1999.
16 KLÜMPER-WESTKAMP, H., HOFFMANN, F., MAYR, P., in: *Proceedings of the 24th Annual Conference of the IEEE Industrial Electronics Society,* Vol. 4; Aachen, 31 August–4 September 1998, p. 2371.
17 KLÜMPER-WESTKAMP, H., HOFFMANN, F., MAYR, P., in: *Proceedings of International Heat Treatment Conference: Equipment and Processes,* Schaumburg, IL, 18–28 April 1994.
18 KLÜMPER-WESTKAMP, H., *Entwicklung eines Sensors zum In-situ Kontrollierten Anlassen von Härtungsgefügen,* Heft 452; Frankfurt am Main: Forschungsvereinigung Antriebstechnik eV, 1995.
19 BERGER, J., *Dissertation;* University of Bremen, Clausthal-Zellerfeld: Papierflieger, 2000.
20 BRIEM, K., *Ger. Pat.* 9421220U1, 1994.
21 NELLE, ST., LÜBBEN, TH., HOFFMANN, F., ZIEGER, H., *Härterei Tech. Mitt.* **49** (1994) 17–25.
22 LISCIC, B., in: *Proceedings of the 18th International Conference on Heat Treatment of Materials;* American Society of Metals, 1980, pp. 51–63.
23 MORDWINKIN, G., *Eur. Pat.* EP 0014729, 1985.
24 HASSELT, P.A., in: *Heat Treating: Equipment and Processes, 1994 Conference Proceedings,* Totten, E., Wallis, R.E. (eds.); Materials Park, OH: ASM International, 1994, pp. 215–218.
25 HASSELT, P.A., MORDWINKIN, G., *Ind. Heating* **12** (1986) 17–20.
26 EDENHOFER, B., *Härterei Tech. Mitt.* **44** (1989) 339–345.
27 KLÜMPER-WESTKAMP, H., MEHNER, A., HOFFMANN, F., MAYR, P., *Mat.-Wiss. Werkstofftech.* **29** (1998) 1–15.
28 EDENHOFER, B., *Heat Treat. Met.* **1** (1997) 7–11.
29 GRABKE, H.J., *Härterei Tech. Mitt.* **44** (1989) 270–277.
30 KLÜMPER-WESTKAMP, H., *Fortschr.-Ber. VDI* Reihe 5, No. 160; *Dissertation,* University of Bremen; Düsseldorf: VDI-Verlag, 1989.
31 KLÜMPER-WESTKAMP, H., *Eur. Pat.* EP 0324809, 1996.
32 WEBER, D., NAU, M., *Elektrische Temperaturmessung,* 7th edn.; Fulda: M.K. Juchheim, 1998.
33 ZISSIS, W., *The Infrared Handbook;* Washington, DC: Office of Naval Research, Department of the Navy, 1978.

5
Developments in Manufacturing and Their Influence on Sensors

5.1
Ultra-precision Machining: Nanometric Displacement Sensors
E. BRINKSMEIER, *Universität Bremen, Bremen, Germany*

Sensors used for ultra-precision machines include position and velocity sensors, which are mostly part of a control loop, and for some special applications acceleration sensors for active vibration control, and thermal sensors for thermal error compensation [1].

An essential component of each positional feedback loop in a machine tool is the displacement measurement system for detecting the actual position of moving machine parts. It is the performance of such measurement systems that limits the accuracy of machine tools and hence directly affects the quality of the machined part. In ultra-precision machining, accuracies of interest are of one part in 10^6 or even 10^7 with a measurement range of up to 1 m. Adequate resolutions are in the range of nanometers [2]. Such demands can only be fulfilled by laser interferometers, optical scales, and linear voltage differential transducers, which will be covered in detail. Although capacitive and piezoresistive sensors are able to offer resolutions in the sub-nanometer range, their measurement range is rather small, which limits their application to measuring macroscopic workpieces [1].

5.1.1
Optical Scales

Optical scales are based on a repetitive pattern of reflective or transmissive material. On scanning along the scale, a periodic signal is obtained which is a direct measure for the traveled path. Conventional optical encoders, also called photoelectric encoders, are based on counting Moiré lines by means of a light source and a photodiode. Beside optical scales, the patterns might also be made of magnetic or conductive material. The principle is then based on magnetic, capacitive, or inductive measurement [2, 3].

The optical set-up for an incremental photoelectric device is shown in Figure 5.1-1. The scale is divided into alternating optically opaque and transparent sectors. A mask with the same grating period and a phase shift of 90° with re-

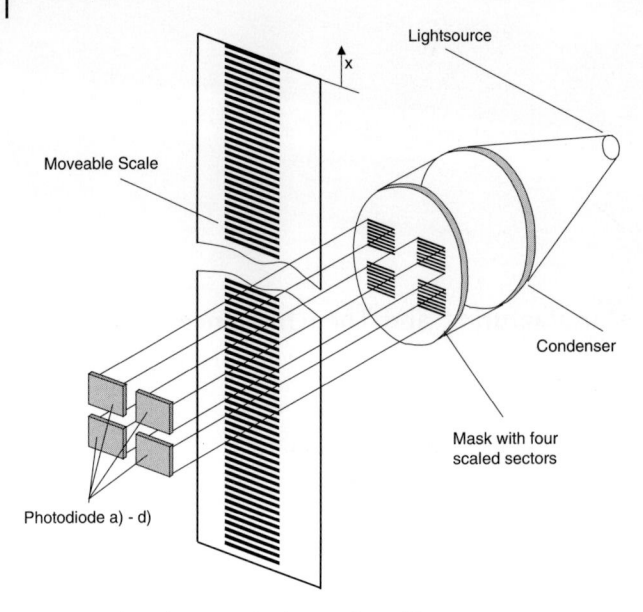

Fig. 5.1-1 Photoelectric scale. Source: Ernst [3]

spect to each other is located between the light source and the scale. The images of the sectors are projected on to the scale by the collimated light from the light source. As the scale moves, four photodiodes on the other side of the mask receive light pulses which pass through the transparent slits of the mask and the scale. Thus, four square sine signals with a relative phase shift of 90° are generated. The determination of the displacement x from these signals is depicted in Figure 5.1-2.

Two of the signals with a phase shift of 180° are subtracted so that two signals, S_1 and S_2, which are symmetrical with respect to the x-axis and have a phase shift of 90°, are generated. These signals are triggered and finally a resolution of ¼ of the grating period g is achieved. By counting the intersections of both functions with the x-axis, the traveled path can be determined. A further increase in resolution can be achieved by means of electronic interpolation. In Figure 5.1-2, the intensities of S_1 and S_2 can be expressed as sine and cosine functions depending on the traveled path:

$$S_1 \sim \sin\left(\frac{2\pi x}{g}\right)$$
$$S_2 \sim \cos\left(\frac{2\pi x}{g}\right).$$

Dividing S_1 by S_2, the path x can be derived by extracting the arctangents from the coefficient:

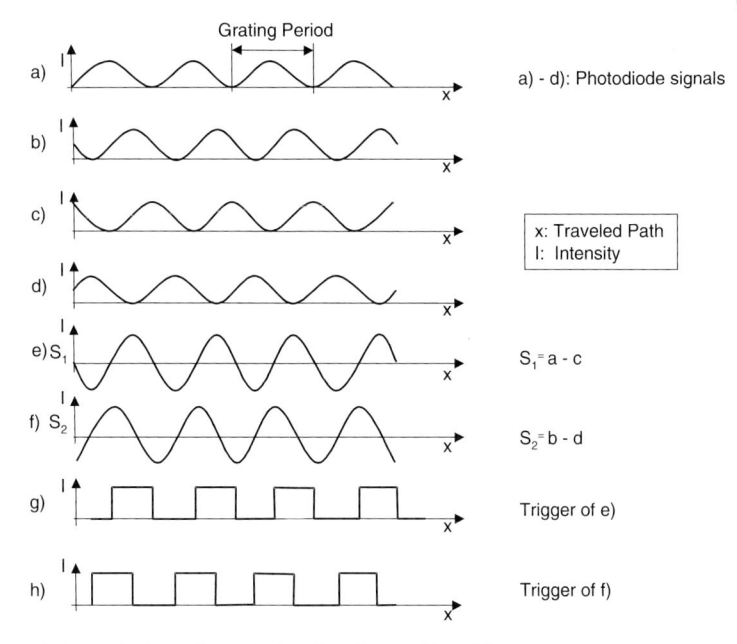

Fig. 5.1-2 Interpolation techniques for optical scales. Source: Ernst [3]

$$x = \frac{g}{2\pi} \arctan\left(\frac{S_1}{S_2}\right).$$

One serious drawback of incremental encoders is that the information about the position is lost when the system is switched off. This disadvantage can be avoided with absolute encoders. Today, most absolute linear encoders are based on the photoelectric principle combined with a coded scale. The linear encoder gives an output which is a binary number that is directly proportional to the distance of the encoder from a fixed point. The simplest code is the binary code consisting of alternating opaque and transparent blocks, as shown in Figure 5.1-3. The number of strips determines the number of binary digits and the total number of increments that can be detected. For example, using 16 bits (and thus strips), the number of resolvable positions is 65 536. The position is measured with a linear array of photodiodes mounted above each strip. One problem that arises with binary coded scales of the above-mentioned type is that reading errors of the photodiodes result in large errors in position. Therefore, another code is used where only a change in one digit occurs when moving from one position to the next. This code is termed the gray code. Transformation from the gray code to the binary code and vice versa can be done using simple logical operations. However, a general drawback of both the gray code and the binary code is the huge number of strips required for high resolutions and large measurement ranges which requires a lot of space within a machine tool. Absolute coding with only a few strips can be

Binary Code

Gray Code

Fig. 5.1-3 Binary and gray code. Source: Sinclair [4]

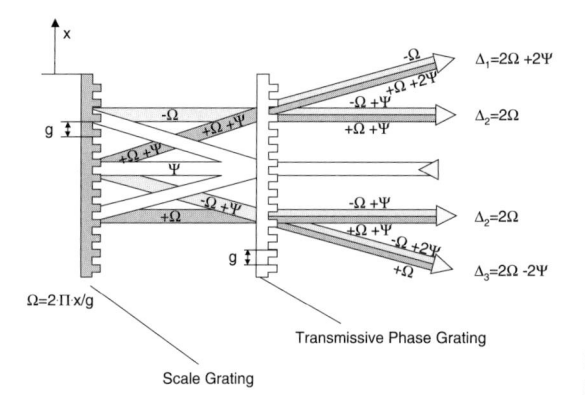

Fig. 5.1-4 Principle of interferometric scale. Source: Spies [5]

achieved with the random code. Here, the strip is coded in a random fashion over the entire length so that a sector of a certain critical length never repeats. If a sensor array with equal critical length scans the coded strip, the information obtained from the array is unique and can be compared with the code stored in the memory. Thus, the absolute position is obtained. Theoretically, one strip is sufficient for absolute encoding, available random coded encoders have three strips for increased resolution [3].

Conventional optical photometric encoders are limited in resolution owing to diffraction effects. On the other hand, so-called interferential scales are based on diffraction [5]. The optical set-up of the interferential encoders (by courtesy of Heidenhain Co.) consist of a movable reflective scale (grating), a transmissive grating with the same grating period as the scale, a light-emitting diode (LED) as light source, a condenser and three photodiodes for scanning the signals. The scale grating consists of 4 µm wide, rectangular plateaus of gold with a height of 0.2 µm, which are located at equal intervals of 8 µm on a gold-plated substrate.

Figure 5.1-4 shows the operating scheme of the Heidenhain linear encoder. On passing through the transmissive grating, a first-order diffraction pattern is created, with three partial waves of nearly the same intensity. The grating is designed in such a way that the zero-order wave is subject to a phase shift ψ, with respect to the first-order diffraction waves [5].

At the reflective scale grating the light is diffracted so that the first-order diffraction has the highest intensity while the zero-order diffraction is negligible. Besides being diffracted, the three waves are also subject to another phase shift Ω at the scale grating. Passing through the transmissive phase grating again, three new interfering waves with different phases are generated. These waves are collimated by the condenser at three different spots to be analyzed by the photodiodes. The phase shift of the three new waves is given as by

$$\Delta_1 = 2\Omega + 2\psi$$
$$\Delta_2 = 2\Omega$$
$$\Delta_3 = 2\Omega - 2\psi \,.$$

However, the phase shift Ω is not a constant like ψ but depends on the movement of the scale grating. When the scale is moving by a distance x, the phase shift is

$$\Omega = \frac{2\pi x}{g}$$

where g denotes the grating period. Hence the phase shifts Δ_1, Δ_2, and Δ_3 can be expressed as a variable depending on the constant ψ and the displacement x. The three photodiodes in the focal plane of the condenser generate three periodic signals with twice the period of the grating and having a phase shift of 2ψ each:

$$\Delta_1 = 2\left(\frac{2\pi x}{g}\right) + 2\psi$$
$$\Delta_2 = 2\left(\frac{2\pi x}{g}\right)$$
$$\Delta_3 = 2\left(\frac{2\pi x}{g}\right) - 2\psi \,.$$

Common to all different interference scales is that the interference phenomenon is used to generate sinusoidal signals which are a function of linear displacement. These signals can further be processed electronically to increase the resolution. Generally, the displacement x can be expressed as [2, 3, 5]:

$$x = \frac{i}{kf}g$$

where i is the number of counts, g the grating period, f the optical multiplication factor, and k the electronic interpolation factor. With electronic interpolation, an increase in resolution by up to the factor $f=400$ is possible, so that nanometric resolution can be achieved.

Since scales are made from glass or steel, they are subject to thermal expansion when the ambient temperature changes. To relate linear displacement measurement to the internationally agreed reference temperature of 20 °C, the thermal coefficient of expansion a and the temperature of the scale must be known. While the temperature can be measured with an accuracy of better than 0.05 °C, careful attention has to be paid to local temperature gradients [2].

5.1.2
Laser Interferometers

The basis of nearly all laser interferometric displacement sensors is the Michelson interferometer and its variations. Modern interferometers for measuring displacements in the nanometer range are based on the Doppler effect, in which two waves of slightly different frequency are made to interfere. As depicted in Figure 5.1-5, a heterodyne He-Ne laser emits two mutually perpendicular polarized beams with the frequencies f_1 and f_2 with a frequency difference $\Delta f = f_2 - f_1 = 20$ MHz, which serves as a reference signal [2, 6].

The reference signal can be interpreted as a single wave with a beat frequency of $\Delta f = f_2 - f_1$. This reference signal is detected in the photodiode P1, after a certain amount of the wave emitted by the laser has been split in the half-reflecting mirror HRM. Next, the wave passing through the half-reflecting mirror is split in a polarizing beamsplitter PBS. The wave with frequency f_2 is led directly to the photodiode P2, whereas the wave f_1 is reflected by the moving reflector MR. Both waves are again made to interfere in the photodiode P2, and this signal serves as measurement channel. If the moving reflector does not move, this P2 likewise detects an amplitude modulation of 20 MHz. However, a moving reflector with velocity v causes a frequency alteration of Δf_1 to the wave f_1 so that the resulting beat fre-

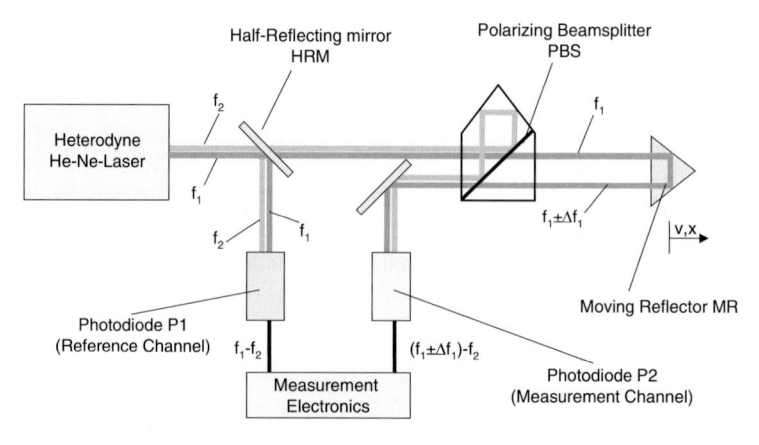

Fig. 5.1-5 Laser interferometer. Source: Sommargren [6]

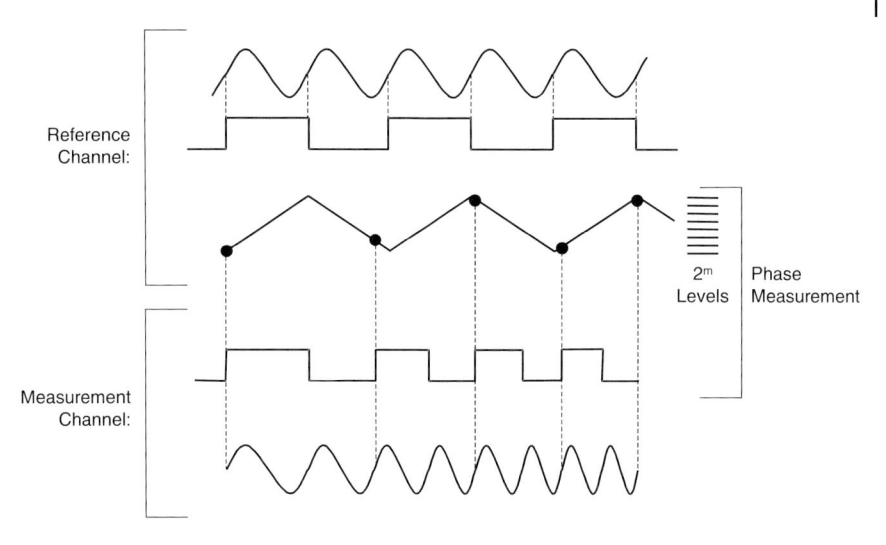

Fig. 5.1-6 Interpolation technique for laser interferometer. Source: Sommargren [6]

quency detected at P2 is $(f_1 \pm \Delta f_1) - f_2$. This wave contains information about the optical path change of the moving reflector in the form of a phase change between the reference signal and the measurement signal. The phase difference between the two signals is measured every cycle and any phase changes are digitally accumulated. Figure 5.1-6 depicts how the phase shift and the optical path change are generated. Both the sinusoidal reference and measurement signal are each converted into square waves. The reference signal is further integrated to produce a triangular wave. At each positive transition of the measurement square wave, the triangular wave is sampled and digitized. Its value is then compared with the previous reading and any difference is added to the memory. Furthermore, the polarity of the reference square wave and the number of transitions of the triangular wave is monitored for correction of the phase measurement. An optical path change of λ, which is equivalent to a movement of $\lambda/2$ of the moving reflector, causes the measurement signal to shift by one cycle with respect to the reference signal. A complete cycle is equivalent to a change of 2^m-1 levels of digitization, where m is the number of available bits in the analog to digital converter. Thus, the optical path change can be resolved with a resolution of $\lambda/(2^m-1)$. This measurement technique achieves both a high resolution and a high rate of optical path change [6].

The accuracy of laser interferometers is generally limited by the stability of the wavelength, which is the measurement reference and is a function of the refractive index of air. This index varies with temperature, humidity, pressure, CO_2 content, and contamination by other gases. Beside these errors, which can be compensated when the properties of the air are monitored permanently, another source of errors is arising from non-linearities in the interpolation process. Misalignment and imperfections of optics increase the systematic errors of laser interferometers [2].

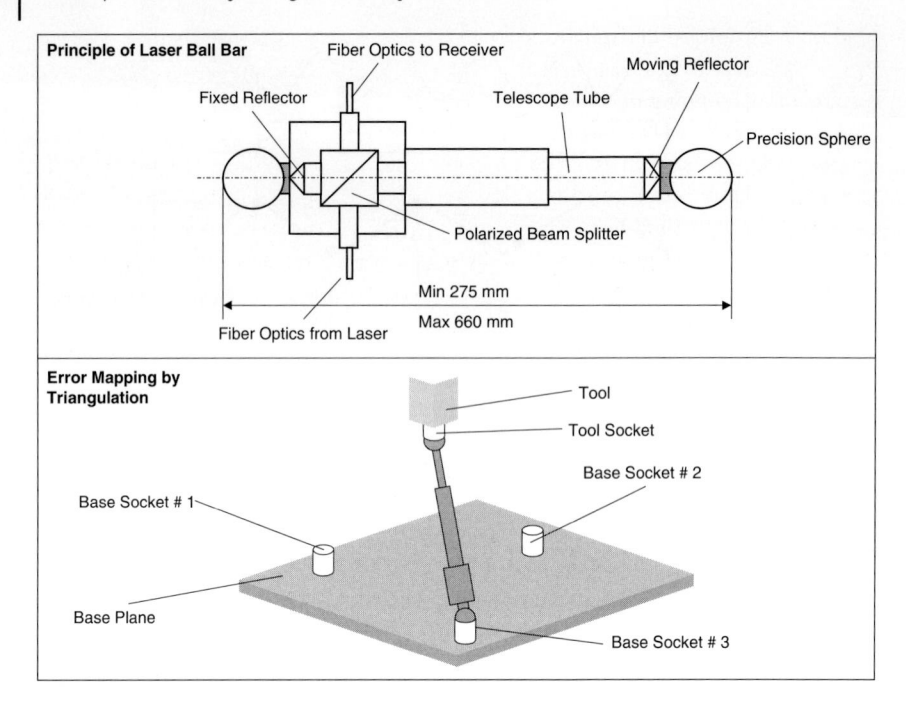

Fig. 5.1-7 Laser ball bar. Source: Ziegert and Mize [7], Srinivasa and Ziegert [8]

To overcome the problems arising from the changing properties of the air, the ultimate solution is to encapsulate the laser interferometer in vacuum; however, this requires huge efforts regarding the design and implementation in a machine structure and is only used for very high demanding applications, eg, calibration for unencapsulated interferometers.

In general, it can be stated that for ultimate accuracy in dimensional metrology, as the first choice vacuum laser interferometers or laser interferometers under extremely stable environmental conditions with very sensitive and calibrated temperature sensors have to be considered. For high accuracy in the range 10^{-7}–10^{-6}/ 10–100 nm, scales and laser interferometers are more or less equivalent. The same applies to accuracies in the range 10^{-6}–10^{-5} but scales are mostly preferred as they are easier to integrate into a machine tool or coordinate measurement machine [2].

A recent application of the heterodyne laser interferometer is the laser ball bar (LBB) for machine tool metrology, whose principle is depicted in Figure 5.1-7 [7]. It is designed to measure volumetric errors rapidly by directly measuring the spatial coordinates of the tool. The LBB consists of a displacement-measuring laser interferometer whose axis is aligned between the centers of two precision spheres as depicted in Figure 5.1-7. Since the LBB is a linear displacement measurement device, it can be used to locate the actual position of a tool relative to a base frame when combined with the technique of triangulation. The tool point positioning error can be derived from the difference between the measured position and the position indicated by the machine. Sequential triangulation leads to a systematic error mapping of the machine tool [8].

Conventionally, error mapping of the volumetric errors of a machine tool, which is the first step in error compensation of a machine tool, is done by measuring parametric error functions for each axis. These parametric functions are combined through rigid body modeling to obtain the volumetric errors of the machine. However, this procedure is difficult and time consuming, eg, a three-axis machine requires a total of 21 error measurements (three translatory and three rotational for each axis and three orthogonality errors for the axes). A useful alternative is to measure directly the spatial displacements of the tool and thus obtain the error map of the machine tool. Here, the LBB serves as a convenient measurement device.

Srinivasa and Ziegert [8] applied LBB triangulation to determine the spindle thermal drift of a two-axis turning center with respect to the machine frame. Thermal errors in machine tools arise from thermal deformations of the machine tools caused by a complex temperature field within the machine. Existing heat sources within the machine tool are the leadscrew bearings and nut, axis drive motors, spindle bearings and drive, heat caused by any friction within the machine, and the cutting process itself. Thermal errors in machine tools, in particular spindle thermal drift, are considered to be the main contributors to overall machine accuracy and are thought to be the main reason for dimensional and geometric errors in workpieces produced on machine tools [8].

5.1.3
Photoelectric Transducers

Both laser interferometers and scales require photoelectric transducers in order to transform the optical signals into electrical information. The most commonly used photoelectric transducers are photodiodes, which are based on the photovoltaic principle. In photovoltaic action, a voltage is generated when light is incident on the photosensitive material. Since photovoltaic cells have an efficiency of only 10–20%, attention has to be paid to ensure that the dissipated thermal power does not cause unacceptable thermal errors to ultra-precision machines. The types of available photodiodes include monolithic and hybrid devices which can produce digital or analog output. Their spectral sensitivity may range from infrared to ultraviolet. Monolithic analog output photodiodes are commonly used in conjunction with LEDs in optical encoders and proximity sensors, and in conjunction with lasers in interferometers. Two-dimensional arrays of photodiodes form the imaging device of most video cameras [9].

For optical position sensors, monolithic photodiodes are most often used which provide an analog voltage output. When photodiodes are used to measure the intensity of incident light, eg, from laser interferometers or optical scales, a bandpass filter, eg, a 630–635 nm filter for the He-Ne laser, is used for excluding ambient light [9].

5.1.4
Inductive Sensors

Inductive sensors are, unlike scales or interferometers, analog position sensors. The most commonly used sensor for measuring displacements in the millimeters and centimeters range is the linear variable differential transducer (LVDT) [1, 10].

The LVDT, depicted in Figure 5.1-8, consists of three coils. An AC voltage, typically 5 kHz, is applied to the primary coil, which is inductively coupled to the secondary coils. For small displacements of the core about the central position, the amplitude of the output voltage across the counterwired secondary windings will be linearly proportional to the displacement. The direction of the core is determined by analyzing the phase of the signal with respect to the reference phase. As the name suggests, the output from the phase-sensitive detector will be linearly proportional to the distance, as resulting non-linearities are subtracted due to the use of two counterwired coils [10].

Short-stroke LVDTs have useful linear range of movement of a few millimeters only. The long-stroke type can provide a displacement range of as large as ±60 mm [4].

5.1.5
Autocollimators

Autocollimators are devices for the precise measurement of small horizontal and/or vertical angular displacements. They are usually not an integral part of a machine sensor system, but are used as inspection and calibration devices. In environmentally controlled laboratories, autocollimators provide a fast and simple method to measure flatness of a machined surface and straightness and orthogonality of moving axes. Both manual and electronic autocollimators are available [9].

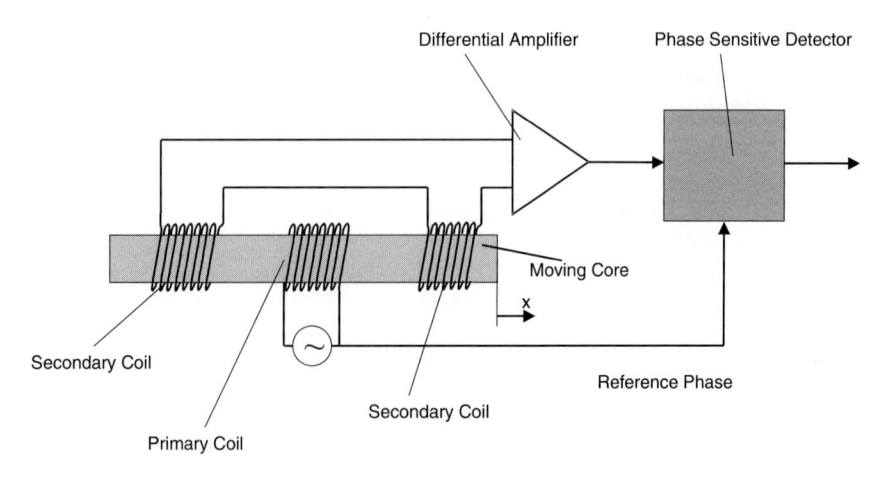

Fig. 5.1-8 Linear variable differential transducer (LVDT). Source: Smith and Chetwynd [10]

Fig. 5.1-9 Autocollimator. Source: Slocum

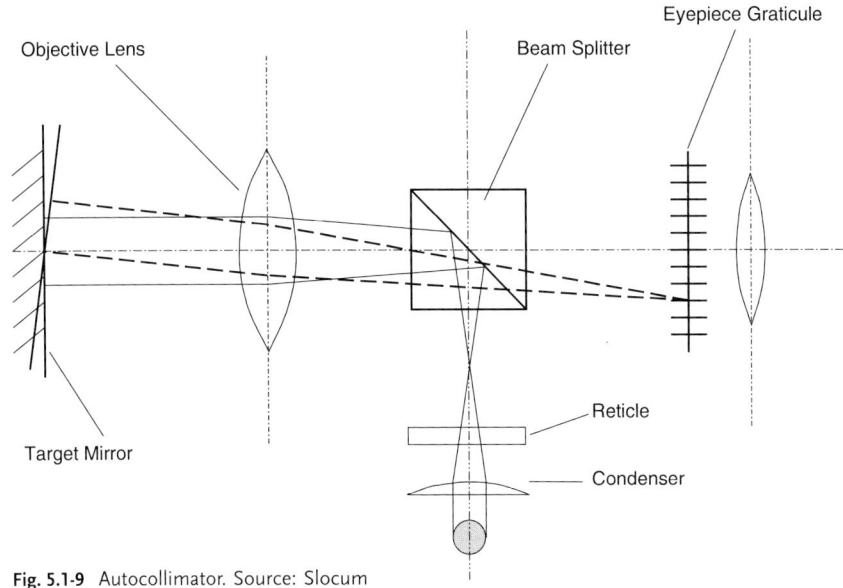

The components of a manual autocollimator are shown in Figure 5.1-9. Light from the light source is focused by the condenser and projected on to the target mirror by a reflecting beamsplitter. The light reflected by the target mirror is observed through the objective lens. Angular displacement can then be measured using the eyepiece graticule [9].

5.1.6
References:

1 Schellekens, P., Rosielle, N., Ann. CIRP **47** (1998) 557–586.
2 Kunzmann, H., Pfeifer, T., Flügge, J., Ann. CIRP **42** (1993) 753–767.
3 Ernst, A., Digitale Längen- und Winkelmessung; Landsberg: Verlag Moderne Industrie, 1989.
4 Sinclair, I.R., Sensors and Transducers; Oxford: Butterworth-Heinemann, 1992.
5 Spies, A., Feinwerktech. Meßtech. **98** (1990) 406–410.
6 Sommargren, G.E., Precision Eng. **9** (1987) 179–184.
7 Ziegert, J.C., Mize, C.D., Precision Eng. **16** (1994) 259–267.
8 Srinivasa, N., Ziegert, J.C., Precision Eng. **18** (1996) 118–128.
9 Slocum, A.H., Precision Machine Design; Englewood Cliffs, NJ: Prentice Hall, 1992.
10 Smith, S.T., Chetwynd, D.G., Foundations of Ultraprecision Mechanism Design; New York: Gordon and Breach, 1992.

5.2
High-speed Machining
H. K. Tönshoff, *Universität Hannover, Hannover, Germany*

Several developments made high-speed cutting (HSC) possible. HSC became an important trend in machining (Figure 5.2-1). The development of tool materials was a prerequisite to higher wear resistance under high temperatures. Up to the 1960s, the dualism of hardness and wear resistance on the one hand and toughness on the other were dominant limiting factors in tool materials. The functional separation of the ability to carry static and dynamic loads and of tribological functions was established by introducing coatings. The development of tougher ceramic materials for cutting purposes with their high-temperature strength was another way to withstand high cutting velocities. Finally, the progress in synthesizing super-hard materials such as diamond, diamond coatings, and cubic boron nitride gave strong impulses from the process side for higher cutting speeds.

Important preconditions had to be established on the machine tool side. High cutting speeds means high spindle rotational frequencies. The bearings of the main spindle had to be enabled:

- to withstand the centrifugal forces increasing with the square of the rotational frequency;
- to generate only small power losses and thus keep heat generation limited;
- to be provided with sufficient dynamic stiffness in the domain of exciting frequencies; and
- to avoid wear even under high thermal and dynamic loading.

Progress has been made by introducing hybrid ball bearings where the bearing balls are made of ceramic with less mass and favorable tribological properties, full ceramic bearings, and active magnetic bearings. This means that the characteristic bearing number $d_m n$ (d_m = average bearing diameter, n = spindle frequency) could increase from 0.8×10^6 to 1.8×10^6 and 4×10^6 mm/min.

tool	spindle	drive/control
coated carbide ceramic super hard material	motor spindle hybrid bearing magnetic bearing	fast control loop high speed ball screw linear motor

- high material removal rate
- fast feed speed
- low forces
- less tool deformation
- low surface roughness
- less thermal impact on surface

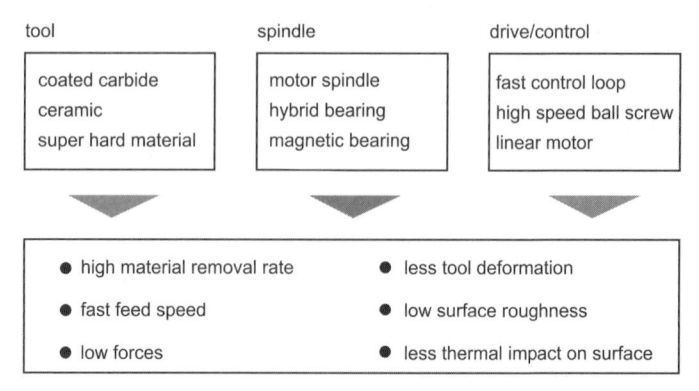

Fig. 5.2-1 Preconditions and advantages of HSC

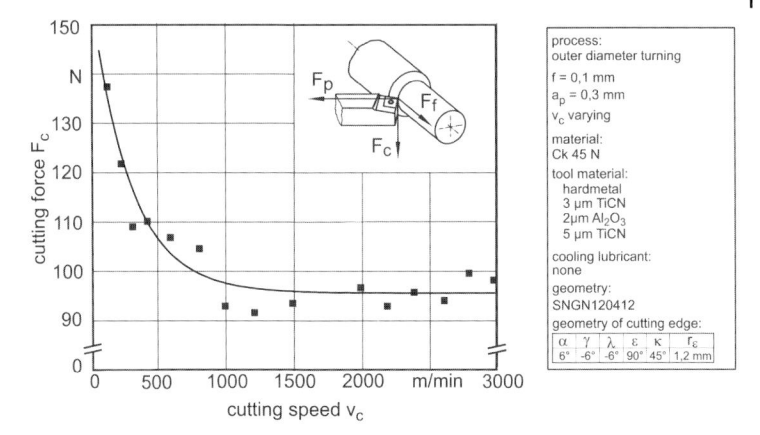

Fig. 5.2-2 Cutting force at varying cutting speeds

In the same way the feed drives of machine tools have also developed considerably. Important innovations are alternating current (AC) rotational servodrives with new magnetic materials, fast frequency converters, and ball screws whose maximum speed limit used to be 30–50 m/min and now reach 100 m/min. This development was probably induced by another competing technology, the direct linear motor, which permits high accelerations and high speeds. For this reason it is superior to rotational drives provided that the driven mass is not too large.

Finally, developments within the control sector have to be mentioned when speaking about HSC. Powerful closed-loop controls with feedforward and crosscoupling abilities were introduced and made the necessary data rates possible with an increase in speeds by factors of 3–5 or even more.

The advantages of HSC which brought the wider dissemination of this technology especially in the aircraft industry, in tool and molds manufacturing, and in the production of gears and drives are manifold. The material removal rates could be increased with the cutting speed because in milling – the main application field of HSC – the feed velocity could be increased proportionally if no further restrictions exist. The surface roughness may be improved by using parts of the speed improvement for shorter feeds per cutting edge. This is especially valuable in those domains of application in which free-form surfaces are generated by ball- or torus-shaped end mills. It was also stated that the physical state of a surface may be improved by HSC because the generated surface and the subsurface layers are less affected by heat. This is mainly a consequence of the time dependence of heat conduction. It is also due to lower energy consumption which some materials show under high cutting speeds (Figure 5.2-2). The specific energy and hence the cutting force decrease with increase in cutting speed. In addition, the forces can be lowered by decreasing the feed per cutting edge. This may be of decisive importance when filigree parts are machined as in the aircraft industry where integral structure parts are very susceptible to elastic deformations during milling and may consequently be incorrectly machined.

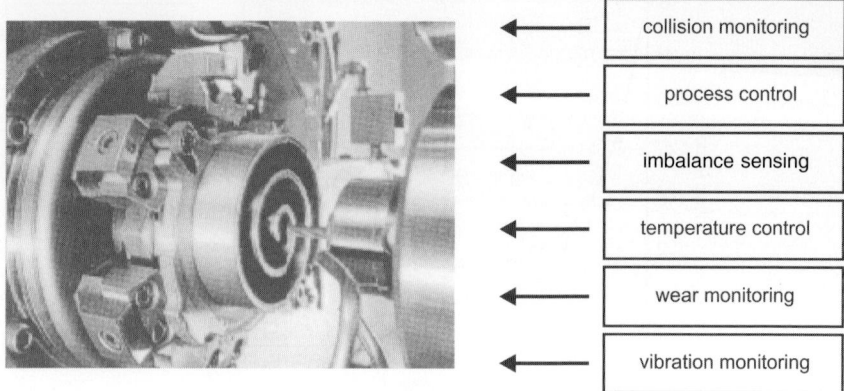

Fig. 5.2-3 Sensor applications of HSC processes

- collision monitoring
- process control
- imbalance sensing
- temperature control
- wear monitoring
- vibration monitoring

Although interesting advantages in connection with HSC can be listed, there are problems, some of which can be avoided or minimized by the use of sensors (Figure 5.2-3). The process cannot be monitored by the operator on-line, and therefore precautions against collisions have to be taken. This is done today by software running off-line or in the background of a numerically controlled (NC) program. For HSC, on-line calculations are normally too time consuming (see Section 2.4). Therefore, sensors are required which can follow all movements in the working area of machining. There have been some developments and investigations in research laboratories, eg, using ultrasonic curtains or infrared radiation, but up to now a solution which is robust enough for practical use, especially in the environment of chips and coolant, has not been found.

Depending on the material, HSC opens up only a narrow process window in which the machining conditions have to be set. This process window is dependent on some influences which are difficult to identify. These are especially the properties of the material to be machined. There may be serious disturbances if the process is implemented outside the window. Therefore, power, torque, and/or force monitoring are important. This is especially true because of the high investment value that HSC machines normally have. Even if such devices cannot prevent collisions, they may limit the consequences and damage which follow such an incident.

The high rotating frequencies in HSC make the run out and the imbalance of the spindle critical. The centrifugal forces grow with the square of the r.p.m. and so does the imbalance. The interfaces of the tool clamping system, that is, the connection between spindle and tool holder and between tool holder and tool, have to be specifically designed. The hollow taper shaft (HSK) was introduced some years ago with good success. Sensorial supervision of its correct fit in the spindle is provided in some machining centers. Monitoring of the run out and imbalance by an accelerometer is another safety feature for HSC.

As is known from the Taylor equation, the cutting speed has a dominant influence on the tool life. HSC means, therefore, that the tools have to be changed fre-

Fig. 5.2-4 Tool life with high cutting speeds.
Source: kindly provided by B. Denkena, University of Hannover

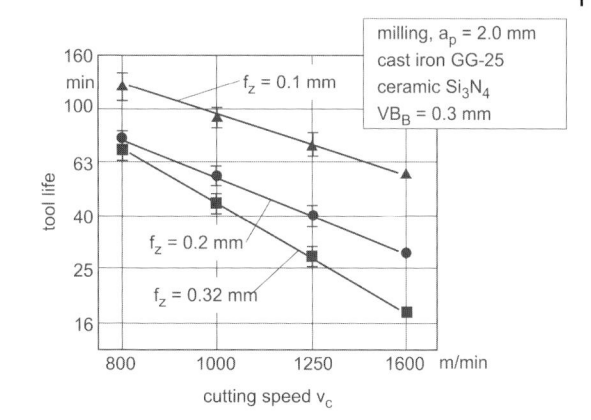

milling, a_p = 2.0 mm
cast iron GG-25
ceramic Si_3N_4
VB_B = 0.3 mm

quently and more often than in conventional cutting (Figure 5.2-4). Wear may be critical and therefore wear sensors are of interest. They should be able to determine the end of tool life reliably. There are several approaches, as discussed in Section 4.3.

One of the main accuracy problems with automated machine tools is derived from the thermal stability. The temperature field in the machine structure changes according to the effect of several heat sources. The most important heat sources are very often the spindle bearings. The monitoring of the bearing temperature is recommended because of the high investment that an HSC machine represents and the critical power losses with high spindle frequencies. The heating of fast-running main spindles can lead to an unstable state: heating increases the pre-stressing of the spindle-bearing system, which increases power losses and heating, etc. Monitoring is therefore advisable. This can be done fairly easily and reliably by thermocouple sensors. Similar measuring devices might be advisable to monitor feed drive components such as spindle-nut systems and direct linear drives to ensure a tolerable increase in temperature.

5.3
Micro-machining
M. Weck, *RWTH Aachen, Aachen, Germany*

The manufacture of micro-components using high-precision machine tools, so-called ultra-precision machines, imposes new demands on integrated sensor systems. In micro-machining, extremely filigree turning, planing, or milling tools are frequently used. In addition, the machining forces are very low, typically in the range below 1 N when natural diamond tools are used.

Very few sensor systems meet the requirements for micro-machining. To determine process forces, piezoelectric force sensors with very high resolutions have to be used to generate a useful measurement signal. Attempts have been made to

Tab. 5.3-1 Precision requirements for tool measuring systems in micro-machining operations

Position	Angular position in relation to the machine axes	<0.1°
	Tool distance from the workpiece (z)	<1 µm
	Tool distance from the rotational axis (y)	<0.1 µm
	Tool distance from rotational axis (x)	<0.1 µm
Geometry	External tool contour (turning)	<1 µm
	Tool radius (milling)	<1 µm
	Wear/spalling	Qualitative

supervise the cutting process via acoustic emission (AE) systems. However, only the tool-workpiece contact can be determined precisely with these methods. An in-process measurement with AE systems has not been developed successfully.

One of the most needed measurement systems for micro-machining are tool measurement systems. If the tool which is clamped in is to be oriented precisely, it is vital that the relative position of the cutting edge to the machine coordinate system, the tool geometry, and the distance between the tool and the workpiece are known. The exact determination of the workpiece geometry permits any centering error in the milling tools, for example, to be corrected. The use of forming tools in micro-turning operations demands that the relative position can be determined precisely and the measurement of the distance between the tool and the workpiece permits defined tool feed motion. The degrees of measuring accuracy required are listed in Table 5.3-1.

Non-contact optical measuring techniques are particularly well suited to this application owing to the small dimensions of the cutting edges of micro-tools and their susceptibility to cutting edge fracture. These techniques entail the use of a charge-coupled device (CCD) camera with a microscope to measure the cutting edge of the diamond tool. The position and geometry of the tool in relation to the machine coordinate system are subsequently measured using image-processing techniques. As an example, a tool measuring system integrating a micro-milling machine is presented below.

The three-axis ultra-precision milling machine and the optical tool measurement system were designed and built at the IPT in Aachen (Figure 5.3-1). This milling machine is equipped with two air bearing spindles and linear air bearing guides. The y-guide is provided with a high-speed milling spindle, permitting ultra-precision milling operations with optical surface quality to be carried out on the machine.

To measure the current position, a laser interferometer system is integrated into the linear slides. In combination with high-precision direct current (DC) torque motors and servo systems, a position accuracy in the submicrometer range has been achieved. The geometry of the rotating milling tool is directly measured using an optical system, mounted on the y-slide.

To achieve a precise tool measurement, telecentric imaging systems, which permit the measuring operation to be performed largely independently of the per-

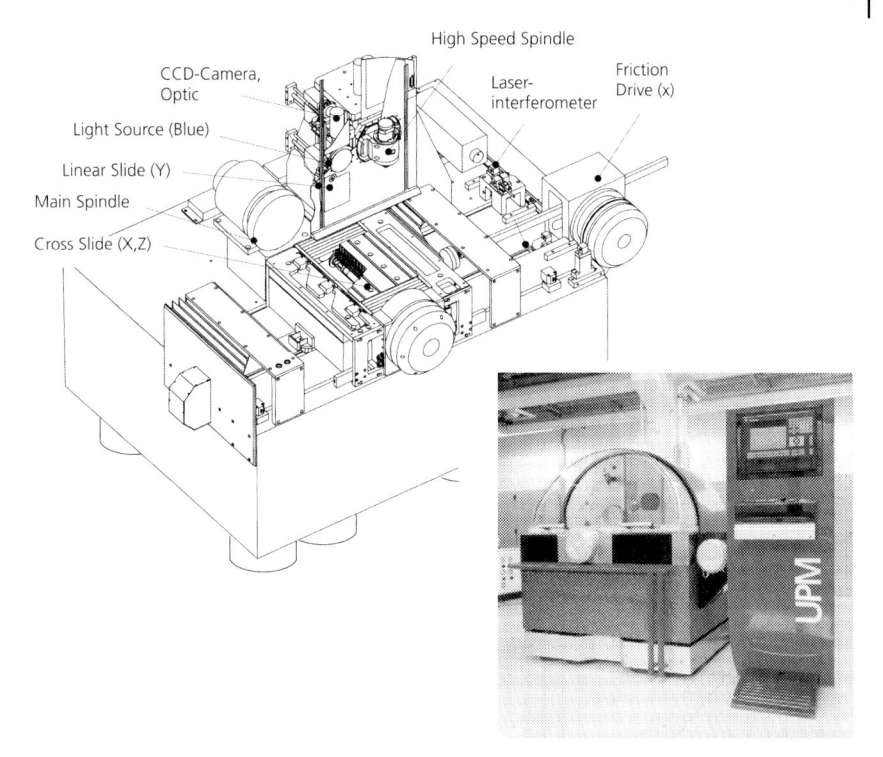

Fig. 5.3-1 Ultra-precision milling machine 'UPM' (Fraunhofer IPT)

spective, were used (Figure 5.3-2). This was the only way of recording high-precision geometric measurements of three-dimensional objects. The illumination unit used is a telecentric, low-wavelength transillumination (blue spectral region). This minimizes diffraction effects at the tool cutting edge, thus increasing the level of absolute measuring accuracy. Because of the low overall dimensions of the micro-machining tools, a measuring range of 0.5×0.5 mm at 100-fold magnification of the optic has proved suitable.

If the exacting requirements in terms of evaluation accuracy are to be met, it is not sufficient to use standardized methods of image processing which draw a conclusion as to the position of the tool on the basis of a straightforward, pixel-oriented evaluation of camera images. The geometric resolution is insufficiently high for this. A combined method of pixel and so-called sub-pixel evaluation has therefore been developed. This technique permits all quantities required to be determined with a resolution lower than 1 µm via high-precision edge detection and exact definition of the tool contour via straight and circular segments.

Appropriate illumination is used in the camera image, which is recorded by a very precise, distortion-free lens system, to contrast the object clearly against the background. Once this object image has been recorded, an edge detection operation is conducted. A number of gradient techniques, which exploit the fact that

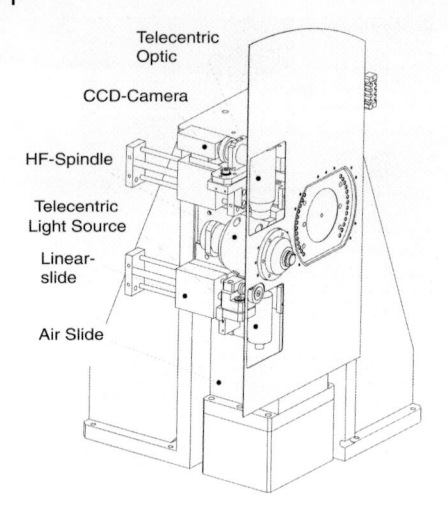

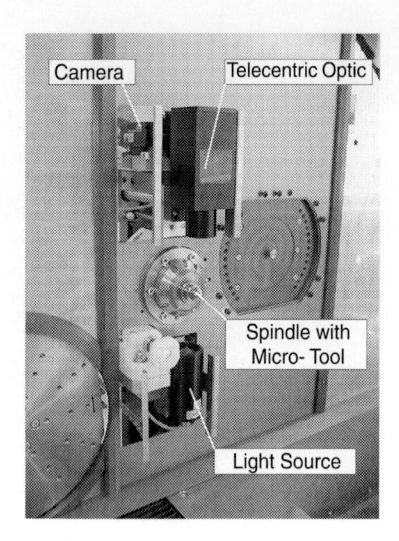

Fig. 5.3-2 Optical tool measuring system

there is a considerable difference in the gray scale value at one edge, are suitable. Good results are recorded particularly by the techniques which use Sobel filters and filters with similar characteristics. The outcome of these techniques is a multi-pixel breadth edge in the image, which is thinned out by mass techniques (non-maximum suppression). The remaining curve, which is the breadth of one pixel, is converted by an edge-following algorithm into a series of coordinates representing the corresponding points on the contour. The result is a list of coordinates which describe the contour in discrete pixel units.

Straightforward evaluation algorithms are stretched to their limits in applications of this nature. They produce a rough grid image of an edge which is actually smooth. Moreover, it is impossible to distinguish between straight lines and circles. In order to achieve a higher degree of accuracy, the contour is approximated in an additional, more extensive step involving the detachment of discrete points. Complex segmentation algorithms can be used to detect corner points. Depending on the course of the contour between the corner points, these sections are allocated an appropriate geometric form (straight or circular segment). The specific parameters of this form are determined and filed in a table. This table contains, for example, data about the starting and end points, the radius, and the mid-points of circular segments and also the tool angles or the angle which is assumed by a straight line in relation to the x-axis.

The openness of the data structure makes it possible to prepare it in such a way as to ensure that the data can be read and processed by CAD programs with suitable interfaces. It is a particularly important feature in this context that all information about the contour of the object can be presented with extreme precision, at a resolution <1 µm.

The approach adopted in the tool measuring operation is described in the following in greater detail, on the basis of the example of an end milling cutter and

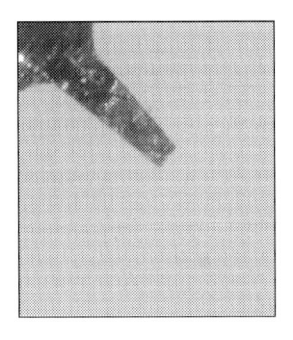

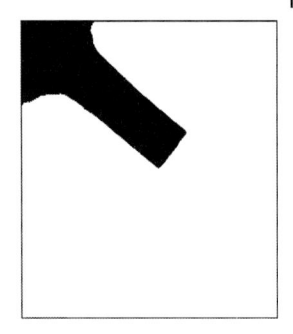

| 1. Camera picture of the non rotating milling tool | 2. Contour of the rotating milling tool | 3. Edge detection |

Fig. 5.3-3 Image recording in optical tool measuring operations

a turning tool. In order to measure a micro-milling cutter, images are recorded continuously while the milling tool rotates at low speed. Each individual image is linked to the result image via minimum formation so as to ensure that after the milling tool has rotated several times, the resultant groove geometry becomes clearly visible (Figure 5.3-3). This is then evaluated with sub-pixel accuracy using an edge detection algorithm.

The resolution achieved is <1 µm. The side walls of the groove are approximated by sections of lines. A geometric evaluation permits the resultant diameter of the groove to be determined subsequently at various distances from the channel base. A centering error of the milling tool which influences the breadth of the groove can be compensated via the machine control system in a later step. Additionally, any angle error which may occur when the milling tool is clamped in, and which results in the course of the machining operation in a V-shaped channel edge geometry, is detected at an early stage. Should this be the case, the tool must be re-clamped and the tool must be measured again.

In addition to measuring end milling cutters, the system can also be used to determine the tool contour of diamond spherical cutters. When a deviation from the spherical reference contour is recorded by the image processing system, the adjustable holder can realign the cutting edge on the machine. Additionally, the optical measuring system permits the exact relative position of turning tools in relation to the machine coordinate system to be determined (Figure 5.3-4). This is of particular importance in the case of forming tools since any alignment error results immediately in a deviation of the workpiece from the reference geometry. When the workpiece to be machined is recorded by the camera, the distance between the tool and the workpiece can also be recorded using image processing techniques (Figure 5.3-4).

This is of particular significance to the machine operator during the so-called 'scraping operation'. When the optical technique of measuring the distance between the tool and the workpiece is applied, a reproducible absolute precision of <1 µm is achieved. When the operation is started in the conventional mode, the

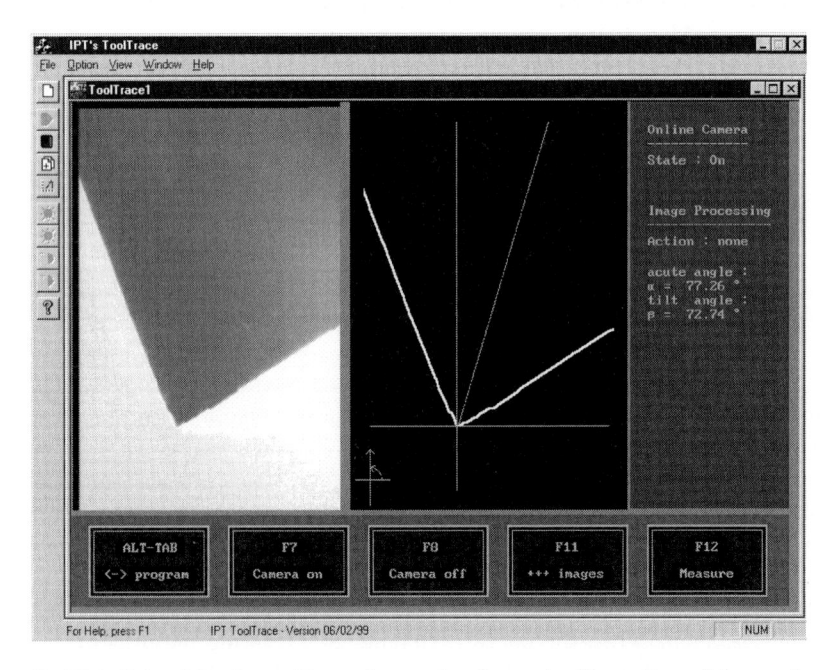

Fig. 5.3-4 Determining the position and geometry of an end milling cutter and the angular position of a turning tool

workpiece is moved at a low feed speed until the first chip formation is visible with the naked eye. The precision and consistency of this feed operation depend heavily on the experience of the machine operator and generally fluctuate by at least 1–2 µm. This, however, is intolerable for demanding applications of the micro-structure technique or for measuring coated sample parts.

5.4
Environmental Awareness
F. Klocke, *RWTH Aachen, Aachen, Germany*

Ecological issues are assuming increasing importance in many areas of the economy as a result of legislation and growing public awareness. Manufacturing, characterized by a chain of resource-intensive processes and by large quantities of waste materials and emissions, is frequently the focus of interest. Government-imposed environmental regulations and increased cost pressure in conjunction with the need to prevent the production of waste materials or to dispose them appropriately are forcing companies to introduce innovative, environmentally compatible manufacturing processes. In the manufacturing environment, the starting points for ecologically oriented improvement lie in the need to prevent the generation of waste materials and pollutants in the first place, or to reduce the volumes produced and re-use them. The advantages which stand to be gained as a result of the application of more environmentally compatible technologies are clear: reduced levels of energy consumption, waste, and disposal costs, together with higher employee motivation and lower rates of absenteeism due to illness [1].

In the successful, practical application of process monitoring systems and components, monitoring- and sensor-related solutions are adapted to meet the specific requirements of the machining task concerned. This demands precise knowledge of the machining operation, ie, the manufacturing environment, the machining process, and any potential process malfunctions [2–4]. The requirements relating to the monitoring system may, however, differ considerably in terms of the objectives and the implementation of the monitoring system.

A reduction in the quantity of cooling lubricant used in machining operations is a good example of the specific demands imposed on process monitoring and sensor systems by manufacturing processes which have been optimized in terms of environmental compatibility. On the one hand, the application of sensors is simplified since the requirements relating to the robustness and coolant resistance of the sensors are lower in this case. On the other hand, a reduction in the amount of cooling lubricant used frequently increases the degree of thermal load to which the parts are subjected. It therefore becomes more important to monitor any temperature-related change in dimensional and form accuracy or in the structure of the material of the finished parts. The influence exerted on the structure is critical, particularly in machining operations conducted on hardened materials.

Demands made on the monitoring system and on measurement engineering, which arise from the specific boundary conditions of environmentally compatible manufacturing processes, are analyzed and discussed below. The priorities will be the reduction in the volume of cooling lubricant used, extending to minimal lubrication and dry machining, as well as the working surroundings and the risk to which the machine operator is exposed by the emissions released during the machining operation. Examples of some approaches to the problems posed by the need to apply metrological techniques in order to measure relevant process variable are presented here for machining operations conducted using both geometrically defined and undefined cutting edges.

5.4.1
Measurement of Emissions in the Work Environment

Manufacturing is characterized by the combination of an extensive range of different types of substance and material flows. The substance flows are composed particularly of emissions, which are released during the manufacture of a product, depending on the processes and materials used. In the specific case of dry machining, the emissions are in the form of particles, which require the application of measurement techniques in order to gage their impact on the working environment and to be able to take appropriate measures, if necessary [5, 6].

5.4.1.1 Requirements Relating to Emission Measuring Techniques for Dry Machining
The type and volume of emissions which occur in dry machining operations depend on the machining operation used, the process control system and the machining parameters. In principle, however, a distinction can be drawn between aerosols (solid and liquid particles), gases, and vapors. Certain characteristics must be established before any conclusive data relating to the impact of emissions on the working environment can be released. The effects on the human organism depend on the characteristics of the material in question, particle geometry, the concentration, and the reaction time. Additionally, small particles released in the course of the cutting process can have an adverse effect on the process or can increase the level of wear sustained by machines and facilities [7–10].

5.4.1.2 Sensor Principles
One of the prerequisites for the reliable determination of emission characteristics is the selection of an appropriate measuring technique and sensor. The two fundamental procedures in any measurement are sampling and analysis. The function of sampling is to take a sample of the air at the measuring location and to ensure that it is available for analysis. A further distinction can be drawn between continuous sampling without enrichment and sampling the materials which do not belong in the air, on a sample carrier. Passive sampling can be performed on gases and vapors by enrichment, diffusion, or permeation. In active sampling operations which can be conducted on gases, vapors, and aerosols, the air containing pollutants is sucked

in and the pollutant is separated off using a sample carrier. This type of sampling operation is followed up by a chemical or physical laboratory analysis of the materials measured, which do not occur naturally in the air. For many areas of application for continuous sampling, there is a wide range of measuring instruments, the majority of which use electrical and optical measuring principles [11–14].

There are various collections of recognized measuring operations and directories listing external measuring centers which are useful sources of information to assist in the selection of suitable measuring techniques. Information and catalogues can be obtained from national employer's liability insurance associations and from institutions for engineering safety standards at the workplace (eg, BIA (Berufsgenossenschaftliches Institut für Arbeitssicherheit, Germany), NIOSH (National Institute for Safety and Health, USA), OSHA (Occupational Safety and Health Administration, USA)).

5.4.1.3 Description of Selected Measuring Techniques

Selected analysis and monitoring techniques used for aerosols in the working environment around dry machining operations are described in the following.

The instruments used most frequently to measure aerosol presence are scattered light-measuring devices. This technique is based on the principle of directing light waves away from their original direction by refraction, reflection, and diffraction caused by small particles. The diameter of the particles can be measured directly by evaluating the intensity, frequency, or phase of the scattered light. A distinction is drawn between instruments which detect the scattered light of a group of particles (photometer) and those which measure the particles individually (optical particle counter) [13–17].

The phase Doppler anemometer technique permits the dynamics of the particle to be measured, ie, simultaneous measurement of the size, speed, and concentration. As soon as small particles exceed a measured volume consisting of a system of plane-parallel interference bands, scattered light is produced which is amplitude modulated due to the interference phenomenon. Initially, this scattered light is used to measure the particle speed (laser Doppler anemometer). Particle size can be determined by evaluating the additional information provided by the phase position of the scattered light. The application of phase Doppler anemometers is limited largely to virtually spherical, transparent particles [18].

Numerous measuring instruments used to determine the particle size distribution exploit the principle that the aerodynamic diameter of a particle can be determined from its acceleration. A nozzle is used to accelerate the aerosol and the aerodynamic particle diameter is determined by measuring the time required by the particles to travel between two points. This technique can also be used to collect further information about other aerosol characteristics such as the distribution of the number of particles, their surface dimensions, or mass concentration [13].

The measuring techniques previously listed generally assume virtually spherical particle geometries and are unsuitable for measuring fibers. Some manufacturers also produce fiber measuring instruments capable of displaying the results immediately. These use the characteristics of the light signals which are scattered by the fibers. The problems in connection with the optical measurement of fiber dust arise from the random orientation of these particles in space, which causes irreproducible measurement results. The orientation of the fibers with their axis in one direction can, however, be achieved when a directed electrical field is used. However, the accuracy of the instruments currently available has been insufficiently high to warrant replacing the microscopic techniques traditionally applied to evaluate fiber dust [13, 17, 19].

In comparison with the particle measuring techniques in which aerosol particles are separated from a sample carrier, the optical particle measuring instruments all operate in non-contact mode, thus avoiding most of the faults resulting from the sampling procedure itself [16].

5.4.1.4 Example of Application

Cutting operations conducted on fiber-reinforced plastics are associated with the well-known problem of the release of dust, which is why an extraction system must normally be used during the machining process. The objective was to measure the efficiency of emission recording in the case of a machine design, which had been adapted to meet the needs of a concrete machining situation (Figure 5.4-1). A measuring technique in which scattered light photometry is combined with gravimetric particle analysis permitted comparison of the progressions with time of the alveolar mass concentration with and without extraction. As demonstrated by a comparison of the caliper gage, a proportion of fine dust is recorded very rapidly and reverts to the starting level about 100 s after conclusion of the milling operation [20].

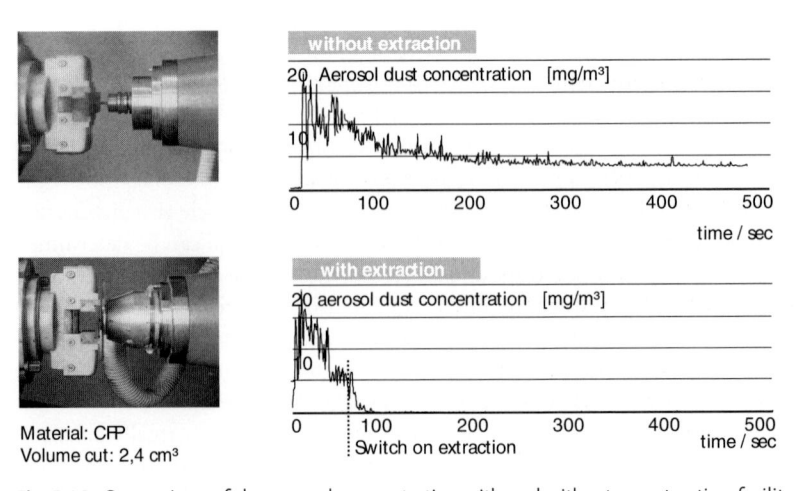

Material: CFP
Volume cut: 2,4 cm^3

Fig. 5.4-1 Comparison of the aerosol concentration with and without an extraction facility

5.4.2
Dry Machining and Minimum Lubrication

Cooling lubricants have come to be accepted as essential elements in production engineering. In many cases, they guarantee the quality of the machining outcome in terms of tool life, surface quality, and part accuracy. However, in recent years, the steadily increasing pressure on costs in manufacturing industries coupled with more pressing questions relating to environmental compatibility and disposal of waste materials have caused many users to reexamine their use of cooling lubricant [21, 22]. Against this background, it is easy to understand why industry and research are putting so much effort into prolonging the service life of cooling lubricants and reducing the amount required to zero – dry machining – if possible [23].

Generally, dry machining tends to facilitate the use of sensors to record and monitor process variables. New or at least extended tasks arise, however, in the field of temperature measurement. Process temperatures rise as a result of the elimination of the cooling lubricant functions of cooling, lubrication, and chip transport. This makes it all the more important to be able to determine the temperatures of tools and workpieces.

5.4.2.1 Measuring Temperatures in Dry Machining Operations

The measurement of temperature during machining operations has been the subject of a number of investigations over many years. A comprehensive overview of temperature measuring techniques was given by Lowack [24] and Kassbaum and Löffler [25].

Owing to the process characteristics, not all of the methods are suitable for in-process monitoring but can be used in basic investigations, eg, to determine the input quantities for modeling. The methods which are suitable for process monitoring are presented in the following.

The use of thermal converters to measure temperatures in tools and workpieces is one of the contact methods of determining temperature. This type of temperature measurement shows not the temperature in the object being measured, but the temperature of the sensor. Ideally, the sensor and the object to be measured are at the same temperature. In reality, this applies only in some cases. Heat sources and heat sinks result in permanent heat transport [26].

The temperature field in tools has also been determined using a series of optical temperature measuring techniques [24]. Thermal imaging is a non-contact measuring technique based on heat radiation and is therefore particularly suitable for measuring the temperature of a rotating tool [27]. The measuring point, however, must be on a visible surface in order to ensure that in a drilling operation, for example, the tool temperature can be measured at the point of exit of the tip of the drill bit from a through hole.

A further feature of this technique is the considerable requirement for calibration. The direct correlation between the level of heat radiated and the absolute temperature applies only in the case of an ideal emitter, the so-called 'full radiator' (Stefan-Boltzmann law) [28]. Given full radiators, it is possible to deduce the absolute temperature on the basis of the level of energy emitted. When determining the temperature of real parts, account must be taken of the emission factor, ie, the deviation from the behavior of an ideal full radiator. The emission factor ε is defined as the ratio of the radiation energy emitted by an object at a given temperature to the radiation energy of a full radiator at the same temperature and depends on the surface structure and temperature range to be measured and also on the material concerned. The emission factor must be known if the temperature of a real object is to be determined reliably. In practice, however, this is no easy matter. This severely restricts the field of application of this method to in-process measuring operations [29, 30].

5.4.2.2 Measuring Droplets in Minimal Lubrication Mode

The application of techniques of machining in the minimal lubrication mode opens up additional fields of application for sensor technology. The term minimal lubrication is used here to refer to the supply to the machining point or to the tool of minimal quantities of a lubricant mixed with compressed air. The amount of lubricant required is considerably lower than 20 mL per process hour in the case of optimally adjusted systems. Certain adjustable parameters, eg, for volume and pressure, should be capable of being monitored as a function of the technique, material, and supply system concerned, in order to ensure that the systems operate smoothly in the minimal lubrication mode. It is particularly important to monitor the operability of the minimal lubrication mode system in large-scale manufacturing series in order to ensure process reliability. Care must be taken to ensure that the machining point is supplied continuously with the mixture of compressed air and lubricant and that there is no incidence of 'intermittent supply'.

The machining outcome depends both on the orientation of the supply unit and on the spray pattern that it produces. It is important that the mixture is directed accurately to the cutting area and that excessive misting is avoided. In external supply, dispersion tests must be conducted to examine the spray pattern of the nozzles. In these tests, a patch of lubricant produced under specified conditions is measured. In-process spray pattern monitoring is a further potential use for the sensor system.

However, the current status of the technology does not permit the quantities listed above, such as spray pattern and droplet size, to be determined in-process during manufacturing operations; instead, they are measured primarily in the laboratory. The spray pattern can be assessed in general terms, eg, with regard to the development of droplets and their distribution, using high-speed cameras with a short exposure time.

The laser Doppler anemometer, described in the previous section relating to the measurement of emissions within the working environment [31], is used to determine the droplet speed.

There are two methods which can be used to determine the distribution of the droplet sizes of the atomized lubricant: first, the measurement of the droplet volume under the microscope, and second, comparative measurements in the range 0–16 µm using a cascade impacter. This entails dividing the area into eight size categories and selecting the particles by their size-dependent flow characteristics.

5.4.3
Turning of Hardened Materials

Hard machining with a geometrically defined cutting edge has gained substantially in importance owing to improvements in the performance of modern cutting materials. The high process temperatures and large specific forces involved in hard machining result primarily in a demand for good high-temperature hardness of the cutting material. This requirement is met especially by ceramic and cubic boron nitride. The use of such materials often permits turning to be substituted for grinding as a hard finishing process. This allows a more flexible, less cost-intensive and more environmentally compatible manufacturing process. The ecological advantages are obtained principally through the avoidance of cutting fluids coupled with smaller quantities of production waste to be disposed of. Additionally, the grinding swarfs that occur in the production of gear and roller bearing components can be reduced substantially.

5.4.3.1 Criteria for Process and Part Quality

Hard turning technology is a machining process at the end of the production chain. Process monitoring can therefore make an important contribution to ensuring reliable production of workpieces with the required quality. In addition to high demands on form accuracy and exacting tolerances, the surface quality and the structure of the surface zone are important criteria for the quality evaluation of hard turned parts.

The surface roughness in hard turning is determined to a considerable extent and is directly influenced by the corner radius of the cutting edge and the feed rate. The occurrence of various forms of tool wear has an adverse effect on surface quality [32–34] and generally results in higher surface roughness values (Figure 5.4-2). Tool wear also influences the structure of the surface zone (Figure 5.4-3).

Owing to the low ductility of martensite, the technology of hard machining is different from the machining of unhardened material. This results in different chip formation mechanisms, which do not rely on the formation of a shear plane or shear zone. The phenomenon is documented by the occurrence of sawtooth chips, already discussed in numerous publications [32–36]. Chip formation influences the surface zone of the machined workpiece and leads to changes in the residual stress state and the microstructure of the surface zone. According to Goldstein [37], mechanical stressing of the workpiece comparable to Hertzian stress occurs at the contact between the tool flank and the workpiece surface, inducing compressive residual stresses in the workpiece surface zone. Tensile residual stresses are super-

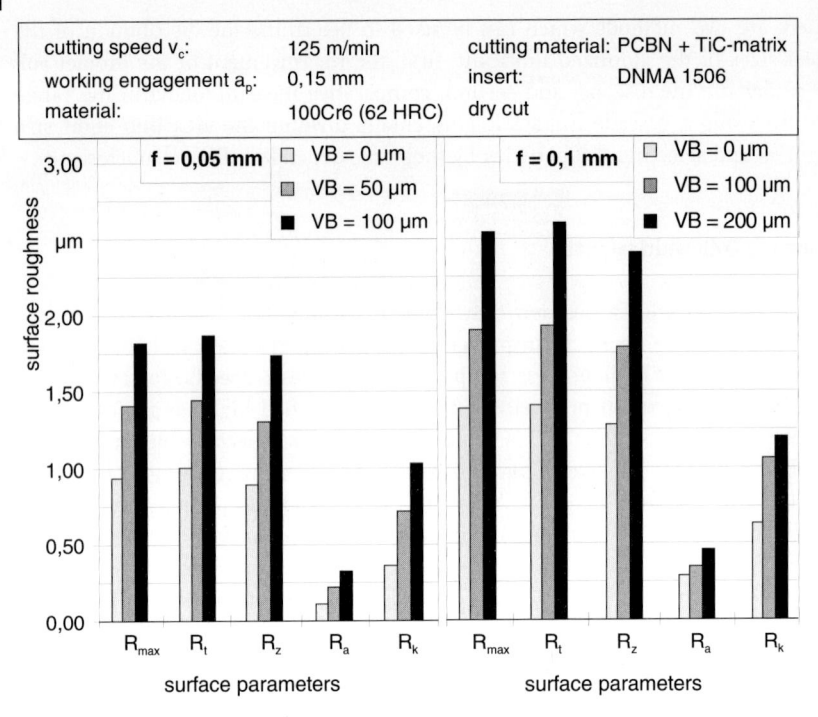

Fig. 5.4-2 Surface roughness as a function of wear

imposed on these compressive residual stresses as tool wear increases. These tensile residual stresses occur as a result of thermal stress caused by flank friction and by deformation phenomena in the work material. The maximum hardness is invariably greater than that of the basic structure and depends on the level of tool wear (Figure 5.4-3). The higher temperatures with increasing wear are also responsible for the structural transformations and the occurrence of 'white layers'.

Deviations from the ideal geometric form of a component following the soft turning and hardening operations result in cutting force fluctuations. These force fluctuations have an adverse effect on form accuracy. It is not only radial allowance fluctuations which lead to deviations from the ideal geometric form of the finished component. Fluctuations in allowances along the feed path cause relative displacements between the workpiece and the tool, due to differing depths of cut. These axial allowance fluctuations show up as inadequate straightness of the contour line.

Apart from cutting forces, a second main variable affecting the form and dimensional accuracy of hard turned parts is the change in temperature during the process. As already noted above, one great advantage of hard turning over the competing grinding process is the ability to machine without cooling lubricant. In studies of surface zone effects and surface quality in hard turning operations, the products showed neither positive nor negative effects resulting from the use of cooling lubricant [33, 38]. Macrogeometrically, however, machining without a cool-

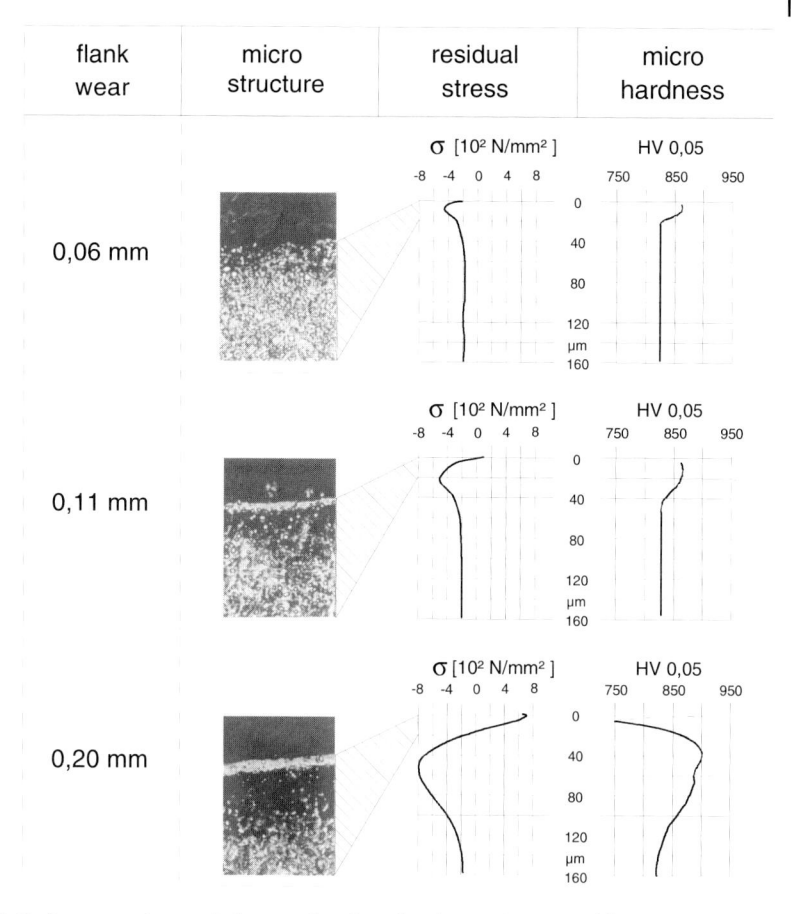

Fig. 5.4-3 Surface zone characteristics as a function of tool wear. Source: Goldstein [37]

ing lubricant has very noticeable effects. The thermal energy of the machining process heats the part and the tool and hence causes shape deviations.

5.4.3.2 Sensing and Monitoring Approaches

There is a correlation between tool wear, passive force, and resulting surface structure [39]. Increasing wear land at the flank face is associated with a significant rise in passive forces [34, 37] and therefore influences the form and dimensional accuracy of the workpiece. The increasing friction between tool and workpiece also leads to higher temperatures with a corresponding impact on surface structure and form deviations, as described above.

As a result, cutting forces and cutting temperature are the most important process quantities for monitoring. Both are influenced by tool wear and change accordingly in the course of the process. Additionally, tool wear has a direct impact on the surface roughness and is itself a target for monitoring.

There are various principles for direct and indirect force measurement in cutting processes. Owing to the low level of cutting forces in the finishing operation, a very sensitive measuring technique is needed for application in hard turning. Force measurements, based on piezoelectric elements mounted in the main force flux, are the most promising approach to provide signals at the high level of sensitivity required. Since cutting force measurements at the rotating workpiece require wireless signal transmission, sensor mounting at the tool site, for example in or below the cutting tool holder, is more suitable. It is impossible to identify any one standard appropriate mounting location, since this depends to a large extent on the machine tool design.

The possibilities for temperature measurements in cutting operations have already been described in the section on dry machining. Owing to the less complex kinematics of the turning operation, it is easier to apply some of the turning techniques than to use the rotating tools used in milling and drilling operations.

For monitoring the cutting tool wear state, the acoustic emission (AE) signal is another promising approach that has been successfully applied in various machining processes for wear monitoring [40–44]. The main advantage over force measurements is the easier sensor mounting and the better signal-to-noise ratio with low chip thickness in precision machining operations [41, 45]. The AE signal has been found to be sufficiently sensitive to monitor wear in hard turning operations [39]. The variations in the AE signal with increasing flank wear depend on the frequency range under examination and require an analysis of the signal characteristic prior to developing a monitoring algorithm [39].

The sensing solutions described so far adopt the approach usually applied in process monitoring to measure process manifestations such as cutting forces or AE signals indirectly and to correlate the measured quantities with process disturbances. New developments in the field of cutting tool coating and micro-system technology permit the wear and temperature at the rake face of a turning tool to be measured directly on-line. Such an approach is still restricted to the field of research, but has considerable potential for application in various fields [46, 47]. Sensors for temperature and wear are applied within the wear-resistant coating of a carbide cutting insert (Figure 5.4-4) and allow the thermal load and the wear geometry at the cutting edge to be determined. Temperature monitoring is particularly beneficial to hard turning applications, since thermal load is the main indicator for changes in the surface zone.

5.4.4
Using Acoustic Emission to Detect Grinding Burn

Much of the energy used in grinding operations is converted into heat. The reason for this is the friction between the grinding grit and the workpiece and the large number of kinematic cutting edges involved in the process. The heat generated is dissipated via the chips, the tool, and the cooling lubricant.

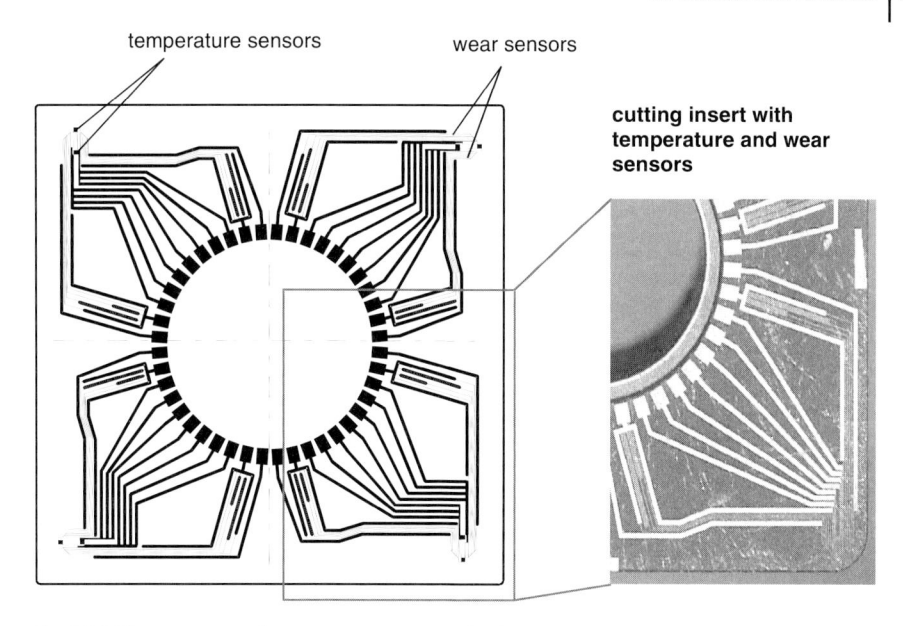

Fig. 5.4-4 Temperature and wear sensors at a cutting insert

From an ecological point of view, it is highly desirable to reduce the volume flow of cooling lubricant, extending if possible to dry machining, in grinding operations also. However, since this entails the loss of an important carrier for the removal of energy, the risk of subjecting the part to be ground to thermal overload increases. This can cause damage in the form of subsurface damage or of burn marks, dimensional deviations, structural changes, loss of hardness, conversion of residual compressive stress into residual tensile stress, and the formation of inter-crystalline cracks in the subsurface. These types of damage result in the rejection of the part, which may represent a considerable financial loss since grinding, which is a finishing operation, is at the end of the process chain.

Until now, it has not been possible to detect grinding burn during the grinding process. In industrial practice, workpieces are picked up in a random sampling operation carried out after the process. The surface and subsurface integrity of the workpieces is analyzed by etching the workpiece structure. In order to ensure 100% monitoring for thermal damage, a system capable of on-line grinding burn monitoring is required which will detect part damage immediately, thus preventing the manufacture of more faulty parts.

5.4.4.1 **Objective**
The AE signal is described in the literature as a reliable signal for process monitoring and control [41, 45]. The AE signal technique is frequently used particularly in precision machining operations because of its good signal-to-noise ratio [45]. Since the increase in workpiece temperature and the friction which occur when

grinding burn develops affect the AE signal generated, this signal can profitably be used for on-line detection of grinding burn in all grinding operations, not only those conducted with a reduced volume flow of lubricant.

5.4.4.2 Sensor System

The signal is influenced by a number of very different factors on its way from the point at which the noise is generated in the grinding gap until it reaches the AE sensor. Mechanical interfaces within the signal transmission path between cutting process and sensor location, for example, cause significant damping, thus weakening the signal. Additionally, the geometry of the parts which conduct the signal influences the AE signal through reflection and through the development of stationary waves. The location at which the sensors are mounted has therefore assumed an important role. Guidelines for mounting the sensors correctly have been developed by Saxler [48].

In grinding operations, the AE sensor can, in principle, be mounted at two locations. From the point of view of the grinding gap, the sensor can be either on the side of the grinding wheel or on that of the workpiece. The grinding wheel-mounted sensor is compared with a workpiece-mounted sensor in Figure 5.4-5.

The sensor on the grinding wheel can be used in more applications since it permits both the grinding signal and the dressing operation to be monitored. The frequency spectra plotted show, however, that the workpiece sensor provides an additional width of the signal frequency band at higher intensity. The reason for this is the shorter distance from the machining point, which is where the sound is produced and the resultant lower level of attenuation. Despite this, the wheel sensors are ideally

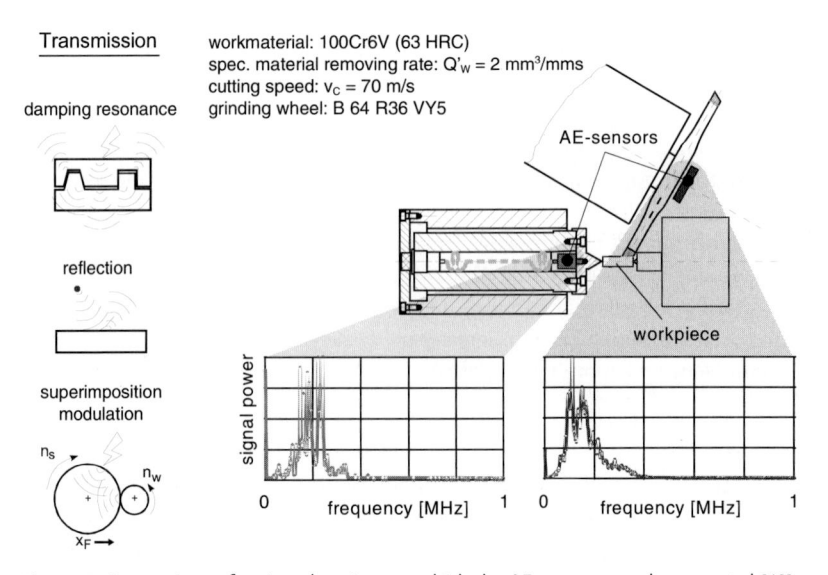

Fig. 5.4-5 Comparison of various locations at which the AE sensors can be mounted [49]

suited to process control [50] or contact detection. The workpiece sensor is, however, by far the better option for the detection of grinding burn resulting from the failure of the lubrication system since the thermal processes which result in the occurrence of grinding burn take place in the workpiece and not in the grinding wheel [48].

5.4.4.3 Signal Evaluation

The AE signals in the frequency range are evaluated in order to detect grinding burn using the AE technique. This initially requires the signals to be digitized in a high-frequency operation in which the scanning rate must be several times that of the highest frequency evaluated. Frequently, only a time-limited area of the grinding signal is taken into account, in order to limit the volume of data involved. A subsequent fast Fourier transformation provides the amplitude frequency spectrum of the signal as shown in Figure 5.4-5.

The frequency spectrum thus obtained depends largely on the type of process control used and on chip formation. The grinding parameters, the grinding wheel used, and the cooling lubricant all exert a particularly strong influence on the AE frequency spectrum. Since the geometry, the material, and, of course, the structural condition of the workpiece affect the frequency spectrum, a comparison of the frequency spectra of undamaged workpieces with those of damaged workpieces must be examined for this specific application in order to detect grinding burn using AE.

5.4.5
References

1 KLOCKE, F., KETTELER, G., REHSE, M., REUBER, M., in: *Proceedings of the 1st International Manufacturing Engineering Conference (IMEC'96)*, 7.–9. August 1996, Storrs, Connecticut, USA, 394–397.
2 KLOCKE, F., REUBER, M., REHSE, M., in: *Proceedings of the 1997 ASME International Mechanical Engineering Conference and Exhibition, Symposium on Sensor Fused Intelligent Machining Systems*, 21 November 1997, Dallas, TX, USA, 83–90.
3 REHSE, M., *Dissertation*; RWTH Aachen, 1998.
4 SEIDEL, D., *Werkstatttechnik* **81** (1991) 161–164.
5 EVERSHEIM, W., *Organisation in der Produktionstechnik*, 2nd edn; Düsseldorf: VDI Verlag, 1990.
6 KLOCKE, F., *VDI-Nachr.* **36** (1996) 24.
7 Luftbeschaffenheit am Arbeitsplatz, Minderung der Exposition durch Luftfremde Stoffe – Allgemeine Anforderungen, VDI-Richtlinie 2262; Blatt 1 (Entwurf), 1990.
8 *Messen von Partikeln, Kennzeichnung von Partikeldispersionen in Gasen, Begriffe und Definitionen*; VDI-Richtlinie 3491, Blatt 1, 1980.
9 *Maximale Arbeitsplatzkonzentrationen und Biologische Arbeitsstofftoleranzwerte*; Weinheim: VCH, 1998.
10 *Staubbrände und Staubexplosionen*; DIN 2263, 1992.
11 SIEKMANN, H., BLOME, H., in: *Gefahrstoffe am Arbeitsplatz*, Sankt Augustin: Berufsgenossenschaftliches Institut für Arbeitssicherheit – BIA, 1987.
12 LENZ, K., et al., *Luftanalytik: Emissions, Immissions- und Arbeitsplatzmessungen*; Ehningen: expert-Verlag, 1993.
13 BARON, P.A., *Analyst* **119** (1994).
14 FISSAN, H., JERKOVIC, A., *Techn. Mess.* **56**, (1989).
15 Ullmanns Encyklopädie der Technischen Chemie, Band 5: *Analyse- und Meßverfahren*; Weinheim: Verlag Chemie, 1980.
16 RÜCKAUER, C., *Tech. Mess.* **56** (1989).

17 BIA-Arbeitsmappe Messung von Gefahrstoffen; Sankt Augustin: *Berufsgenossenschaftliches Institut für Arbeitssicherheit – BIA*; Bielefeld: Erich Schmidt Verlag, 1994.
18 STAFFMANN, M., *Tech. Mess.* **56** (1989).
19 LILIENFELD, P., *J. Aerosol Sci.* **18** (1987).
20 RUMMENHÖLLER, S., *Dissertation*; RWTH Aachen, 1996.
21 KÖNIG, W., OSTERHAUS, G., GERSCHWILER, K., et al., in: *Wettbewerbsfaktor Produktionstechnik – Aachener Perspektiven*, AWK Aachener Werkzeugmaschinen-Kolloquium, Düsseldorf: VDI-Verlag, 1993, 5/1–5/48.
22 KLOCKE, F., SCHULZ, A., GERSCHWILER, K., et al., in: *Wettbewerbsfaktor Produktionstechnik; Aachener Perspektiven*. AWK Aachener Werkzeugmaschinen-Kolloquium,. Düsseldorf: VDI-Verlag, 1996, 4/35–4/105.
23 KLOCKE, F., EISENBLÄTTER, G., *Ann. of CIRP*, **46** (1997) 1–8.
24 LOWACK, H., *Dissertation*; RWTH Aachen, 1967.
25 KASSBAUM, D., LÖFFLER, N, *Dissertation*; TU Dresden, 1982.
26 FAY, E., in: *VDI-Lehrgang BW36-11-16 Technische Temperaturmessung*, 8.–10. Oktober 1990, Stuttgart (Vortrag).
27 MESTER, U., *Temperaturstrahlung und Strahlungsthermometer*, Wiesbaden: Heimann (pers. Kommunikation).
28 BEITZ, W., KÜTTNER, K.-H., *Dubbel – Taschenbuch für den Maschinenbau*, 18th edn; Berlin: Springer, 1995, D32–D33.
29 SCHULTE, D., in: 11th International Colloquium on Industrial and Automotive Lubricatio, 13–15 January 1998, *Technische Akademie Esslingen*, 109–123.
30 MÜLLER-HUMMEL, P., LAHRES, M., *Temperature Measurement on Diamond-Coated Tools During Machining*; IDR 2/95, 1995, 8–12.
31 EYERER, P., *Minimalschmiersysteme für die Zerspantechnik – Anwendung, technische Optimierung und wirtschaftlich – umweltliche Bewertung*, Abschlußbericht des Verbundvorhabens des Wirtschaftsministeriums Baden-Württemberg, 1997.
32 ACKERSCHOTT, G., *Dissertation*; RWTH Aachen, 1989.
33 KOCH, K. F., *Dissertation*; RWTH Aachen, 1996.
34 WINANDS, N., *Dissertation*; RWTH Aachen, 1996.
35 BERKTOLD, A., *Dissertation*; RWTH Aachen, 1992.
36 NAKAYAMA, K., ARAI, M., KANDA, T., *Ann. CIRP* **28** (1988) 89–92.
37 GOLDSTEIN, M., *Dissertation*; RWTH Aachen, 1991.
38 BRANDT, D., *Dissertation*; Universität Hannover, 1995.
39 SCHMIDT, J., *Dissertation*; Universität Hannover, 1999.
40 DIEI, E. N., DORNFELD, D. A., *J. Eng. Ind. Transaction ASME* **109** (1987), 234–240.
41 KÖNIG, W., KLUMPEN, T., presented at the *5th International Grinding Conference*, 26–28 October, 1993, Cincinnati, OH, paper MR93-358-1–MR93-358-23.
42 INASAKI, I., BLUM, T., SUZUKI, H., ITAGAKI, H., SATO, M., *Adv. Flexible Autom. Robot.* **2** (1988) 1017–1024.
43 KETTELER, G., *Dissertation*; RWTH Aachen, 1996.
44 LAN, M. S., DORNFELD, D. A., in: *Proceedings of the 10th NAMRC, SME*, McMaster University, Hamilton, Ontario, 1982, 305–311.
45 DORNFELD, D., in: CIRP/VDI Conference on High-Performance Tools', 3–4 November 1998, Düsseldorf, 215–233.
46 KLOCKE, F., REHSE, M., *Ann. Ger. Soc. Prod. Eng.* **2** (1997).
47 LÜTHJE, H., LÖHKEN, T., BÖTTCHER, R., *ITG-Fachber.* **148** (1998) 571–578.
48 SAXLER, W., *Doctoral Thesis*; University of Aachen, 1997.
49 KLOCKE, F., MUCKLI, J., presented at the *3rd International Machining and Grinding Conference*, 4–7 October 1999, Cincinnati, OH, Technical Paper SME 1999 MR99-232-1–MR99-232-23.
50 KLOCKE, F., ALTINTAS, Y., MEMIS, F., *Int. J. Machine Tools Manuf. Vol. 35* **10** (1995) 1445–1457.

List of Symbols and Abbreviations

Symbol	Designation	Section
a	linear acceleration	2
a_C	carbon activity	4.9
a_e	depth of cut	4.4
A	cross section	2
A	Sievert constant	4.1.1
A	uncertain event	1.3
B	magnetic flux density	2
B	Sievert constant	4.1.1
C	auto-correlation function	1.3
C_H	concentration of hydrogen dissolved in aluminium	4.1.1
d	distance	3.1
d	spindle length	3.1
d_m	average bearing diameter	5.2
d_{theor}	resolution	3.2
D	shadow diameter	3.1
E	measured potential	4.9
E_{kin}	kinetic energy	2
f	focal length	3.1, 3.2
f	optical multiplication factor	5.1
f	sine or cosine deviation	3.1
f_{max}	upper limit frequency	1.3
F	reaction force	2
F_N	normal force component	4.3
F_p	back force	3.3
F_T	tangential force component	4.3
g	grating period	5.1
H_{cM}	coercivity	3.3
i	integer	1.3
i	number of counts	5.1
$I_T(x,y)$	tunneling current	3.2
j	unit number	1.3

k	coupling factor of transformer	2
k	electronic interpolation factor	5.1
k_c	specific cutting force	4.3
K	constant	3.1
K_N	nitriding potential	4.9
l	length of measurement	2
l	length of windings	2
l	arm length	3.1
l_k	contact length between tool and workpiece	3.3
L	layer number	1.3
m	mass	2
m	number of nodes	1.3
m_F	coating mass	4.8
M	number of digital data	1.3
M_{max}	maximum amplitude of Barkhausen noise	3.3
MA	moving average	1.3
n	number of bits	2
n	number of modules	2
n	refractive index	3.1
n	spindle frequency	5.2
O	output	1.3
p_{H2}	partial pressure of segregated hydrogen	4.1.1
P	partial pressure	4.9
P_α	friction power at flank face	3.3
P'_α	specific friction power	3.3
P''_c	specific grinding power	3.3
P_c	grinding power	3.3
P_j	availability	2
P_k	power spectrum	1.3
Q'_w	material removal rate	4.4
Q'_w	specific material removal rate	3.3
r	radius of circle	2
R	ideal output	1.3
R_a, R_z	surface roughness	4.4
R_a, R_z, R_q	roughness	3.2
R_b	resistance	2
R_{pk}	reduced peak height	3.3
s	length displacement	3.1
S	signal intensity	5.1
S_N	scatter value	3.2, 4.4
t	time	1.3, 2
T	observation period of signal	1.3
T	temperature	4.9
$T_{\text{Cold Junction}}$	temperature of cold junction	4.2
$T_{\text{Hot Junction}}$	temperature of hot junction	4.2

List of Symbols and Abbreviations

u	deviation of coordinate	2
U	DC voltage	4.8
U	electrical signal	2
U	induced voltage	2
U_a	inductive voltage	3.1
U_d	dressing overlap	4.4
U_{PD}	potential difference	4.2
U_x	voltage	2
v	deviation of coordinate	2
v	speed	2
v	velocity of reflector	5.1
v_f	feed speed	2
V'_w	material removal	3.3
VB_c	width of flank wear land	3.3
w	deviation of coordinate	2
w	number of windings	2
W	weight	1.3
x	coordinate	2
x	shift of ocular scale image	2
x	traveled path	5.1
$x(t)$	signal time series	1.3
X	input	1.3
y	coordinate	2
y	distance	3.1
z	coordinate	2
$Z_T(x,y)$	height of needle tip	3.2
α	angle of inclination of mirror	2
α	rotation angle	2
α	thermal coefficient of expansion	5.1
β	Scheimpflug angle	3.1
δ	sampling slot width	2
Δ	phase shift	5.1
ε	emission factor	5.4
θ	angle of incidence	3.1
θ	angle of triangulation	3.1
Θ_i	moments of inertia of components i	2
λ	division period	3.1
λ	wavelength	1.3, 3.1, 3.2, 4.4
λ_0	wavelength of vacuum	3.1
ρ	specific resistance	2
τ	incremental width	2
φ	angular deviation	3.1
ψ	phase shift	5.1
ω	angular speed	2
ω_0	diameter of surface illuminated area	3.2
Ω	phase shift	5.1

List of Symbols and Abbreviations

Abbreviation	Explanation	Section
AC	adaptive control	1.3, 4.3, 4.4, 5.2, 5.4
AC	alternating current	1.2, 1.3, 3.1, 3.2, 4.2, 4.3, 4.6, 4.8, 5.1
ACC	adaptive control constraint	4.3
ACO	adaptive control optimization	4.3
AD	analog-to-digital	1.3
AE	acoustic emission	1.3, 3.3, 4.3, 4.4, 4.6, 5.3
AFM	atomic force microscopy	3.2
AI	artificial intelligence	1.3
AMB	active magnetic bearing	4.4
ASK	angular speckle correlation	3.2
c.w.	continuous wave	3.3
C-AFM	contact atomic force microscopy	3.2
CCD	charge-coupled device	3.1, 3.2, 4.3, 4.4, 4.5, 4.7, 5.3
CD	continuous dressing	4.4
CIM	computer integrated manufacturing	4.1.1
CMM	coordinate measuring machine	3.1
CT	computed tomography	4.1.1
CT	current transformer	4.9
CVD	chemical vapor deposition	4.8
DBB	double ball bar	2
DC	direct current	1.2, 1.3, 2, 3.2, 4.6, 4.8, 5.3
DFT	digital Fourier transform	1.3
DMD	digital mirror device	3.1
EDM	electrical discharge machining	4.6
EFM	electric force microscopy	3.2
FET	field effect transistor	1.2
FMM	force modulation microscopy	3.2
FPW	flexural plate wave	1.2
FWHM	full width at half maximum	3.2
HF	high frequency	4.6
HIP	hot isostatic pressing	4.1.2
HSC	high-speed cutting	5.2
HSK	hollow taper shaft	5.2
I/O	input/output	1.3
IC-AFC	intermittent-contact atomic force microscopy	3.2
ID	inner diameter	4.4

IR	infrared	4.7, 4.8
LBB	laser ball bar	5.1
LCD	liquid crystal display	3.1
LED	light-emitting diode	3.2, 4.3, 5.1
LFM	lateral force microscopy	3.2
LiMCA	liquid metal cleanliness analyzer	4.1.1
LVDT	linear variable differential transformer	1.2, 3.1, 5.1
M & D	monitoring and diagnosis	2
MEMS	microelectromechanical systems	1.2
MF	medium frequency	4.6
MFM	magnetic force microscopy	3.2
MQL	minimum quantity lubrication	4.4
MTTR	mean time to repair	2
NC	numerical control	2
NC-AFM	non-contact atomic force microscopy	3.2
OD	outer diameter	4.4
OMM	on-the-machine measurement	4.6
PDM	phase detection microscopy	3.2
PM	powder metallurgy	4.1.2
PSD	position-sensitive detector	3.3
PSPD	position-sensitive photodetector	3.2
PT	potential transformer	4.9
PVD	physical vapor deposition	4.8
QCM	quartz crystal microbalance	1.2
SAW	surface acoustic wave	1.2
SCM	scanning capacitance microscopy	3.2
SEM	scanning electron microscopy	3.2
SNAM	scanning near-field acoustic microscopy	3.2
SNOM	scanning near-field optical microscopy	3.2
SOS	silicon-on-sapphire	1.2
SPM	scanning probe microscopy	3.2
SSK	spectral speckle correlation	3.2
SThM	scanning thermal microscopy	3.2
STM	scanning tunneling microscopy	3.2
TFT	thin-film thickness	4.8
TQM	total quality management	4.1.1
TSM	thickness shear mode	1.2
VHF	very high frequency	4.6
VS	vibroscanning	4.6

Index

Abbé principle   49, 72
abrasive processes   123, 236 ff, 262
absolute measurement methods   52, 89
accelerometers   13, 154
– cutting   214
– grinding   239
– piezoelectric   58
accuracy   2, 8
– electrical discharge machining   283
acetone   160
acoustic emission (AE) sensors   6, 16, 31
– chip control   228
– cutting   204 f, 214 ff
– grinding burn   372
– grinding   237 ff, 245 f
– laser processing   275
– machine tools   65
– micromachining   358
– punching processes   173 f
– workpieces   131
acoustic radiation, electrical discharge machining   283
active magnetic bearings (AMBs)   246
active sampling, dry machining   364
actuators, mechanical manufacturing   25
adaptive control   231, 249
– abrasive processes   265
– coating   310
– cutting   231
– high-speed machining   355
aerosols   365
air classification sensors   162
alcohol   160
aliasing   35
aluminum alloys   184
aluminum oxide ceramics   264
analog signal processing   32 f
analog-to-digital conversion   34
angle measurement, potentiometric   77

angular deviation, workpieces   73
angular speckle correlation (ASK)   104
annealing   336
arc discharge, transient   277
arc welding   286, 295 f
artificial intelligence   40
artificial neural networks   301
atmosphere control
– heat treatments   329
– sintering   167
atomic energy, conversion   11
atomic force microscopy (AFM)   98, 113 f
austenites   336
autocollimators   58, 352
autocorrelation, mechanical manufacturing   37
autofocus methods   96
automatic monitoring   3 f

band reject filters   33
Baratron   311, 314
barium titanate   155
Barkhausen noise   138
batch furnaces   167, 171
beam bounce detection   115
beam monitoring
– laser processing   274
– welding   298
bearing balls, high-speed machining   354
binary code, absolute measurements   89
binder metals   160, 166
binding energy   11
blackbody coating   320, 327
blankholders, deep drawing   184
blending, metal powders   159
Bloch wall motion   137
bolometers   311, 320 f

384 | Index

boron nitride   354
breakage
– chips   228
– punching processes   173, 180
– tools   8
bright-field microscopy   98
broaching   203
burst type signals, acoustic
   emission   131

calibration
– deep drawing   185
– machine tools   60
calipers
– oil-proof   99
– workpieces   72
camera-based monitoring   90, 156, 225
   see also: charge coupled device
   cameras
cantilevers   16, 118
capacitive sensors   14
– chip control   229
– coating   311, 314
– displacement   76 f
– grinding wheels   245
– incremental   81
– laser processing   274
carbides   160
carburizing   329
cassette systems, modular   177
casting   143 ff
CBN, grinding   258
ceramics   144, 354
cermets   206
characteristic variables, workpieces   71
charge coupled device (CCD) cameras
   222 ff, 247, 358
– interferometry   102
– laser processing   275
– welding   292
– workpieces   90
charge determination   17
chatter vibrations
– cutting   204, 209

– grinding   236
– loose abrasive processes   264
chemical loads, cutting   220
chemical properties, powder
   metallurgy   166
chemical sensors   4, 10, 18
chemical vapor deposition (CVD)
   308 f
chip breakage   131
chip control systems   228 f
chip formation
– cutting/grinding   123, 203
– hard turning   369
chromel-alumel thermocouples   169
classification, sensors   6, 10 f, 143
closed-loop control systems   310
coating   182, 307 ff
cold cathode method, coating   316
cold junctions   193
cold welding   184 f
collision detection, machine tools   62
combination vacuum gages   316
communication techniques   24, 43
compacting, metal powders   160 ff
comperator principle   72 ff
compositions, melts   144
compression, metal powders   159
computer-integrated manufacturing
   (CIM)   143
conditioning, grinding   256
conductivity   112, 146
construction principle   see: working
   principle
contact atomic force microscopy
   (C-AFM)   114
contact electrodes   148
contact sensors   220, 225
– grinding   249
– welding   287
contact stylus method   99
contactless measurements,
   workpieces   71
continuous dressing, grinding   247
continuous emission, acoustic   131
continuous measuring systems   7, 84 f

control
- electrical discharge machining   279 f
- powder metallurgy   165
control loops   48, 343
control parameters   7
convective heat flux   340
conversion, signals   32 f
conversion processes   3, 11
coolant supply, grinding   259
cooling lubricants   363
cooling time, process monitoring   200
coordinate measuring machine (CMM)   83 f
correlation curves, scattered light sensors   254
correlation function, mechanical manufacturing   37
correlation statistics, deep drawing   183
corrosion resistance   16
corrosive effect, melts   144
corrosive load, heat treatments   326
corundum grinding wheels   246
cosine deviation, calipers   74
cover slide monitors, welding   299
crack formation
- acoustic emission   131, 136
- cutting   220
- metal forming   184
crankshafts
- grinding   249
- punching processes   173
cubic boron nitride   354
Curie temperature   334
current control   17, 149
current modes, microscopy   109
current-through-gap, electrical discharge machining   281
current transformer, heat treatments   340
cutting   2 f, 26, 203 ff
- edges   123
- high-speed   354 ff

damage, thermal   126
damped mass spring element   251
dark-field microscopy   98
decision making   19, 39
deep drawing   182 f
deep penetration laser welding   302
deflection, cantilevers   118
deformations
- cutting   204, 214
- forging   192
- welding   286
- workpieces   126
deposition temperature, coating   308
depth-of-focus, point triangulation   91
deviations, workpieces   73 f
dewaxing   170
diagnosis, machine tools   65
dial comperators   76
dial gages   75
diameter measuring systems   226
diamonds   354
diaphragms   16, 314
differential plunger coil   79
diffraction
- optical scales   346
- scanning methods   107
digital increment sensor   54
digital mirror device (DMD)   93
digital signal processing   36 f
direct current level, electrical discharge machining   283
direct measuring systems   7, 48
direction discriminator   57
disparate systems   5
displacement transducers   76, 154, 225 f
distance-coded reference marks   88
distance gages   194
distance measurements
- interferometric   94
- welding   288
disturbances, punching processes   173
docking stations   178

dosage control   148, 153
double ball bar (DBB) device   60
double skidded systems   101
dowel pins, force sensors   208
dressing, grinding   247, 257
drilling   26, 203 ff
– torque sensors   209
dry machining   364 ff
dynamic measurements   27, 85
dynamometers
– coolant supply   261
– cutting   207
– piezoelectric   125, 237

eddy current sensors   80, 136, 154
– heat treatments   334 ff
– welding   292
ejector force-time courses, forging   199
elastic flexural plate wave (FPW)   16
electric force microscopy (EFM)   121
electrical discharge machining (EDM)   277
electrical energy conversion   11
electrical properties, iron/compounds   334
electrical sensors   17, 71, 76 f
electrical signals   4, 10
electrochemical cell, oxygen   329
electrode contact sensors   290
electromagnetic radiation   283
electromagnetic sensors   225, 292
electromechanical measurements   83 ff
electromotive force   146, 327
encoders, optical scales   343
energy conversion   11
environmental awareness   363 ff
errors, workpieces   72
exactness, mechanical manufacturing   27
extension sensors   208
extrusion, metal powders   159

failures, machine/tools   4, 63 ff
Faraday rotation   18
feature selection process   19
feed force collision sensor   65
ferrites   336
fiber optic sensors   220, 328
fiber-reinforced plastics   366
film thickness   311
filters, mechanical manufacturing   32
flange insertion, metal forming   182, 186
flexural plate wave (FPW)   16
flow behavior, grinding   259
flow meters   14
fluid quench sensors   338
flux sensors, heat treatments   339
focal position, laser processing   274
folding   35
force control
– powder metallurgy   166
– process monitoring   146, 155
force modulation microscopy (FMM)   120
force sensors   6
– chip control   229
– cutting   204 f
– forging   192, 195, 198
– grinding   237 f
– punching processes   173 f
– workpieces   125
forging   191 ff
form testing   95
forming, metals   172
forms, workpieces   71
Fourier transform   37
fracture, single grain   132
frequency domain
– fusion   22
– signal processing   36
frequency ranges, electrical discharge machining   283
frequency response   27 f
friction
– cutting   214
– metal forming   182

– workpieces 118, 125, 131
fringe projection 92
front illumination 86 f
full radiators, dry machining 368
full width-at-half maximum (FWHM) 112
furnace tracker systems 328
furnaces, sintering 167
fusion 5, 21 f
fuzzy logics 42, 301

galling 184
gap conditions, electrical discharge machining 279 f
gas analyzers 321, 331
gas content, melts 144
gas dosing systems, coating 324
gas mixtures, vapor deposition 308 ff
gas pressure, laser processing 274
geometric properties
– grinding wheels 245
– workpiece sensors 225, 278
geometry-oriented sensors 287 f
GMA welding 295 f
grain boundaries 167
grain refining agents 144
gravitational energy 11
Gray code 89, 93
grazing incidence X-ray reflectometry 105
green products 159, 163, 170
grid-based sensors 299
grinding 4, 123, 236 ff
grinding burn 132 ff, 236
– acoustic emission sensors 372
grinding wheels 244 f
grooves, welding 286
GTA welding 301
gyroscopes 15

half-bridge probe 79
half-reflecting mirrors 348
Hall probe 18, 138, 282

heat flow 17
heat radiation
– chip control 230
– infrared films 213
– workpieces 130
heat transfer, hollow wire sensors 339
heat treatments 326 ff
Heidenhain linear decoder 347
height modes, microscopy 109
helium leak detector 318
heuristic signal-based systems 66
hexanes 160
high-frequency receivers, electrical discharge machining 283
high-pass filters 32
high-speed cameras 228
high-speed machining 354 ff
hollow taper shaft (HSK) 356
hollow wire sensor 338
holographic interferometry 95
honing processes 236, 251
hot isostatic pressing (HIP) 167
hot junctions, forging 193
hot welding 184 f
human monitoring 19
human-machine interfaces 44
hybrid ball bearing, high-speed machining 354
hydrogen 144
hysteresis, mechanical manufacturing 27

identification techniques, mechanical manufacturing 39
illumination, workpieces 86 ff
image sensing
– laser processing 273
– micromachining 361
– welding 301
image velocimetry 260
in-process sensors 7
– deep drawing 186 f
incremental sensors 81, 86
indentation cracks 132

indirect measuring systems 7, 48
induction heating 339
inductive sensors 148, 352
– deep drawing 186
– displacement 76 f
– grinding wheels 245 f
– incremental 82
inductosyn sensor 53, 82
inert gases, sintering 167
infrared absorption 331
infrared chip measurement 229
infrared films 213
infrared spectroscopy 185
infrared thermography 263
injection molding 159
inner diameter, grinding 238
input/output (I/O) devices 44
input-output relation, transducers 27
intelligent systems
– cutting 233
– grinding 236, 268 ff
interfaces 24
interferometric distance measurements 88, 94
interferometric scales 346
interferometry, whitelight 102
intermittent contact atomic force microscopy (IC-AFM) 117
intermittent systems 7
interpolation techniques 345, 349
interprocess measurements 7
iron group metals, powder metallurgy 160

junctions
– forging 193
– thermocouples 326

KiNit sensor 334

lapping processes 262
large-scale systems 4

laser ball bar (LBB) 350
laser beam welding 289
laser Doppler anemometers, aerosols 365, 368
laser interferometers 348 f
laser level measurements 156
laser processing 272
laser scanning 91, 252
laser scattering 225
laser triangulation system 135, 245
laser variants, tool sensors 220
laser welding 291
lateral force microscopy (LFM) 118
leak detectors, coating 317
lever-type test indicators 75
light emitting diode (LED) 102, 222, 346
light scattering systems, powder metallurgy 162
light section method, workpieces 92
linear movement, machine tools 47
linear variable differential transformer (LVDT) 13, 78 f, 352
linearity, mechanical manufacturing 27
liquid crystal displays (LCD) 93, 185
liquid metal cleanliness analyzer (LiMCA) 147
liquid-phase sintering, powder metallurgy 166
load analyzer, heat treatments 340
loads, cutting 203, 220
loose abrasive processes 262 f
Lorenz force 18
low-pass filters 32
lubrication
– environmental awareness 363, 367
– grinding 259
– metal forming 182 ff
– sintering 167 f

machine tools 47–70
macrogeometric quantities 71 ff, 246, 249

magnetic bearings, high-speed machining   354
magnetic energy   11
magnetic field measurements   153
magnetic force microscopy (MFM)   117
magnetic incremental sensors   81
magnetic properties
– deep drawing   183
– iron/compounds   334
magnetic sensors   4, 10, 18
magneto-inductive signals   183
magnetostrictive effect   209
malfunctions   5, 30
man-machine interfaces   24
martensites   336, 369
mass determination, slug   192, 195
mass spectrometry   311, 315
material removal, cutting   203
mean time-to-repair (MTTR)   66
measurands   11
mechanical energy   11
mechanical impact   124
mechanical loads
– cutting   203, 220
– heat treatments   326
mechanical manufacturing   24, 31 f
mechanical properties
– deep drawing   183
– powder metallurgy   166
mechanical sensors   13, 71 f
mechanical signals   4, 10
medium frequency, electrical discharge machining   283
melt contact, process monitoring   143 ff, 149 ff
membranes   314
metal forming   172
metal transfer modes, welding   296
metallurgy   143, 159 ff
Michelson interferometers   95
microelectrical discharge machining   285
microelectromechanical sensing systems (SENS)   15

microgeometric quantities   98 ff, 247, 251
micromachining   357 f
micromagnetic sensors   137
micrometer gages   73
milling   203 ff, 26
– powder metallurgy   160
– torque sensors   209
– ultraprecision   359
miniaturization, workpieces   98
minimum quantity lubrication (MQL)   238, 244
mixing, metal powders   159
model-based monitoring systems   39, 66
modular cassette system, metal forming   177
Moiré effect, position measurements   55
Moiré lines, optical scales   343
molecular energy   11
molten pools, welding   287, 300
monitoring
– abrasive processes   236 ff
– coating   307 ff
– cutting   203
– grinding   128, 236 ff
– heat treatments   326 ff
– laser processing   272
– machine tools   65
– powder production   161
– sintering processes   169
– unit processes   3 f
motion sensors   25, 47
multiple regression   183
multipoint cutting   203
multisensor approach   10, 177

Nd:YAG laser welding   299
necks, sintering   167
Nernst law   147
neural networks   41, 301
NiCr-Ni thermocouples   326
niobium carbides   160

nitriding 331
noise free data 19
noncontact atomic force microscopy (NC-AFM) 116
nozzle bounce plate principle, grinding 246, 256
nuclear energy 11
numerical controlled machine tools 48, 62, 67

ohmic resistance
– arc welding 295
– platinum 212
oil films, metal surfaces 184 f
oil-proof calipers 99
on-the-machine measurement (OMM) 284
opaque sectors 343
open-loop control system 310
operating principles *see:* working princiles
optical fibre sensors 225 f, 229
optical measuring methods 90, 98, 101
optical pyrometer, heat treatments 327
optical scales, ultraprecision machining 343 ff
optical sensors
– grinding 247, 253
– laser processing 275
– machine tools 58 f, 221
– micromachining 360
– welding 291
– workpieces 71
optoelectronic sensors 8, 71, 86
orientation sensors 58
oscillation frequencies, welding 297
output types 12
overload, metal forming 180
oxide layers 290
oxidizing 332
oxygen 144
oxygen probes 329 ff

parasitic induction 197
partial pressure measurements 144 f
particle image velocimetry 260
passive sampling 364
path signals, forging 198
pattern recognition 19 ff, 40
PCNB 206
penetration depth, welding 287
Penning gages 311, 315
peripheral systems, grinding 256
permeability
– iron/compounds 334
– magnetic films 209
phase detection microscopy (PDM) 119
phase Doppler anemometers 365
phase shifts
– optical scales 347
– signals 56
photodiodes 18
– grinding 247
– optical scales 343
– welding 302
photoelectric encoders 343
photoelectric transducers 351
photogrammetry 94
photosignals, position measurements 55
phototransistors 18
physical properties
– heat treatments 326
– powder metallurgy 166
– workpieces 71, 99 f, 123 ff, 225
physical vapor deposition (PVD) 308 ff
piezoelectric crystals 13
piezoelectric dynamometers 125
– chip control 228
– grinding 237
piezoelectric force sensors 155, 174
piezoelectric pressure gages 194
piezoelectric quartz force transducers 206
piezoresistive sensors 14
Pirani gages 311 ff

Planck law 327
plasma environment, vapor deposition 308
plasma processes 341
plastic deformation, cutting 204
platinum 328
platinum-rhodium alloys 169, 326
plunge grinding 238, 248
plural in/output, mechanical manufacturing 39
pneumatic sensors 96, 153, 221, 225
– grinding wheels 245, 253
– welding 291
point triangulation 91
polarizing beam splitters 348
pollutants 365
position sensitive detectors (PSD) 134
position sensitive photodetectors (PSPD) 115
position sensors
– cutting 205
– deep drawing 186
– machine tools 47, 71
– ultraprecision machining 343
postprocess measurements 7
potential difference 17
potential transformers, heat treatments 340
potentiometers 13, 50, 77
powder metallurgy 143, 159 ff
power sensors 6, 128
– cutting 211
– grinding 237 f
precision 27
– forging 196
preoxidation, heat treatments 333
preprocess measurements 7
press load, punching processes 173
pressforming 184
pressing, metal powders 159, 163
pressure gages, piezoelectric 194
pressure measurements 145
– coating 312
primers 290

probe measuring methods 106
probe tips 83, 99
process monitoring 143–342
process oriented sensors 288, 295
process parameters 5 ff, 32, 125
– coating 309 f
– grinding 237
– laser processing 272
production sequence, powder metallurgy 160
productivity, unit processes 2
protecting tubes, melt contact 144
protractors, universal 73 ff
proximeters 28
pulsed-arc welding 301
punching processes 172 ff
purity, melts 144
pyrometers 148, 152
– coating 309. 318
– grinding 244
– heat treatments 327

quadrupole mass spectrometer 311, 321
quality control
– coating 309
– laser processing 273 ff
– ultraprecision machining 343
– workpieces 226
quartz crystal microbalance (QCM) 16
quartz crystal sensors 323
quartz force transducers 125, 155, 238
quenching 337

radar sensors 245
radiant energy 11
radiant signals 4, 10
radiation, black body 327
radiation monitoring, welding 302
radiation pyrometry 213
radioactive methods 220

ram path, forging   194, 197
ratio thermometers   328
raw materials   1
rear illumination   86 f
redundancy, fusion   21
reference surface tactile probing   100
reflection sensors, grinding wheels   245
reflective materials, optical scales   343
refraction   88, 95
refraction index, X-rays   105
relaxation type generator, electrical discharge machining   281
reliability, mechanical manufacturing   30
repeatability
– infrared analyzers   185
– mechanical manufacturing   27 f
replicated systems   5
residual stress determination   125
resistance, electrical   147, 334
resistance thermometers   328
resistive displacement sensors   76
resolution
– camera methods   90
– mechanical manufacturing   27
– scanning microscopes   112
resolver, position measurements   52
resonance type sensors   219
resonant frequency   29
response   8, 27
robots   47–70
roller ball sensor   188
rolling, cold/hot   185
root mean square (RMS) converters   17
rotating roughness sensors   251
rotational movement, machine tools   47
roughness sensors   251, 255

safety   4
sampling   35
– dry machining   364
– position measurement   55

sapphire wafers   16
scanning, grinding   252
scanning capacitance microscopy (SCM)   111
scanning electron microscopy (SEM)   98, 107 f
scanning near field optical microscopy (SNOM)   110 f
scanning near-field acoustic microscopy (SNAM)   122
scanning probe microscopy (SPM)   106
scanning thermal microscopy (SThM)   111
scanning tunneling microscopy (STM)   108
scaper mirrors   303
scattered light sensors
– aerosols   365
– grinding wheels   245
– workpieces   103
Scheimpflug condition   91
scintillation meter   149
SCOUT imaging system   273
screening, powder production   161
seam tracking system, welding   287
sedimentation   161
Seebeck effect   17
self-arc-lightening   301
self-oscillations   251
semiconducting oxide powder-pressed pellet (Taguchi sensors)   19
semiconductor sensors   6
sensivity, mechanical manufacturing   27 f
set-ups   *see:* working principles
shadow casting methods   91
Shannon sampling theorem   35
sheet metal forming   181 ff
shrinkage, powder metallurgy   167
shunt resistor   281
Sievert law   145
signal processing   19, 24
– analog   32 f
– digital   36 f

signal transmission   31 f
signals   4, 10 ff
– metal forming   182
– punching processes   173 f
signal-to-noise ratio   22, 27
silicon etching   320
silicon microsensors   10 f, 14
silicon-on-sapphire (SOS)   16 ff
silicon photodiodes   247
sine deviations, workpieces   73
single grain fracture   132
single in/output, mechanical manufacturing   39
single waveband thermometers   328
sintering   159, 166 f
size analysis, powder production   161
skidded systems   100
slide path, punching processes   173
slide position, metal forming   180
slot, force sensors   176
slug temperature   192, 195
solid phase sintering   166
solidification control   149
sound monitoring   8
spatial resolution   112
speckle interferometry   95, 104
spectral analysis, welding   304
spring elements   251
stability, mechanical manufacturing   27
stainless steel   184
static measurements   85
Stefan–Boltzmann law   368
strain gages   13, 175
– cutting   206
– grinding   238
strain gages forging   196
strain rates, von Mise criterion   190 ff
structural changes control   334
structural deformation   126, 132
stylus method   99
substrates, coating   308
superabrasives   258
superconductor magnetic sensors   18
superhard materials   354

surface acoustic waves (SAW)   16
surface coating   307 ff
surface grinding,   131, 237
surface integrity
– acoustic emission analysis   133
– characterization   125
– grinding   252
surface locations, process monitoring   175
surface oil films   184
surface roughness
– grinding   236, 240
– hard turning   369
– workpieces   99, 106
surface texture, metal forming   185
switch type sensors   288
switching probe system   84

table slot force sensor   176
tactile measuring methods
– grinding wheels   245
– workpieces   99
tactile seam tracking system   274
Taguchi sensors   19
tantalum carbides   160
tapping   26
Taylor series   72
temperature control   147, 150 f
– coating   318
– dry machining   367
– forging   192
– heat treatments   326 f
temperature resolution, scanning microscopes   112
temperature sensors
– cutting   211 f
– grinding   241
– workpieces   129
– sintering   168
tensile strength   182
theodolite measuring systems   93
thermal conductivity   146
thermal conductivity gage   312
thermal damage, workpieces   126

thermal energy, conversion 11
thermal imaging 152
thermal impact 124
thermal loads, cutting 203, 220
thermal sensors 4, 10, 17
– welding 304
thermocouples 17, 28, 147
– coating 319 f
– grinding 242
– heat treatments 326
– sintering 168
thermoelectric couples, forging 194
thermoelectric potential, heat treatments 326
thermography, infrared 263
thermometers 328
thermophile based sensors 299
thickness shear-mode sensor (TSM) 16
thin film resistance 220
thin film sensors 213
thin film thermocouples 243
thin film thickness (TFT) control 322
thixo billet 153
TIG welding 296 f
time deviation 57
time domain
– fusion 22
– signal processing 36
time monitoring, punching processes 173
time resolution measurements 91
TiO_2 333
titanium carbides 160
titanium diaphragms 16
tool breakage
– cutting 203, 209
– milling 218
– punching processes 173, 180
tool sensors 220
tool system, integrated 195
tool temperatures, forging 193, 197
tools 134
tools materials 2

topography 99
– grinding 248
torch orientation 289
torque sensors 209
total quality management (TQM) 143
tracer particles, velocimetry 260
transmission band, acoustic sensors 174
transmission techniques 31, 43
transparent sectors 343
transport mechanisms, powder metallurgy 166
triangulation sensors 91
– grinding wheels 245
– lasers 135
tribological load 326
tribological properties, high-speed machining 354
triggering 185
tungsten carbides
– cutting 206
– grinding 246
– powder metallurgy 160
tunneling effect 109
turbine blades 247
Twyman-Green interferometers 95

ultraprecision machining 4, 343 ff, 357
ultrasonic assisted lapping 263
ultrasonic sensors 225, 293
unique memory triggers 19
unit processes 2 ff
universal protractor 73

vacuum coating processes 308 ff
valves, coating 324
vapor deposition 308 ff
velocimeters
– grinding 260
– ultraprecision machining 343
– electrodynamic 58

venturi meters   14
Vernier scales   72
vibrations
– cutting processes   214
– grinding   236
– loose abrasive processes   264
vibroscanning   284
visual-based sensing   275
VLSI patterning   15
voltage divider   50
von Mises criterion   190

water jet machining, abrasive   264
wear determination   125, 220
– cutting   217, 220
wear land width   8
wear resistance   354
welding   184, 286 ff
Wheatstone bridge   16, 196, 313
whitelight interferometry   102
wide-area measurements   71
wire sensors   330
working principles
– acoustic emmission sensors
   174, 237
– adaptive control   231
– autocollimator   353
– beam bounce detection   115
– contact sensors   250, 288
– dynamometers   207
– eddy current sensors   155, 292
– electrical discharge machining   278
– face lapping   262
– fluid quench sensors   338
– forging   191
– gas dosing system   324
– incremental sensors   81
– infrared analyzers   185
– intelligent grinding   270
– interferometric scale   346
– KiNit sensor   334
– laser interferometers   348
– laser level measurements   156
– laser wear sensors   224

– laser-stripe sensors   291
– linear variable differential
   transformers   352
– magnetic film sensors   210
– magnetic force microscope   117
– nitriding gas sensors   333
– oxygen probe   330
– Penning gage   316
– phase detection   119
– photodiode detectors   162
– photoelectric scale   344
– Pirani gages   313
– powder pressing   165
– quadrupole mass spectrometer   322
– resolvers   52
– roller ball sensors   188
– scanning microscopes   110 f, 122
– scattered light sensors   103
– sintering   168
– speckle correlation   104
– temperature sensors   212, 242
– thermocouples   147, 194, 197, 313
– ultrasonic sensor   294
– vibroscanning   284
– weld pool monitoring   300
– whitelight interferometers   102
– wrinkle sensors   186
– X-ray imaging   150
workpiece coatings   290
workpiece geometry
– laser processing   273
– welding   286
workpiece sensors   225 ff, 249 ff
workpieces   71–142
wrinkles
– deep drawing   187
– metal forming   182 ff

X-ray imaging   149
X-ray reflectometry   105

zirconium oxide   333